Emerging Organic Contaminants in Soil

Emerging organic contaminants (EOCs) are either newly identified or newly detected contaminants that are toxic or potentially hazardous to the ecosystem. The EOCs include, but are not limited to, pharmaceuticals and personal care products (PPCPs), antibiotics, perfluoroalkyl sulfonate (PFAS), flame retardants, endocrine disruptors, etc., presenting new challenges for scientists, policymakers, and the public. Soil is one of the most important sinks of EOCs arising from industrial emissions, incidental discharge, wastewater irrigation, or atmospheric deposition. The presence of EOCs in soil may threaten the soil ecosystem's health and subsequently transfer to humans via food and water consumption. This book provides comprehensive knowledge on the behavior and translocation of EOCs in soil and the related environmental medium and introduces remediation methods and technologies. From fundamentals to future risks, it explains the importance of understanding EOCs.

Features:

- Provides a comprehensive overview of EOCs in environmental systems.
- Covers comprehensively the physical, chemical, and biological processes of EOCs in soil.
- Addresses innovative remediation and management approaches of EOC-contaminated sites.
- Is the first well-organized book on EOC written by international experts with long engagement in EOC studies.

This book is an excellent foundational text for upper-level undergraduate and graduate students taking courses in soil science, environmental science, environmental chemistry, social ecology, and waste management. It is also essential for those who work with environmental hazards such as environmental engineers, ecologists, environmental professionals, and managers.

Emergent Environmental Pollution

Series Editor:
Jörg Rinklebe
University of Wuppertal, Germany

The nature of pollutants in the environment depends largely on their interactions in the soil, water, air, and organisms. Understanding the fate and transport of pollutants in the environment is essential for humans and animals, plants, water bodies, air, soils, and sediments. This series provides timely information on the biological, chemical, and physical processes governing the interactions of pollutants in the environment. Such knowledge is needed for minimizing the negative impact of pollutants on human health and the health of ecosystems. Books from this series are of value to professionals and researchers, students, industrial and mining engineers, government regulators, authorities, and consultants in the environmental arena.

Phosphate in Soils: Interaction with Micronutrients, Radionuclides and Heavy Metals
Edited by H. Magdi Selim

Permeable Reactive Barrier: Sustainable Groundwater Remediation
Edited by Ravi Naidu and Volker Birke

Nitrate Handbook: Environmental, Agricultural, and Health Effects
Edited by Christos Tsadilas

Vanadium in Soils and Plants
Edited by Jörg Rinklebe

Maritime Accidents and Environmental Pollution – The X-Press Pearl Disaster: Causes, Consequences, and Lessons Learned
Edited by Meththika Vithanage, Ajith Priyal de Alwis, and Deshai Botheju

Methylmercury Accumulation in Rice: Process and Regulation
Edited by Xinbin Feng, Jianxu Wang, and Jörg Rinklebe

Green Technologies for Pollutants Remediation and Recovery from Water and Soil
Edited by Jakub Zdarta, Teofil Jesionowski, and Jörg Rinklebe

Emerging Organic Contaminants in Soil: Properties, Translocation, and Remediation
Edited by Ming Zhang and Mahtab Ahmad

For more information on this series, please visit: https://www.routledge.com/Emergent-Environmental-Pollution/book-series/CRCEEP

Emerging Organic Contaminants in Soil

Properties, Translocation, and Remediation

Edited by

Ming Zhang and Mahtab Ahmad

CRC Press
Taylor & Francis Group
Boca Raton London New York

CRC Press is an imprint of the
Taylor & Francis Group, an **informa** business

Designed cover image: Shutterstock

First edition published 2026
by CRC Press
2385 NW Executive Center Drive, Suite 320, Boca Raton FL 33431

and by CRC Press
4 Park Square, Milton Park, Abingdon, Oxon, OX14 4RN

CRC Press is an imprint of Taylor & Francis Group, LLC

ISBN:9781032834771 (hbk)
ISBN: 9781032834788 (pbk)
ISBN: 9781003509523 (ebk)

DOI: 10.1201/9781003509523

Typeset in Times
by codeMantra

Emerging Organic Contaminants in Soil

Properties, Translocation, and Remediation

Edited by

Ming Zhang and Mahtab Ahmad

CRC Press
Taylor & Francis Group
Boca Raton London New York

CRC Press is an imprint of the
Taylor & Francis Group, an **informa** business

Designed cover image: Shutterstock

First edition published 2026
by CRC Press
2385 NW Executive Center Drive, Suite 320, Boca Raton FL 33431

and by CRC Press
4 Park Square, Milton Park, Abingdon, Oxon, OX14 4RN

CRC Press is an imprint of Taylor & Francis Group, LLC

ISBN:9781032834771 (hbk)
ISBN: 9781032834788 (pbk)
ISBN: 9781003509523 (ebk)

DOI: 10.1201/9781003509523

Typeset in Times
by codeMantra

Contents

Preface.....vii
About the Editors.....viii
List of Contributors.....ix

Chapter 1 Emerging Organic Contaminants.....1

Innocent Tayari Mwizerwa, Mansuur Husein, Shahzad Afzal, Joseline Kateita, Susan Praise, and Muhammad Ibrahim

Chapter 2 Environmental Behaviors of Emerging Organic Contaminants.....32

Yajing Wu, Ming Zhang, and Jianjie Fu

Chapter 3 Occurrence of Emerging Organic Contaminants in Soil.....57

Areej Tahir, Anam Khalid, Zarish Imtiaz, Faiqa Faryal, Mehmood Aslam, Maosen Zhao, and Mujtaba Baqar

Chapter 4 Sorption of Emerging Organic Contaminants (EOCs) in Soil.....90

Qandeel Zainab, Iqra Muhammad, Shahid Mahmood, Muzammil Anjum, Qaiser Hussain, and Azeem Khalid

Chapter 5 Transport of Emerging Organic Contaminants (EOCs) in the Soil-Groundwater System.....107

Noshin Masood, Nisbah Mushtaq, Mahtab Ahmad, and Abida Farooqi

Chapter 6 The Uptake of EOCs from Soil by Plants is Essential for Agricultural Food Safety and Phytoremediation.....130

Saher Shahid and Mahtab Ahmad

Chapter 7 Emerging Organic Contaminants in Soil Biota.....148

Ruwandi Nilakshi, Nalin Madusanka Gunathilaka, Sahan Nishada Perera, Meththika Vithanage, and Anushka Upamali Rajapaksha

Chapter 8 Global Regulations on Emerging Contaminants.............................. 169

Fang Tao

Chapter 9 Physical Remediation of EOCs in Soil.. 198

Qaiser Hussain, Summaya Sajjad, Cheng Liu, Azeem Khalid, Shahzada Sohail Ijaz, Muhammad Akmal, Shahid Mahmood, and Muhammad Ahtisham Ali

Chapter 10 Chemical Remediation of EOCs in Soil..220

Hina Imtiaz, Lena Madden, and Mahtab Ahmad

Chapter 11 Bioremediation and Related Interfacial Processes of EOCs in Soil 240

Dong Zhang, Hongting Zhao, Zhengfeng Hu, and Ming Zhang

Chapter 12 Remediation Approaches and Cases for EOC-Contaminated Soils..268

Yuchi Zhong, Dong Zhang, and Ming Zhang

Index..293

Preface

The lovely planet, where we live from generation to generation, provides us with everything we need to survive. Soil in the terrestrial ecosystem plays an extremely important role in the matter cycling, linking water, air, plants, animals, and microbes. Maintaining the purity, quality, and safety of soil is the guarantee to human beings for what we eat, drink, and breathe.

However, over the past decades, the purity of the soil has been spoiled by the soaring introduction of organic contaminants from the ever-growing industries, which are thriving to meet the demands of human life. Some of these organic contaminants are of emerging concern, named "emerging organic contaminants (EOCs)." These are either newly identified or newly detected contaminants that are toxic or potentially hazardous to the ecosystem. The EOCs include, but are not limited to, pharmaceuticals and personal care products (PPCPs), antibiotics, perfluoroalkyl sulfonate (PFAS), flame retardants, endocrine disruptors, etc., presenting new challenges for scientists, policymakers, and the public. Soil is one of the most important sinks of EOCs arising from industrial emissions, incidental discharge, wastewater irrigation, or even wet and dry atmospheric deposition. The presence of EOCs in soil may threaten the soil ecosystem's health by altering microbial diversity, being taken up by crops/plants, leaching into groundwater, and subsequently transferring to humans via food and water consumption.

This book aims to reveal the complex world of EOCs in the soil. One section of this book explores the occurrence and environmental behavior of EOCs in the soil, especially the sources and translocations of EOCs in the soil system and the factors that influence their distribution. By understanding where these contaminants come from and how they spread, we can better assess their potential risks and develop targeted interventions. Besides, the regulations and standards of EOCs in the soil from different countries and regions are also summarized. Finally, this book addresses the related innovative remediation and restoration technologies for EOCs in the soil, including physical, chemical, and biological approaches. By bringing together the research, policies, technologies, and innovations, we expect to provide comprehensive resources for scientists, students, policymakers, and the public.

This book is co-edited by Prof. Ming Zhang from China Jiliang University and Prof. Mahtab Ahmad from Quaid-i-Azam University. Great thanks to our contributors who shared their expertise and insights in each chapter. We also extend our gratitude to our readers for their acceptance of this book.

Ming Zhang & Mahtab Ahmad
March, 2025

About the Editors

Ming Zhang is a professor in the Department of Environmental Engineering, College of Energy, Environment and Safety Engineering at China Jiliang University. His research interests include soil remediation of organic contaminants, biochar and other novel sorbents for environmental application, and the environmental behaviors of microplastics. He has published over 80 SCI-indexed papers. He was recognized as a Clarivate Highly Cited Researcher in 2022. He was a recognized scientist listed in the World's Top 2% Scientists (2023–2024).

Mahtab Ahmad is an associate professor at Quaid-i-Azam University. He is a recognized scientist listed in the World's Top 2% Scientists. His research encompasses environmental chemistry, remediation of contaminated soils and waters, revitalization of abandoned soils, production and application of biochar from agricultural and solid waste materials, and carbon sequestration. Being actively engaged in research, Dr. Ahmad published research papers in reputed journals. He was awarded the Best Performance Award 2022–2023 as the Best Cited Scientist by Quaid-i-Azam University. He is leading the Waste Management Group working on environment-friendly utilization of waste resources with a team of PhD, MPhil, and BS scholars.

Contributors

Shahzad Afzal
Department of Environmental Engineering
China Jiliang University
Hangzhou, Zhejiang, China

Muhammad Akmal
Institute of Soil and Environmental Sciences
PMAS-Arid Agriculture University
Rawalpindi, Punjab, Pakistan

Muhammad Ahtisham Ali
Institute of Soil and Environmental Sciences
PMAS-Arid Agriculture University
Rawalpindi, Punjab, Pakistan

Muzammil Anjum
Department of Environmental Sciences, Institute of Soil and Environmental Sciences
PMAS Arid Agriculture University
Rawalpindi, Punjab, Pakistan

Mehmood Aslam
MOE Key Laboratory of Pollution Processes and Environmental Criteria, College of Environmental Science and Engineering
Nankai University
Tianjin, China

Mujtaba Baqar
Sustainable Development Study Centre
Government College University
Lahore, Pakistan

Abida Farooqi
Department of Environmental Sciences, Faculty of Biological Sciences
Quaid-i-Azam University
Islamabad, Pakistan

Faiqa Faryal
Institute of Environmental Sciences and Engineering
National University of Sciences and Technology
Islamabad, Pakistan

Jianjie Fu
State Key Laboratory of Environmental Chemistry and Ecotoxicology, Research Center for Eco-Environmental Sciences, Chinese Academy of Sciences, School of Environment, Hangzhou Institute for Advanced Study
University of Chinese Academy of Sciences
Hangzhou, Zhejiang, China

Nalin Madusanka Gunathilaka
Faculty of Graduate Studies
University of Kelaniya
Dalugama, Kelaniya, Sri Lanka

Zhengfeng Hu
Key Laboratory of Environmental Pollution Control Technology Research of Zhejiang Province
Ecological and Environmental Science and Research Institute of Zhejiang Province
Zhejiang Environmental Science & Design Institute
Hangzhou, Zhejiang, China

Mansuur Husein
School of the Environment and Safety Engineering
Jiangsu University
Zhenjiang, Jiangsu, China

Qaiser Hussain
Institute of Soil and Environmental Sciences
PMAS Arid Agriculture University
Rawalpindi, Punjab, Pakistan

Muhammad Ibrahim
College of Environment
Hohai University
Nanjing, Jiangsu, China

Shahzada Sohail Ijaz
Institute of Soil and Environmental Sciences
PMAS-Arid Agriculture University
Rawalpindi, Punjab, Pakistan

Hina Imtiaz
Department of Environmental Sciences, Faculty of Biological Sciences
Quaid-i-Azam University
Islamabad, Pakistan
and
Technological University of the Shannon
Shannon ABC
Limerick, Ireland

Zarish Imtiaz
College of Earth and Environmental Sciences
University of the Punjab
Lahore, Pakistan

Joseline Kateita
National Water & Sewerage Corporation
Kampala, Uganda

Anam Khalid
College of Earth and Environmental Sciences
University of the Punjab
Lahore, Pakistan

Azeem Khalid
Department of Environmental Sciences, Institute of Soil and Environmental Sciences
PMAS Arid Agriculture University
Rawalpindi, Punjab, Pakistan

Cheng Liu
Institute of Eco-Environmental Research
Zhejiang University of Science and Technology
Hangzhou, Zhejiang, China

Lena Madden
Technological University of the Shannon
Shannon ABC
Limerick, Ireland

Shahid Mahmood
Department of Environmental Sciences, Institute of Soil and Environmental Sciences
PMAS Arid Agriculture University
Rawalpindi, Punjab, Pakistan

Noshin Masood
Department of Environmental Sciences, Faculty of Biological Sciences
Quaid-i-Azam University
Islamabad, Pakistan

Iqra Muhammad
Department of Environmental Sciences, Institute of Soil and Environmental Sciences
PMAS Arid Agriculture University
Rawalpindi, Punjab, Pakistan

Nisbah Mushtaq
Department of Environmental Sciences, Faculty of Biological Sciences
Quaid-i-Azam University
Islamabad, Pakistan

Innocent Tayari Mwizerwa
College of Environmental Science & Engineering
Donghua University
Shanghai, China

Ruwandi Nilakshi
Faculty of Graduate Studies
University of Kelaniya
Dalugama, Kelaniya, Sri Lanka

Sahan Nishada Perera
Faculty of Graduate Studies
University of Kelaniya
Dalugama, Kelaniya, Sri Lanka

Susan Praise
Faculty of Agriculture
Yamagata University
Tsuruoka, Yamagata, Japan

Anushka Upamali Rajapaksha
Ecosphere Resilience Research Centre & Instrument Center, Faculty of Applied Sciences
University of Sri Jayewardenepura
Nugegoda, Sri Lanka

Summaya Sajjad
Institute of Soil and Environmental Sciences
PMAS-Arid Agriculture University
Rawalpindi, Punjab, Pakistan

Saher Shahid
Department of Environmental Sciences, Faculty of Biological Sciences
Quaid-i-Azam University
Islamabad, Pakistan

Areej Tahir
College of Earth and Environmental Sciences
University of the Punjab
Lahore, Pakistan

Fang Tao
Department of Environmental Engineering
China Jiliang University
Hangzhou, Zhejiang, China

Meththika Vithanage
Ecosphere Resilience Research Centre, Faculty of Applied Sciences
University of Sri Jayewardenepura
Nugegoda, Sri Lanka

Yajing Wu
Chinese Academy of Sciences, School of Environment, Hangzhou Institute for Advanced Study
University of Chinese Academy of Sciences
Hangzhou, Zhejiang, China

Qandeel Zainab
Department of Environmental Sciences, Institute of Soil and Environmental Sciences
PMAS Arid Agriculture University
Rawalpindi, Punjab, Pakistan

Dong Zhang
Department of Environmental Engineering
College of Materials and Environmental Engineering
Hangzhou Dianzi University
Hangzhou, Zhejiang, China

Hongting Zhao
Department of Environmental Engineering
College of Materials and Environmental Engineering
Hangzhou Dianzi University
Hangzhou, Zhejiang, China

Maosen Zhao
MOE Key Laboratory of Pollution Processes and Environmental Criteria, College of Environmental Science and Engineering
Nankai University
Tianjin, China

Yuchi Zhong
Zhejiang Taohuayuan Environmental Protection Technology Co. Ltd
Hangzhou, Zhejiang, China

1 Emerging Organic Contaminants

Innocent Tayari Mwizerwa, Mansuur Husein, Shahzad Afzal, Joseline Kateita, Susan Praise, and Muhammad Ibrahim

1.1 INTRODUCTION

With an increasing concern about environmental pollution emanating from various anthropogenic activities and natural processes, discovering and understanding emerging organic contaminants (EOCs) have become crucial for safeguarding public health and ecological systems. EOCs are diverse chemicals detected in the environment but are not usually monitored or regulated. They include pharmaceuticals, personal care products, industrial chemicals, and agricultural compounds (Kumar, 2023). EOCs are anthropogenic or natural substances discharged from both point and nonpoint sources and end up in environmental spheres at small concentrations that cause adverse impacts on flora and fauna. A broad category of substances are found in the environment but not frequently controlled or monitored are known as EOCs (Dhangar and Kumar, 2020). Pharmaceuticals, personal care products, industrial chemicals, and agricultural substances, among others, are a few examples of these pollutants as presented in Table 1.1 which might pose threats to human health as well as ecosystems. The EOCs are manufactured or naturally occurring compounds released into the environment in low quantities that harm plant and animal life. They are released from point and nonpoint sources. Their presence in environmental samples signals a pressing need for accurate toxicity testing, understanding of degradation pathways, and assessing their environmental impact (Rodrigues, 2022). Therefore, with their capacity to persist in the environment and potentially disrupt endocrine systems among humans and wildlife, studying these organic contaminants signifies an essential advancement in environmental science and public health policy.

EOCs are not widely monitored and measured, and as a result, they have negative consequences for various ecosystems. Pharmaceuticals, personal care products, detergents, steroid hormones, industrial chemicals, pesticides, and other contaminants end up in different ecosystems, thus causing quality concerns. Even though the impacts of EOCs in different ecosystems are not well-appraised there are significant indicators regarding their acute and chronic effects on ecosystems. Therefore, efforts to promote soil and water quality from contamination have been spearheaded through technical advancements for convenient use, sustainability, and economic feasibility. Furthermore, there is little knowledge about such contaminant accumulation in the

DOI: 10.1201/9781003509523-1

TABLE 1.1
Major Types of Emerging Organic Contaminants, Uses, and Examples

Emerging Organic Contaminant	Examples	Origin	Characteristics
Pharmaceuticals (Loganathan et al., 2023)	Hormones and antibiotics	Human and veterinary drugs, such as contraceptives, are used in aquaculture and released from animals at pasture.	i. The parent neutral compound and associated salts form a polymorphic solid ii. Enter into the ecosystems and then into human metabolism iii. Have large, chemically complex structures iv. Ionizable with active sites around the whole molecule
Persistent organic pollutants (POPs) (Kumar et al., 2023)	Insecticides, pesticides, polychlorinated biphenyls (PCBs), PFAS, polyhaogenated carbazoles (PHCZ)	Firefighting retardants and cleaning foams	v. Toxic; persistent, resists normal processes that break down contaminants vi. They accumulate in living organisms
Personal care products (Loganathan et al., 2023)	Dyes, shampoos, soaps, toothpaste, shower gels, lotions, perfumes, and makeup.	Beauty salons and laundry wastewaters	vii. Body irritation, hormonal disruption, viii. Environmentally free radicals ix. Aquatic animals' migration x. They accumulate in crop tissues
Endocrine-disrupting compounds (EDCs) (Lin et al., 2020)	BPA, phthalates, phenol, brominated flame retardants, PCBs, phthalates, parabens, triclosan	Dyes, hormones, foams, laboratory chemicals	xi. Activates hormone receptors; xii. Antagonizes hormone receptors; xiii. Alters hormone receptor expression; xiv. Changes in signal transduction among hormone-responsive cells xv. Triggers epigenetic in hormone-production xvi. Change hormone synthesis xvii. Disrupts hormone transport xviii. Interrupts hormone distribution or circulating levels of hormones xix. Hormonal metabolism disruption xx. Changes the fate of hormone-producing or hormone-responsive cells.

soil, particularly after a long period of partially treated wastewater for irrigation and uptake, including bioaccumulation of contaminants triggered by the plant tissues (Kathi and El Din Mahmoud, 2024).

EOCs can be pharmaceuticals, personal care products, pesticides, veterinary medicines, industrial products/by-products, food additives, and engineered nanomaterials. These enter the soil ecosystem via landfill sites through percolation into the soil layers. The generation and disposal of animal and domestic waste, hospital and industrial effluent, septic tanks, and wastewater treatment plants also contribute to the sources of EOCs. They are thus associated with side effects such as endocrine disruption, attributed to exposure to sewage sludge and sewage sludge-amended soils during sewage management and through the food web and farming activities (Loganathan et al., 2023). This has triggered a couple of scientific communities tailored to dive into studies to discover the trends of emerging contaminant concerns. The findings have been reported in scientific journals and academic publications in different countries for a considerable time such as that portrayed in Figure 1.1, from (1999 to 2024), in the illustrations. Thus this book chapter gives a snapshot of EOCs and further highlights how they disseminate to other ecosystems, including their effects on mankind's health and the ecosystem safety.

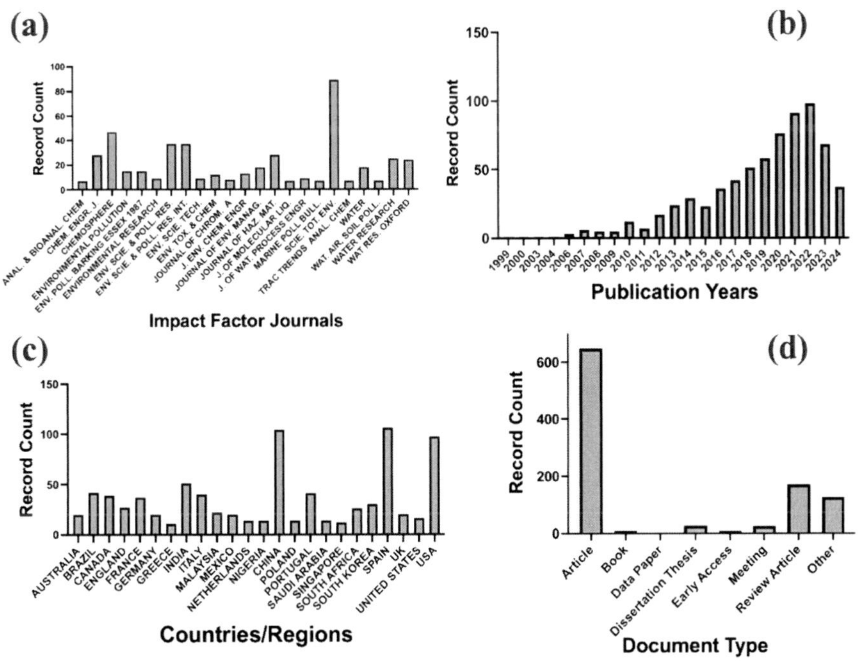

FIGURE 1.1 Publication trends of emerging organic contaminants (EOCs). (a) The number of annual impact factors; (b) trend of publications year in subject of EOCs as the keywords; (c) publishing country; (d) published articles, review articles, and proceedings/editorial materials per year (retrieved on December 30, 2024), sourced from the search engine of Web of Science on annual publications.

1.2 SOURCES AND PATHWAYS OF EOCs

EOCs originate from various sources that can be divided into natural and anthropogenic origins (Cerar, 2016). Their environmental burden horizon has been presented globally distributed in different quantities as represented in Figure 1.2. Each type of EOC has unique properties and ways of interacting with the environment, making their study critical for assessing the full scope of their impact. These contaminants' persistence and potential toxicity necessitate ongoing research to understand their behavior, effects, and environmental degradation pathways. This section emphasizes human activities that contribute significantly to the occurrence of EOCs in ecosystems. They include the following;

a. **Industrial Effluents:** Fabrication plants, hospital establishments, and food and beverage processing plants, among others, discharge waste containing EOCs into aquatic systems. These point-source pollutants have spatially discrete impacts but are significant due to the concentrated nature of their emissions.
b. **Municipal Sewage Treatment Plants and Integrated Sewage-Storm-Water Overflows:** Municipal sewage and integrated sewage storm wastewater contribute to the EOCs. This is attributed to inefficient treatment

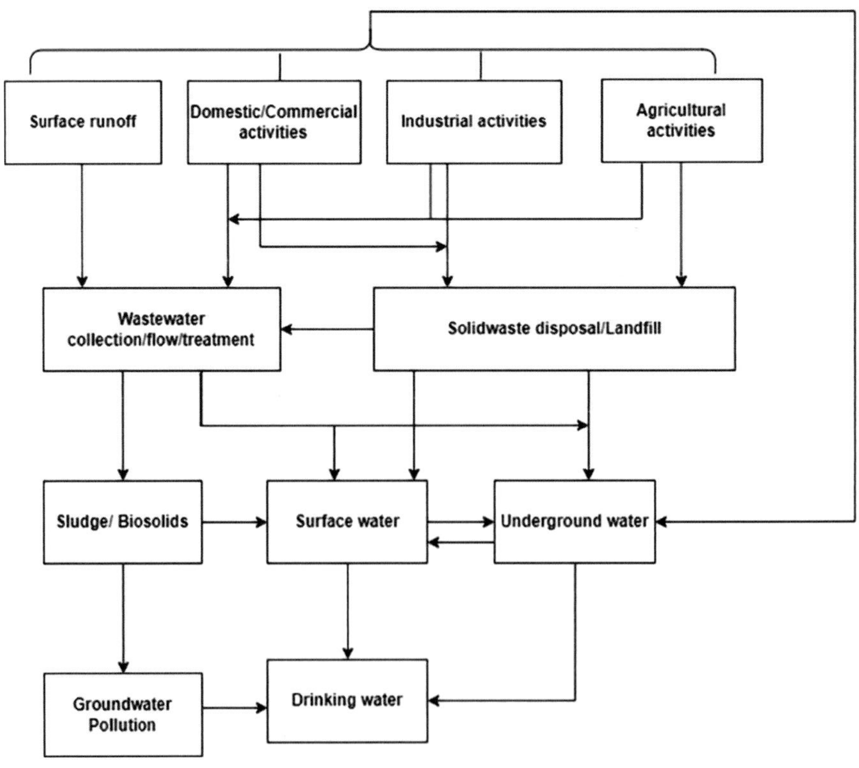

FIGURE 1.2 Sources and pathways of EOCs.

systems and careless discharge of solid and liquid wastes, especially from landfills and domestic sanitation facilities. Such contaminants may include hormones and endocrine disruptors, which end up in the food chain through drinking water and eating food items. Therefore, these facilities are critical points of concern as they handle large volumes of waste, including EOCs, which can sometimes bypass treatment processes and enter the natural water cycle.

c. **Resource Extraction and Waste Disposal Sites:** Economic activities, such as mining and the operation of landfill sites and industrial impoundments, contribute to EOC contamination. These activities leave tailing and residuals in the ecosystem, thus affecting the safety of both biota and aquation resources. Notably, older landfill sites continue to impact groundwater due to historical practices that were less stringent than current standards, especially when they settle in the groundwater aquifers, threatening both aquatic and human mankind (Mwizerwa et al., 2020a, b).
d. **Septic Systems:** The construction of septic tanks presents another critical significant source of groundwater quality deterioration by EOCs, mainly in both rural and urban areas set up homesteads, particularly in regions with high water tables and permeable aquifer (Li et al., 2021a). This is because the soil layers are not stable and easily broken because of the poor soil structure of such treatment facilities. Therefore, the diffusing nature of this pollution makes it challenging to monitor and manage.
e. **Agricultural Runoff:** Using veterinary antibiotics and hormones, for example, in the sex reversal in fish farming and aquaculture hatcheries, as well used in the seed protection against spoilage in store to increase the shelf lives, acaricides, and herbicides. All these leave residues in the aquatic and biotic ecosystems. Therefore, these concentrated animal feeding operations, manure, and biosolids on agricultural land are prominent sources of EOCs. These practices contribute to the deterioration of soil quality because of inefficient removal during soil and wastewater treatment.

1.2.1 Pharmaceutical and Personal Care Products

These are global consumables for humans, termed pharmaceutical and personal care products (PPCPs), for health care and beautification. The PPCPs are increasingly popular in environmental compartments where the traditional wastewater treatment facilities cannot effectively remove them after they discharge into the natural environment in small doses, thus gradually affecting the soil. The emerging contaminants can also be discharged into aquatic and soil environments through waste from animal farm yards and sewage treatment facilities. Sewage sludge generated from wastewater treatment plants is usually applied in agriculture as a fertilizer. Nevertheless, newly discovered pollutants may also be present in treated sewage sludge, including chemicals, pharmaceuticals, and engineering nanomaterials (Chen et al., 2024; Koumaki et al., 2021; Loganathan et al., 2023). PPCPs are among the newly discovered toxins in sewage sludge which, if released into ecosystems, could negatively affect the environment (Van et al., 2021).

There is limited knowledge about the likely environmental effects of PPCPs (Vasilachi et al., 2021). Other sources of PPCPs in the soil have been discovered, such as inefficiently treated wastewater, surface run-offs, or contaminated river water for irrigation (Gallego et al., 2021). Soils are usually contaminated with EOCs with low hydrophobicity, which are then accumulated in the soil via organic substance interactions (Beltrán et al., 2020). Similarly, research in Spain on 166 emerging pollutants and heavy metals (such as Cd, Ni, Pb, and Hg) discovered 38 pharmaceuticals in wastewater that had undergone tertiary treatment, albeit in minimal amounts. These results show that wastewater treatment technologies cannot effectively remove some developing pollutants. The primary cause of the recent detection of EOCs in the water bodies utilized to discharge treated wastewater is the ineffective attenuation of these contaminants.

The study carried out by Bai and colleagues clarified the condition of emerging contaminants in surface water located in Denver and Colorado (Bai, 2018). This investigation covered sampling points where the discharge from many wastewater treatment plants dispersed around the area and snow melting in the springtime affected surface water flow. The results revealed that 76% of detected pharmaceuticals were registered in water samples. Notably, despite the treated wastewater being significantly diluted by the naturally flowing water, a high percentage of emerging pollutants has been detected in water samples. Similarly, Petrie et al. provided a comprehensive overview of the levels of various chemical contaminants in surface water and influent and effluent water from UK wastewater treatment plants (Petrie et al., 2015). In surface water samples exposed to treated wastewater containing high quantities of effluent with a discernible pattern of newly detected pollutants.

The term PPCPs is defined as a wide variety of products consumers use for their health care and appearance. These include but are not limited to toothpaste, skincare products, fragrances, antibiotics, and various veterinary medications utilized in the agroindustry. These PPCPs are becoming increasingly prevalent in areas where conventional wastewater treatment facilities cannot adequately remove them after they are released into the environment in tiny amounts, progressively altering the soil ecosystem. Animal farms and sewage treatment plants' effluent can potentially release emerging pollutants into soil and aquatic environments. Agriculture frequently uses wastewater treatment plant sludge as a fertilizer. For instance, potassium, nitrogen, manganese, and iron are among its many nutrients. However, treated sewage sludge may also comprise a range of recently identified contaminants, such as chemicals, medications, and engineering nanomaterials. One of the recently identified toxins in sewage sludge is PPCPs, which could harm humans or the environment if discharged into ecosystems.

PPCPs are regarded as EOCs of concern because there is limited knowledge about the potential environmental effects of these substances. Treated wastewater and polluted river water used for irrigation are the other sources of PPCPs in the soil. Low hydrophobicity chemicals of emerging concern (CECs) are frequently found contaminating soils. Then, through interactions with organic materials, they might accumulate in the soil and can end up in other aquatic ecosystems. For instance, in studies targeting emerging contaminants and heavy metals (including cadmium, nickel, lead, and mercury) carried out in Spain, 38 pharmaceuticals were found, albeit in minimal concentrations, in wastewater that had received tertiary treatment. These findings demonstrate that wastewater treatment systems cannot successfully remove some emerging contaminants. These are especially pharmaceuticals, among others.

1.2.2 Pesticides

A pesticide is a substance or chemical that kills or protects crops from damage. For instance, pesticides are mainly used for controlling/eradicating plant pathogens. Some are also toxic to humans and other living organisms, such as soil fauna or microorganisms, when they enter the food web or are exposed to human life. It has been investigated that the surface water of 38 streams in the US had the top 10 most commonly detected anthropogenic contaminants. These included eight pesticides such as 2-chloro-4-isopropylamino-6-amino-*s*-triazine, chlorpyrifos, aminomethylphosphonic acid, metolachlor, dieldrin, atrazine, de-sulfinyl fipronil, and glyphosate from degradation of aminomethylphosphonic acid from farming and household refuse, as well as two pharmaceutical drugs from caffeine and metformin containing elevated detection rates. Therefore, pesticides are among the most persistent chemicals in soil. Their excessive use can contribute to soil microorganism reduction. The use of pesticides is significantly increasing in some regions despite being banned. Examples of such pollutants are the organochlorine pesticides, also called OCPs. They have been cited to elevate toxicity, bioaccumulation, and environmental biomagnification.

1.2.3 Phthalate Esters

Phthalate esters (PAEs) are among the constituents of domestic consumables as additives to enhance the flexibility of specific polyvinyl chloride resins, cellulose, vinyl acetate, and polyurethanes in industrial manufacturing. PAEs can also be utilized as enteric coating agents for various applications. It is worth noting that PAEs are poorly water-soluble chemicals that can end up in the soil horizons. Although PAEs are sparingly water-soluble, they can be readily absorbed by organic residue and solid surfaces. Owing to the extensive use of PAEs, their ubiquity can lead to the contamination of food chains.

For example, high levels of soil pollutants are PAEs from their source materials via evaporation, leaching, and abrasion. It is, therefore, paramount to note that phthalates are transformed from the soils to plants. However, they are at significantly lower concentrations than the recommended soil remediation efforts in some regions. For example, the concentration of phthalates in top soils during winter wheat harvest was extremely high compared with summer. This evidenced that wet deposition scavenges the phthalates more smoothly from their source materials.

Furthermore, phthalates penetrate soil horizons using agricultural plastic films, irrigation, sewage sludge, organic fertilizer applications, and biosolids. Additionally, e-waste can pose significant soil pollution due to the chlorinated chemicals, especially polychlorinated biphenyls (PCBs), used to fabricate it. Although stringent regulations have been advocated and bans have been imposed, biphenyls remain notorious contaminants in soils and sediments.

PAEs are found in some household commodities. They are also utilized in cellulose, vinyl acetate, and polyurethanes, among other resins. In various applications, PAEs can also be used as enteric coating agents. As PAEs are weakly soluble in water, they may enter the soil horizons. Another essential aspect that affects a chemical's aquatic toxicity and biodegradability is its solubility in water. It has an impact on these compounds' distribution as well. Despite their recognized limited solubility in water, PAEs are readily absorbed by organic waste and solid surfaces. The slow and steady buildup and leakage of these substances may impact the natural conditions of water systems. Contamination of plants and food chains may result from the buildup of PAEs in agricultural soils. Additionally, it might expose people directly or indirectly.

For instance, the extensive use of materials such as plastic products, pesticides, cosmetics, personal care products, building materials, automotive parts, medical devices, food packaging, toys, teethers, adhesives, paints, and lubricants is attributed to high levels of additives in soil pollutants, such as PAEs. According to research, soil pollution is caused by the materials' transportation from the source by evaporation, leaching, and abrasion. It is crucial to understand that phthalates are transferred from soil to crop plants, even if their concentrations are far lower than those at which soil remediation activities are advised to reach threshold limit values, over which the soil becomes contaminated—seasonal differences in the soil's phthalate deposition. For instance, phthalate concentration levels in top soils after the winter wheat harvest were much more significant than during the summer wheat harvest. This demonstrated that the phthalates are more easily scavenged by wet deposition from the source materials responsible for their transportation into soil media. Moreover, sewage sludge, irrigation, biosolids, organic fertilizer applications, and agricultural plastic films allow phthalates to seep into soil horizons. Furthermore, because they are made of chlorinated chemical class materials, e-waste can seriously pollute soil. Biphenyls remain well-known pollutants in soils and sediments despite the advocacy for strict controls and a ban implementation.

1.2.4 Hormones

Worldwide, there has been a revolutionary use of hormones to boost agriculture, aquaculture, and fisheries productivity among others. These hormones are used in sex reversal in fish breeding or genetic engineering in animal production. For example, during worldwide industrial growth, steroidal estrogen was regarded as an emergent problem. Many emerging contaminants, including synthetic or natural hormones, are referred to as hormone disrupters. It has been estimated that the human population discharges tons of natural steroidal estrogens and synthetic estrogens solely contraceptives each year. The release of estrogens from livestock is also estimated to be relatively high. Similarly, natural estrogens discharged from animal and human wastes are also a severe environmental threat. This is alarming since animal manure has been widely adopted in modern agriculture.

Using hormones to increase productivity in aquaculture, agriculture, and fisheries has proven revolutionary worldwide. These hormones are employed in genetic engineering for the creation of animals or in fish breeding for sex reversal. Since steroidal estrogen has been seen as an emerging problem due to the world's industrial growth, it has been established that it negatively impacts soil fertility and aquatic life. The adrenal cortex and other regions of the human and animal bodies release steroidal hormones, which are manufactured or naturally occurring variants of estrogen. Hormone disruptors are a group of new pollutants that include both natural and synthetic hormones. The use of birth control pills is the only source of the tons of manufactured and natural steroidal estrogens that the human population releases annually. Similarly, cattle can release a significant number of estrogens. The estimated annual output of estrogens is higher in the United States and the European Union. Likewise, natural estrogens released from human and animal waste are significant environmental hazards, especially in the soil ecosystem. This environmental issue is particularly concerning since organic farming increasingly relies on biosolids.

1.2.5 Perfluorinated Compounds

Perfluorinated compounds (PFCs) have various functional groups, such as perfluoroalkyl and perfluorocarboxylic acids. Per- and polyfluoroalkyl substances (PFAS) are anthropogenic chemicals of worldwide attention. This is because most of them are believed to be bioaccumulative in aquatic species and plants. The wide use of these substances ends up in the soil, which later enters the food web, resulting in bioaccumulation. The data used to incorporate exposure via the ingestion of fish and shellfish by humans and aquatic-dependent species when developing ambient water quality criteria for PFASs have also been recorded.

The data quality, gaps, and limits in bioaccumulation data were assessed. Carbonyl and sulfonyl PFAS classes have a lot of data, but phosphate, fluorotelomer, and other PFAS classes have very little evidence for fish and none for most other taxonomic taxa. The PFCs are generally used in textile, pesticide, and refrigeration industries. The Perfluorinated alkyl acids are believed to contaminate the soil in many ways, such as when the reclaimed wastewater is used for irrigation or when biosolids are added as a fertilizer. In the United States, around 60% of the land used for farming is devoted to applying these materials that inflict biosolids into the environment as residues. Unfortunately, there are no regulations for perfluoroalkyl acids (PFAAs) in bio solid discharge and use in the agriculture sector, thus leading to soil ecosystem contamination.

It is known that PFCs contain a variety of functional groups, including perfluoroalkyl and perfluorocarboxylic acids. Worldwide attention is focused on anthropogenic chemicals known as PFAS. This is because most of them bio accumulate in aquatic and plant species. The extensive usage of these materials leaves them in the soil, where they eventually find their way into the food chain and cause bioaccumulation. When creating ambient water quality criteria for PFASs, the data are also utilized to account for exposure through human consumption of fish and shellfish and aquatic-dependent species. Gaps and limitations in the bioaccumulation data and the overall quality of the data were evaluated. The phosphate, fluorotelomer, and ether PFAS classes contain little evidence for fish and none for most other taxonomic groups.

In contrast, the carbonyl and sulfonyl PFAS classes have abundant data. Because of their strong surface activity, resilience to chemicals, and heat in many different industries, including textiles, insecticides, and refrigeration, PFCs are often employed as industrial chemicals. One kind of perfluoroalkyl carboxylic acid that persists in the environment is PFAA. PFAAs are regarded as a potentially dangerous new contaminants that risk the environment and human health.

PFAAs can contaminate soil through various means, such as using reclaimed wastewater for irrigation or applying biosolids as fertilizer. However, biosolids, organic materials derived from treated sewage sludge, are typically processed according to governmental regulations. For example, studies have indicated that agricultural land utilizes these materials, introducing biosolids into the environment. Unfortunately, there are no specific regulations for PFAAs in the discharge and use of biosolids in agriculture, leading to soil contamination. This implies that the repeated application of biosolids poses a risk of potential soil contamination.

1.2.6 Micro- and Nano-Plastics

Materials known to range in size from 0.001 to 1.0 mm and in length from less than 1.0 mm are employed in the industrial sector as micro- and nano-plastics (MNPs), respectively. The terrestrial pollution caused by MNPs has received little attention. Similarly, two further categories of plastics have also surfaced: bioplastics and Oxo-degradable plastics. However, these new plastics are less environmentally beneficial than anticipated. A significant amount of plastic waste has accumulated in landfills. These ultimately wind up in soil environment compartments, exacerbating the pollution linked to the severity of MNPs in agricultural soils. The examination of processed food samples revealed that adults and children are exposed to an average of particles annually through ingestion and inhalation. Thus, MNP hotspots worldwide are emerging in agricultural areas.

1.2.7 Novel Flame Retardants

Polybrominated diphenyl ethers (PBDEs) are flame retardants (FRs). The most widely used PBDE worldwide is c-decay BDE, a BDE-209 congener. For example, the BDE-209 congeners are among the intensively prevailed PBDEs in soils. There were no apparent variations in the concentrations of PBDEs in soils posing significant threats via the trophic transfer of PBDEs for food quality. This included BDE47 through food, significantly higher than the reference dose. The organophosphorus esters (OPEs)-based FRs have replaced the brominated FRs (BFRs). But, no differences were determined between the environmental risks BFRs and OPEs pose. For example, deplorably, tris(2-chloroethyl) phosphate (TCEP), an FR, is ubiquitous in plasticizers, preservatives, and fragrances (Maddela et al., 2020), which are linked to the highly volatilized TCEP.

Notably, the burden of EOCs in soils, especially perfluorooctane sulfonate (PFOS), is dominant in Norway and the USA, with the least in Uganda, China, and Antarctica. These are in origin like additives, which are most prevalent in Kenya, with the least prevalent in China, Turkey, Ethiopia, and the USA. Chile and China

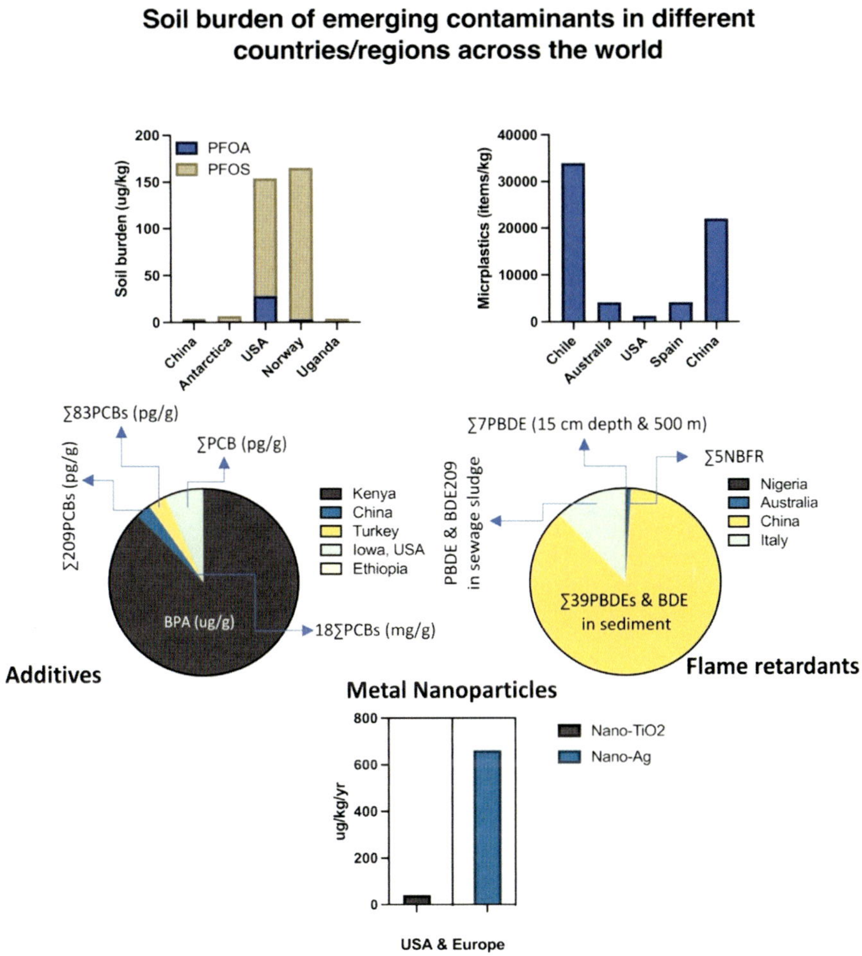

FIGURE 1.3 Burden of EOCs in the soil of different countries/regions.

presented high microplastics with the least in the USA. Finally, FRs were prevalent in China, followed by Italy, and had less value in Nigeria and Australia. The prevalence of EOCs in the soil of different regions has been summarized in Fig. 1.3 from the above-mentioned literature in section 1.2 and Oppeltová et al. (2024).

1.3 PROPERTIES OF EOCs

Properties of emerging contaminants are paramount in the detection, analysis, remediation and degradation in different matrices. EOCs are naturally hydrophobic with various degrees of polarity and different physicochemical properties such as pK_a, water solubility, partition coefficient, and vapor pressure (Leung et al., 2023). Understanding partition coefficients is essential to elucidate the chemical transfer in the environment and critical properties of vapor pressure (P)

TABLE 1.2
Some Properties of Emerging Organic Contaminants in Soil Samples

Contaminant	pK_a Values	Solubility (mg/L)	Neutral log K_{ow}	log K_{oa}	log K_{aw}	K_{oc} (L/kg)	f_n
Carbamazepine (CBZ)	NA	112	2.25	10.81	−8.36	168.6	1.000
Diethyl phthalate (DEP)	NA	1080	2.65	7.02	−4.60	135.70	1.000
Ethyl paraben (EPB)	8.5 [0/−]	885	2.49	9.18	6.71	246.9	0.666
5-Methyl-1Hbenzotriazole (5-TTri)	0.77 [+/0] 8.85 [0/−]	3069	1.71	6.89	−5.18	87.87	0.817
Primidone (PMD)	2.36 [+/0] 3.94 [0/−] 5.42 [−/2−]	500	0.73	9.01	−8.10	23.84	0.003
Surfynol 104 (S104)	13.15 [0/−] 13.83 [−/2−]	26.35	3.61	8.61	−5.00	125.60	1.000
Tris(2-chloroethyl) phosphate (TCP)	NA	7000	1.63	5.31	−3.87	66.83	1.000
Bisphenol A (BPA)	9.8 [0/−] 10.4 [−/2−]	120	3.64	12.75	−9.43	1245	0.974

Source: Compiled by authors from references (Maddela et al., 2020, Li et al., 2021, Leung et al., 2023). [0], neutral; [+], cationic; [−], anionic; NA, not applicable; K_{ow} (L/L), the partition coefficient octanol to water for the neutral molecule, K_{oa} (L/L), the partition coefficient octanol to air, K_{aw} (L/L), the partition coefficient air to water for neutral molecules, K_{oc} (L/kg), the soil organic carbon-water partition coefficient, f_n, the neutral fraction at soil pH (8.1), pK_a, dissociation reaction (the values are reported using the EPI-Chemaxon software).

and solubility of a compound (S). It is crucial to determine the partition coefficients to reveal a chemical's fate, including air/water (K_{aw}), n-octanol/water (K_{ow}), and n-octanol/air (K_{oa}). Therefore, a compound's aqueous solubility and vapor pressure determine the partitioning of a chemical from pure liquid or solid into the air and soil ecosystem (Table 1.2).

1.4 POTENTIAL ENVIRONMENTAL TOXICITY AND HEALTH IMPACTS OF EOCs

Environmental toxicity and health impacts of the already detected emerging contaminants have grown exponentially in both mankind and ecosystems (Morin-Crini et al., 2022; Yadav et al., 2021). This is majorly because they easily persist in groundwater due to groundwater's relatively long residence time and structural arrangement. This persistence threatens freshwater bodies and necessitates a comprehensive understanding of their spatial and temporal variations. Therefore, the scientific community and policymakers design appropriate guidelines to regulate their distribution and removal from the production chains of products and materials, especially consumables (Gibson et al., 2023). This can also include regulations on discharge and strict and efficient treatment systems.

1.4.1 Effects on Water Quality

The discharge of EOCs has vast effects on water quality. The previously mentioned pharmaceuticals, including antibiotics, personal care products such as dyes, and other industrial chemicals, are continuously introduced into aquatic environments (Häder et al., 2020). Their presence is often pseudo-persistent, maintained by constant inputs from sources such as sewage treatment plants. This steady-state concentration of EOCs in water bodies can disrupt the ecological balance, affecting aquatic and terrestrial biota. Long-term exposure to these contaminants and their metabolites can lead to significant environmental risks, such as disruption of the endocrine systems of organisms and changes in microbiological resistance, which can cascade up the food chain, causing biomagnification of micro pollutants and adversely impacting organisms at higher trophic levels (Babuji et al., 2023).

1.4.2 Effects on Soil and Air

The impact of EOCs extends beyond water quality to soil and air. PPCPs, often introduced into the soil segments through irrigation with treated effluents or reclaimed water, gradually accumulate, contaminating soil and groundwater. This contamination can inhibit the growth and biomass of crops, affecting various plant functions such as stem, root, and shoot growth. It can also disrupt the activity and structure of soil microorganisms. The unique characteristics of these contaminants, such as polarity and semi-volatility, make them particularly challenging to manage and remediate. Additionally, the presence of EOCs in the vadose zones and soils has been a concern for several decades. Therefore, it highlights the need for targeted attention to their environmental behavior and their potential risks to soil health and air quality.

The environmental impact of EOCs is a pressing issue, primarily driven by human activities that introduce these chemicals into various ecosystems. As more efforts continue to uncover the broad scope of their effects, it becomes increasingly important to develop strategies to mitigate their presence and to protect human health and the stability of environmental systems (Jain et al., 2023).

1.4.3 Human Health Risks Associated with EOCs

EOCs' human health risks are classified according to the degree of the effect they inflict on human beings as follows:

Short-Term Health Effects: Exposure to specific EOCs can lead to acute health effects, particularly impacting respiratory and cardiovascular systems. For instance, air pollution, a carrier of various organic pollutants, has been linked to exacerbations of respiratory conditions and cardiovascular diseases. The causal relationship between air pollution and these health outcomes is well-documented, with acute exposure potentially increasing susceptibility to infections and other short-term health issues (Roche et al., 2024). Specific contaminants such as methyl tert-butyl ether have been associated with symptoms such as sicchasia, headaches, and nasal stimulation, reflecting the direct impact of short-term exposure to EOCs.

Long-Term Health Effects: The long-term health risks associated with EOCs are significant and include various forms of cancer and reproductive issues. For example, exposure to ethylene oxide (EtO) has been recorded as a risk factor for breast cancer and various types of leukemia. Workers in facilities that use or produce EtO, as well as residents living near such facilities, are at high risk, with a potential for 100 or more cases of cancer per million exposed over a lifetime (Jones et al., 2023). Furthermore, PFOS and PFOA have been studied for their potential to reduce fecundity and increase the risk of hormone-related cancers, such as breast cancer, in specific populations (Chen et al., 2023; Wee and Aris, 2023). Long-term exposure to trihalomethanes, a byproduct of water chlorination, has been associated with increased risks of bladder and colon cancer. The persistent nature of these contaminants in the environment and their ability to bioaccumulate pose ongoing risks to human health, necessitating continuous monitoring and rigorous health risk assessments to manage their impact effectively (Li et al., 2024).

1.4.4 PFAS Toxicity

PFAS are a group of synthetic chemicals used in industry and consumer products (Chen et al., 2023). Animal study experiments have revealed PFASs with a high possibility of endocrine-disrupting effects (DeWitt et al., 2019). For example, the exposure of zebrafish embryos to PFOSs led to a reduced islet area structure of tissues; however, the impact of PFOSs was neutralized at higher concentrations (Chen et al., 2014; Jantzen et al., 2016). In human pluripotent stem cells, PFASs disrupt the creation of human pancreatic progenitor cells and affect pancreatic and endocrine differentiation. Pancreatic progenitor cells were exposed to low, human-relevant doses of PFOA and PFOS (Liu et al., 2019), which evidenced their endocrine toxicity. Health risks are extreme when there is chronic exposure to PFASs due to the mean estimated half-lives of PFOS and PFOA in the human body being relatively higher.

1.4.5 PFAS Carcinogenic Risks

PFAS exposure has exponentially increased, especially in consumer materials; its footprint is enormous. These phenomena have suggested that exposure to PFOS and PFOA increases the different forms of cancer risk (Chang et al., 2024; Chen et al., 2023; Fenton et al., 2021). It has been pinpointed that there are frequent cases of mortality from prostate cancer among workers in the PFOA industry. However, the cohort analysis could not show any connection between exposure and cancer risk. A recent occupational cohort study carried out in Italy suggested that the individuals exposed to PFOA showed a seven times higher mortality from liver cancer and fivefold higher lymphatic and hematopoietic cancer mortality than nonchemical factory staff. Many of these studies have demerits, such as differences in data methods, and sample size. Yet, there is a substantial correlation between PFAS exposure and cancer signs and symptoms (kidney, prostate, and testicular) in humans. High concentrations of PFOA escalate pancreatic tumors. Additionally, children with germ cell tumors presented relatively higher serum levels of PFAS than the tumor-free pediatric patients (Lin et al., 2020). Furthermore, the International Agency for Research on Cancer classified PFOA as the widely investigated PFAS. This also reported a high possibility of human carcinogens based on inadequate epidemiological results (van Gerwen et al., 2023). PFAS are capable of carcinogenicity and presents a substantial maternal link, increasing the cancer risk. Meanwhile, the maximum allowed values for PFOS and PFOA in drinking water have been legislated to be 0.2 and 0.6 mg/L, respectively. This is according to Health Canada guidelines, although other countries are following the same direction to safeguard the lives of mankind and the ecosystem resources. However, there is a dearth of knowledge and understanding of PFAS contamination-related soil health guidelines in other nations (Abunada et al., 2020). This calls for international and inter-governmental investment in research and training about PFAS.

1.4.6 Reproductive Disorders

Brominated retardants can disturb the endocrine system by mimicking the estrogen receptors (Bajard et al., 2021; Legler, 2008). For instance, investigating T3 and T4 hormone levels revealed a decline in an experimental rats' survey. This is evidence for the use of emerging contaminants and hormonal interference. Therefore, the gene assay confirmed the modulation efficiencies and associated hydroxylated analogs on estrogen/thyroid/androgen receptors (ER/TR/AR) (Arowolo et al., 2022). Interestingly, hydroxylated DE congeners possessed higher antiestrogenic potencies. In a similar investigation, rats' gestation and lactation exposure to BDE47 at conducive environmental concentrations significantly influenced the male reproductive system (Li et al., 2021b).

1.4.7 Carcinogenic Risk

The research investigations have been tailored toward the relationship between cancer risk and PBDE exposure in turmoil and uncertainty (Renzelli et al., 2023). However, minimal arguments or evidence confirms the cancerous risk of these PBDEs. Typically, some studies show the serum levels of PBDEs and OH-PBDEs are elevated, especially in the thyroid cancer population (Liu et al., 2017). Nonetheless, it is unclear whether the PBDE-induced variations of thyroid hormone regulations are prevalent. However, low cancer risk was traced in humans via PBDEs at e-waste recycling sites and municipal waste burning (Chakraborty et al., 2017; Ohajinwa et al., 2019). Recently, confirmation was traced between the serum levels of PBDE congeners. BDE47, BDE100, and BDE153 were associated with invasive breast cancer risk in women in California (Hurley et al., 2019). Nonetheless, other forms such as BDE47, BDE99, and BDE209 have cytotoxic and genotoxic effects in human bronchial epithelial cells (Hurley et al., 2019). Likewise, the genotoxic properties of PBDE congeners are depicted in human neuroblastoma cells, grass shrimp embryos, and dolphin blast cell lines (Rajput et al., 2021). The OPEs also possess carcinogenic effects in humans and animals (Hurley et al., 2019). The OPE congeners such as 2-ethylhexyl diphenyl phosphate, tri-*N*-butyl phosphate, 2,4,6-trimethoxyphenyl phosphine, and triphenyl phosphate substantially pose more risks for breast and cervical cancers. In a related investigation, a rare Chinese minnow showed the presence of TDCIPP exposure (Feng et al., 2024; Wu et al., 2023). These could adversely affect the cell cycle, DNA replication, Fanconi anemia pathway, p53 signaling pathway, and DNA repair pathways (Avila-Barnard et al., 2022).

1.4.8 Cellular Toxicity

There is extensive research on EOC toxicology, but research on the endocrine disruptive characteristics of nanoparticles is limited (Xuan et al., 2023). The bioaccumulation and thyroid disruption via tetrabromobisphenol A in zebrafish larvae increased when SiO_2 nanoparticles were present. Therefore, n-SiO_2 enhanced the uptake of BDEs by zebra fish embryos owing to the development of n-SiO_2-BDE in the environment (Zhu et al., 2021).

1.4.9 Hormonal and Developmental Disorders

Hormonal and development disorders are endocrine-disrupting chemicals that influence the endocrine system even at low threshold limits. This di-2-ethylhexyl phthalate (DEHP) is a PAE congener that reduces steroidogenesis, interferes with the activity of peroxisome proliferator-activated receptors, and depicts agonistic influence on endocrine receptors. The research studies on Sprague-Dawley rats fed with DEHP showed histopathologic variations in the thyroid and liver (Tesic et al., 2023). Similarly, Cystathionine Beta-Synthase (CBS) has shown agonistic influence on endocrine receptors and anti-androgenic properties. They also contribute to the variations in adult males. The testis weight was reduced by 10% in animals fed with DEHP/PCB compared to single-ED dosed groups (Delbes et al., 2022). The DEHP/PCBs trigger collective toxic influence on liver function (Maddela et al., 2022; Rouiller-Fabre et al., 2015).

1.4.10 Thyroid Carcinogenicity

Thyroid carcinogenicity is among many side effects of EOCs.

The study targeting the pair-matching case-control revealed exposure to phthalate congeners that could pose high risks of thyroid papillary cancer. At the same time, the odds ratio represented 5.3, and 95% confidence interval (CI) recorded higher values. Marotta and coworkers reported that to DEHP exposure was increased. There were approximately 14 times higher risks of thyroid cancer in patients having thyroid disorders (Marotta et al., 2019). A similar study of varied approaches found a relationship between phthalate exposure and thyroid cancer cases. This gives a more excellent snapshot of how cancer pose danger to human health. This is especially true for the thyroid cells in other animals such as rats, targeting the hormones and genetic setup. The experimental study with thyroid carcinoma cell lines and rats reported a high possibility of DEHP to induce cell proliferation and DNA damage in the thyroid. The estrogen-mimicking and endocrine disruptors, biphenyl congeners, were linked with carcinogenicity in animals. PCB138 congeners have a substantial association to breast cancer development. Bisphenol A also presented higher thyroid cancer proneness in rats pretreated with *N*-bis-(2-hydroxypropyl)-nitrosamine. This further indicates that the thyroid carcinogenicity from environmental risks of soil exposure to biphenyls/phthalates is so high.

1.4.11 Endocrine and Immune Toxicity

The endocrine system and immunity can be disrupted by consuming EOC food materials from polluted soils (Vázquez-Tapia et al., 2022). The immunity of cells can assimilate MNPs through various points of entry. These may include phagocytosis, micropinocytosis, and clathrin/caveolae-dependent or independent assimilation. The mutual interactions of immune cells and these MNPs trigger various immune disorders (Firdessa et al., 2014). For example, an *in vitro* study scenario reported beads of 20 nm with polystyrene (PS), which induced oxidative stress in human monocytes. The report further revealed that tiny nanoparticles (20 nm) exhibited cellular toxicity. This is probably ascribed to the assimilation of abundant NPs because of their small size. However, the larger NPs induce cell toxicity by producing reactive oxygen species. Likewise MPs have also been reported to result in size-dependent cellular toxicity. MPs also have endocrine-disrupting property with organism behavior changes (Pei et al., 2022). The exposure to PS NPs in the pancreas of zebrafish results in reduced glucose levels, which causes aberrant movement activity. These findings pin the contaminant for great potential for hindering immunity and endocrine functioning at nanomolar concentration. These could even be disastrous upon chronic bioconcentration in mammalian living organisms' tissues and other organs such as lungs, liver, and prostate (Maddela et al., 2022).

1.4.12 Micro- or Nano-Plastic Genotoxicity

There is an increasingly alarming discovery of genotoxicity from MNPs (Resmi and Vaishnavi, 2024; Singh et al., 2024a). These include PS MPs, which present pronounced DNA denaturing in humanity, especially the Hs27 cell line (Demir et al., 2022). The aminated PS nano beads depicted significant DNA damage in pulmonary epithelial cells and human macrophage cell lines (Hu and Palić, 2020). The spherical structure of PS NP genotoxicity in outer skin cell lines by inducing chromosome breakdown and deleting chromosome segregation machinery. Similarly, COOH-altered PS NPs neutralize the genotoxicity. The mechanism of genotoxicity is that MPs such as polyethylene-based MPs may look like clastogen. This causes the misrepairing of double-strand breaks or DNA repair capacity potential. Delayed solutions may consequently promote the DNA strand breakdown. These contaminants

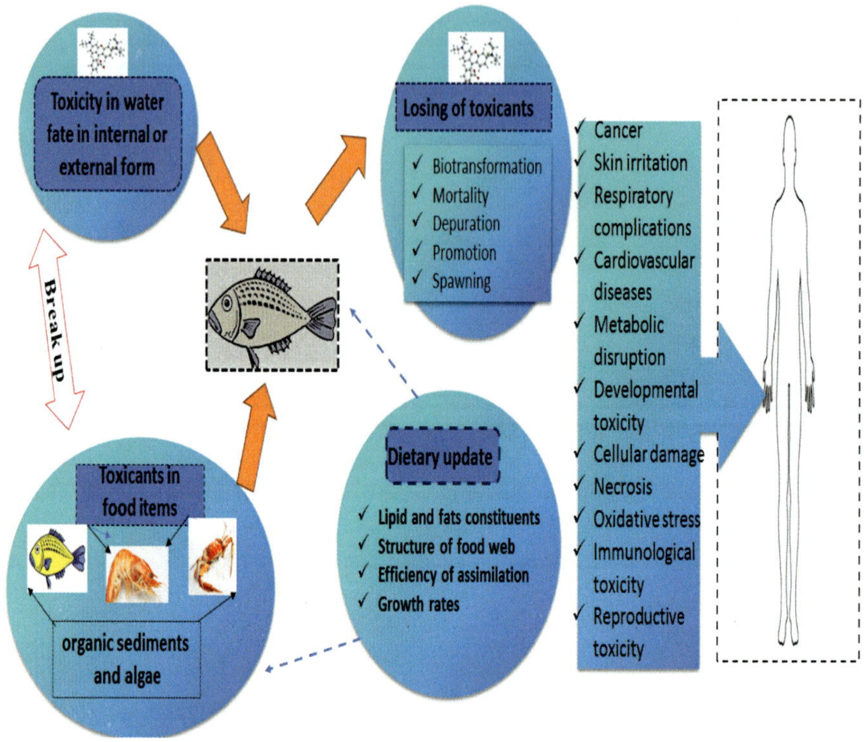

FIGURE 1.4 EOC pathway and potential effects.

can cause genotoxicity by damaging the nuclear membrane. It is worth pointing out that the unrepaired or misrepaired DNA lesions trigger mutagenic effects. This carcinogenesis can happen when mutations are present in the essential genes of genome stability and integrity or cell cycle (Tang, 2025). More studies are required for more snapshots about the genotoxic nature of MPs in varying human cells, particularly with MNPs of various sizes and types (Majeed and Mashaan Rabee, 2025; Tang et al., 2024). Therefore, toxicity investigations with varied pollutants can offer an extended understanding and give the keynote areas required to better analyze health risks in the soil ecosystem. The uptake pathways of EOCs and toxic effects have been summarized in Fig. 1.4 from the above-mentioned literature in section 1.4 and Maddela et al. (2022).

1.5 DETECTION AND ANALYSIS OF EOCs

1.5.1 Sampling Methods

The samples of EOCs and their detection and analysis in the environment require sophisticated and efficient sampling methods. This increases the accuracy and reliability of the results for proper scientific actions and sensitization (Lukač Reberski et al., 2023). The increasing prevalence of EOCs poses significant risks to both human health and ecological systems. A systematic review of existing literature is essential for identifying known EOCs and their sources. Risk assessment helps prioritize EOCs based on their toxicity, prevalence, and potential impact, ensuring that resources are focused on the most harmful contaminants (Masinga et al., 2024). After identification, a strategic sampling plan is implemented considering spatial distribution and temporal variations. Thus proper sample preservation techniques are necessary to prevent degradation during transport and storage, ensuring accurate analysis.

1.5.2 Analytical Techniques

Analyzing EOCs requires appropriate tools, equipment, and procedures, depending on their capacity; for example, the colored and the noncolored ones respond differently during analysis (Alcudia-León et al., 2009), as presented in Figure 1.5. For example, the dyes are colored while antibiotics such as tetracycline and other pesticides like dichlorodiphenyltrichloroethane are not colored; therefore, they have different quantum yields and thus respond differently to modes of quantification like spectrophotometer and high-performance liquid chromatography (Kadadou et al., 2024). Effective detection and analysis of these pollutants are crucial for understanding their fate, transport, and removal (Coronado-Apodaca et al., 2023). High-resolution mass spectrometry and advanced chromatographic techniques (gas chromatography coupled with mass spectrometry and liquid chromatography-tandem mass spectrometry) provide accurate identification and quantification of EOCs. This is because nontargeted screening approaches help identify unknown or unexpected EOCs, expanding the scope of detection. However it should be noted that stringent QA/QC protocols ensure the reliability and accuracy of data. Gas chromatography coupled with mass spectrometry is employed to analyze these concentrated samples, effectively identifying and quantifying EOCs (Maji et al., 2023). The complexity of EOCs in environmental samples necessitates using advanced molecular techniques to detect and interpret these contaminants accurately. This, therefore, requires different equipment for extra high performance, especially for the elute (Banerjee et al., 2023).

Other detection methods include spectroscopy using Fourier transform infrared spectroscopy (FTIR) and surface-enhanced Raman spectroscopy, among others which are employed to identify organic contaminant structures and concentrations (Shipp et al., 2017). FTIR is a fast, accurate, and sensitive method that measures interferograms to determine infrared spectra. It uses nanostructural factors to boost Raman scattering intensity.

The liquid chromatography-tandem mass spectrometry method effectively detects low-level contaminants and quantifies a wide range of ECs (Meher and Zarouri, 2025). Enzyme-linked immunosorbent assay is one of the popular techniques used to detect specific pollutants. Biosensors are a facile biosensor technology developed to enable rapid on-site detection of contaminants. This is more reliable because it captures the values of target samples at the site and thus can be more reliable because of fewer interferences such as temperature drop, among others. Electrochemical

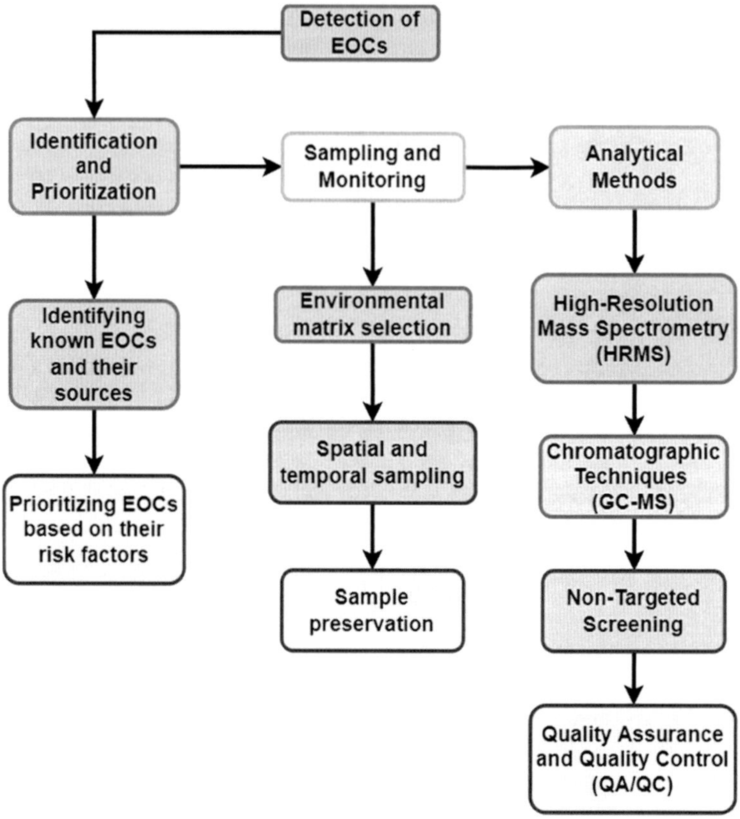

FIGURE 1.5 EOC detection and analysis procedure.

detection methods are emerging, especially for measuring pharmaceuticals, because they are less spectral and detect mainly the ions in the samples. Last but not least is an online solid-phase extraction-liquid chromatography-mass spectrometry. This technique is simple, automated, low-cost, and time-saving. It can detect EOCs in water, atmosphere, and biota. Mass spectrometry and bioanalytical techniques are equipped mainly to effectively analyze specific EOCs that are less opaque and require maximum detection. This has furthermore promoted electrochemical methods targeting emerging contaminants measurement more prevalent (Meher and Zarouri, 2025) (Figure 1.5).

1.6 REMEDIATION TECHNIQUES FOR EOCs

EOCs are prevalent in wastewater treatment facilities because of domestic and industrial effluent emissions (Parida et al., 2021). EOCs' physicochemical and biological properties are crucial in determining their fate in wastewater treatment processes and subsequent discharge into natural environments (Zhang et al., 2023). Therefore gathering scientific data on EOCs' fate and effects is imperative to assess their environmental risks accurately (Kumar et al., 2023). This helps in designing the remediation techniques. In doing so, a combination of physical (adsorption, filtration), chemical (advanced oxidation processes), and biological (bioremediation, phytoremediation) treatment methods have been recorded to offer a comprehensive approach to their removal and remediation (Aparicio et al., 2022; Kathi and El Din Mahmoud, 2024). These hybrid systems that integrate these methods can enhance removal efficiency.

1.6.1 Biological Methods

Biological remediation techniques have gained prominence due to their eco-friendly and cost-effective nature (Tripathi et al., 2023). These methods primarily include bioremediation and phytoremediation. These utilize microorganisms and plants to clean polluted sites. In doing so, bioremediation strategies such as bioaugmentation, biostimulation, bioventing, and bioleaching harness the natural processes of microorganisms to degrade organic pollutants into less harmful substances. Phytoremediation techniques such as phytoextraction, phytostabilization, and phytovolatilization involve plants in absorbing, stabilizing, or volatilizing contaminants from the soil and water (Bartucca et al., 2023). Emphasizing the application in marine systems, biological remediation can be effective in seawater and sediments (Aryal, 2024). This approach addresses contaminants from oil spills to persistent organic pollutants by employing inorganic and organic remediation strategies. Furthermore, advanced methods also explore genetic engineering to enhance the efficiency of bioremediation and phytoremediation (Bala et al., 2022; Hassanien et al., 2022; Kafle et al., 2022).

1.6.2 Chemical Methods

Chemical remediation encompasses a variety of approaches including physical, chemical, and combined physicochemical methods. These techniques are designed to remove or neutralize contaminants through reactions that render them harmless (Kuppan et al., 2024). Traditional chemical methods are widely used due to their immediate effectiveness in reducing pollutant levels. Nanoscience and engineering have introduced innovative solutions in this domain (Singh et al., 2024b). Nanostructured materials, such as nanoparticles, have shown significant potential in removing toxins from effluents (Olawade et al., 2024). Thus, despite the less comprehensive studies, the effectiveness of treating various contaminants especially in wastewater has been thought of from a circular economy perspective (Mwizerwa et al., 2024a).

1.6.3 Physical Treatment Methods

Physical treatment methods for emerging organic pollutants (EOPs) leverage various mechanisms to remove these contaminants from water and soil environments. One primary approach is adsorption, where pollutants get attached to the surface of materials like activated carbon, zeolites, biochar, or advanced nanomaterials (Mahinroosta and Senevirathna, 2020; Mwizerwa et al., 2024b). Activated carbon, in particular, is effective due to its high surface area and porous structure, although it requires frequent regeneration or replacement due to saturation (Larasati et al., 2021; Santos et al., 2022).

Another widely used method is membrane filtration. This has various formats such as microfiltration, ultrafiltration (UF), and nanofiltration, including reverse osmosis. These techniques physically separate contaminants based on size and charge but often face challenges such as membrane fouling and high energy consumption (Shoshaa et al., 2023). Advanced filtration techniques, such as hybrid systems, combine membrane filtration with other processes such as advanced oxidation processes (AOPs) or adsorption to enhance EOP removal efficiency.

Ion exchange involves resins that exchange ions with the solution to remove charged pollutants, although it is limited in capacity for nonionic compounds and requires resin regeneration (Czupryński et al., 2022). Additionally, sedimentation and flotation processes are employed to physically separate particulate-bound contaminants from liquids, which are effective for particulates but not dissolved contaminants. These physical methods provide critical pathways for removing resistant EOCs, especially when combined with chemical or biological treatments for comprehensive remediation (Shahid et al., 2021).

Therefore, remediating EOCs involves a multifaceted approach, incorporating established and novel technologies (Satyam and Patra, 2024). Each method offers unique advantages and, depending on the specific environmental context and type of contaminant, can be deployed effectively to mitigate the impact of these harmful substances on the environment (Khan et al., 2021) (Figure 1.6). The remediation techniques of EOCs have been summarized in Fig. 1.6 from the above-mentioned literature in section 1.6.

1.6.4 EOC Removal Challenges

At present, the removal of EOCs is faced with significant challenges due to their diverse chemical structures and properties (Bhagat et al., 2024; Crini et al., 2022; Naeem et al., 2022). Conventional wastewater treatment processes often fail to eliminate these contaminants (Otálora et al., 2021; Wan Ismail and Mohktar, 2020). Physicochemical methods such as adsorption, ultrafiltration, electrodialysis, and advanced oxidation processes like the Fenton reaction and photocatalysis are employed but with varying degrees of success. Biological methods, including biofilters and algal wastewater treatment ponds, also play a role, yet they face limitations in removing EOCs. While effective at removing certain pollutants such as suspended solids and biodegradable organic matter, constructed wetlands exhibit less efficacy in eliminating recalcitrant organic and bacterial contaminants. These systems face challenges in removing nutrients and pathogens unless combined with other treatment processes. Hybrid systems that integrate different treatment stages show promise in enhancing the removal efficiency of EOCs, particularly when advanced oxidative processes are incorporated (Asheghmoalla and Mehrvar, 2024; Dhangar and Kumar, 2020).

1.6.5 Technology Advancement Trends

Various technologies have been employed to address the removal challenges. These include both stand-alone and combined treatment methods. Stand-alone treatments

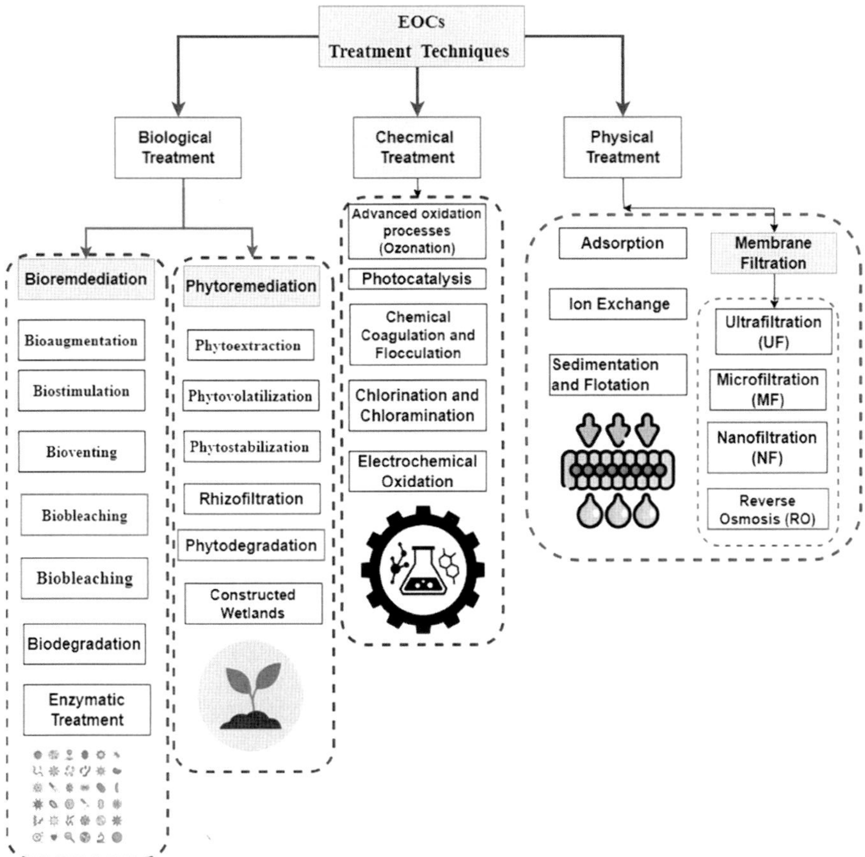

FIGURE 1.6 Remediation techniques of EOCs.

often suffer from operational issues and may not effectively address all EOCs due to their unique physicochemical properties. Combined treatment processes have shown tremendous success, integrating techniques such as adsorption, ozonation, and biological treatments with advanced oxidation processes and membrane-based technologies (Kanafin et al., 2023). These holistic treatment technologies leverage the advantages of individual processes to achieve higher removal efficiencies at sustainable levels. Therefore, membrane processes have emerged as one of the best technologies for effectively and safely eliminating EOCs. These processes are highly adaptable and can be optimized to enhance the overall performance of wastewater treatment, making them a crucial component in the management of EOCs.

1.7 CONCLUSION AND FUTURE OUTLOOK

EOCs are continuously detected in the soil and in other environmental matrices such as air and water. They are not new since their footprint is traced many decades back; the only difference is changes in the chemical composition, especially in mother forms to intermediates, and their effects on the living biota caused by environmental changes. Traditionally, these substances have been identified and detected, including their environmental impact; however, fewer investigations have found new adverse effects that were not previously evaluated, depending on the methods used. Although we have made much more progress to analyze, evaluate, and regulate the conventional persistent pollutants, still more work is required on emerging pollutants. As the research on emerging pollutants is in the juvenile stage, still scientists have paid less attention to them. At present standpoint, the identification of new emerging pollutants is tremendously ongoing with innovations such as artificial intelligence and simulations at play. At the same time, a few emerging pollutants can be moved from the available list due to new scientific evidence. The categories of the emerging pollutants vary widely in properties, making it hard to propose a general analysis strategy. There is also limited access to information regarding their destiny, fate, and biotransformation, thus making it complex and challenging to find ever-expounding facts about tiny intermediate quantities. They might be highly disastrous at specific values. Therefore, the field of emerging pollutant analysis is a happening field of research, such as quantitative structure–activity relationship and density function theory simulations, with more literature published yearly. Various databases with excellent reviews and research papers are available on this subject matter. This chapter, therefore, gives a concise snapshot of all the key concepts of EOCs in the soil ecosystem. Same as other environmental protocols, the analysis of emerging pollutants also involves sampling. Conventional water sampling protocols such as isokinetic and nonisokinetic methods were also employed for sampling EOCs. However, very low concentration of many emerging contaminants and a dearth of knowledge about their release and fate in the ecosystem require novel sampling techniques such as passive sampling. Herein, the analyte is collected in the receiving phase by the difference in chemical potential. Currently the development of new passive samplers catering to both specific and general emerging pollutants is a hot research direction. An austere quality control measure is suggested to ensure the accuracy of sampling which involves analyzing emerging contaminants and various forms of replicates, blanks, and spikes.

This aims to develop methods that suit a broader spectrum of EOCs and direct efforts toward eco-friendly procedures. Therefore, economical and user-friendly on-site analytical methods such as electrochemical methods are under intensive application investigations. In this case, understanding and addressing the issue of EOCs is essential for protecting environmental and public health. Continued research, innovation, and collaboration among scientists, policymakers, and the public are crucial to managing the risks associated with these contaminants based on (i) risk assessment, (ii) technological innovations, and (iii) policy development as follows;

i. Traditional treatment technologies must first be used to generate a liquid free of organics or other materials, which adds to the complexity of water and wastewater. Innovative technologies that deliver destruction have been used to remove EOCs from drinking water and groundwater, but their effectiveness is still limited. The authors recommend new destruction and separation technologies in the book chapter for several EOCs and their precursors and some of the existing technologies for eliminating EOCs from soil ecosystems and wastewater treatment plants. Pretreatment and residual management are two issues with these systems. Residual management is difficult and might result in higher concentrated fractions being transferred to biota.

ii. Remediation is a potential strategy to decontaminate the environmental matrices. However, most of its practice still faces some problems at the implementation level in different communities on a full scale. Improving public awareness on the migration and measures to devoid EOC materials as well as choosing between phasing out existing chemicals containing such substances or replacing chemicals coming up in synthesizes is essential.

iii. The recent advances in toxicological and bioaccumulation research give hope for novel developments. However, research is needed concerning different biota at the individual species level and emerging facile detection options as well as their economic viability and affordability, especially for developing countries.

REFERENCES

Abunada, Z., Alazaiza, M.Y.D. and Bashir, M.J.K. 2020 An overview of per- and polyfluoroalkyl substances (PFAS) in the environment: Source, fate, risk and regulations. *Water* 12(12), 3590.

Alcudia-León, M.C., Lucena, R., Cárdenas, S. and Valcárcel, M. 2009. Stir membrane extraction: A useful approach for liquid sample pretreatment. *Analytical Chemistry* 81, 8957.

Aparicio, J.D., Raimondo, E.E., Saez, J.M., Costa-Gutierrez, S.B., Álvarez, A., Benimeli, C.S. and Polti, M.A. 2022. The current approach to soil remediation: A review of physicochemical and biological technologies, and the potential of their strategic combination. *Journal of Environmental Chemical Engineering* 10(2), 107141.

Arowolo, O., Pilsner, J.R., Sergeyev, O. and Suvorov, A. 2022. Mechanisms of male reproductive toxicity of polybrominated diphenyl ethers. *International Journal of Molecular Sciences* 23(22), 14229.

Aryal, M. 2024. Phytoremediation strategies for mitigating environmental toxicants. *Heliyon* 10(19), e38683.

Asheghmoalla, M. and Mehrvar, M. 2024. Integrated and hybrid processes for the treatment of actual wastewaters containing micropollutants: A review on recent advances. *Processes* 12(2), 339.

Avila-Barnard, S., Dasgupta, S., Cheng, V., Reddam, A., Wiegand, J.L. and Volz, D.C. 2022. Tris(1,3-dichloro-2-propyl) phosphate disrupts the trajectory of cytosine methylation within developing zebrafish embryos. *Environmental Research* 211, 113078.

Babuji, P., Thirumalaisamy, S., Duraisamy, K. and Periyasamy, G. 2023. Human health risks due to exposure to water pollution: A review. *Water* 15(14), 2532.

Bai, X., Lutz, A., Carroll, R., Keteles, K., Dahlin, K., Murphy, M. and Nguyen, D. 2018. Occurrence, distribution, and seasonality of emerging contaminants in urban watersheds. *Chemosphere*, 200, 133–142.

Bajard, L., Negi, C.K., Mustieles, V., Melymuk, L., Jomini, S., Barthelemy-Berneron, J., Fernandez, M.F. and Blaha, L. 2021. Endocrine disrupting potential of replacement flame retardants: Review of current knowledge for nuclear receptors associated with reproductive outcomes. *Environment International* 153, 106550.

Bala, S., Garg, D., Thirumalesh, B.V., Sharma, M., Sridhar, K., Inbaraj, B.S. and Tripathi, M. 2022. Recent strategies for bioremediation of emerging pollutants: A review for a green and sustainable environment. *Toxics* 10(8), 484.

Banerjee, A., Singh, S. and Ghosh, A. 2023. Detection and removal of emerging contaminants from water bodies: A statistical approach. *Frontiers in Analytical Science* 3, 1115540.

Bartucca, M.L., Cerri, M. and Forni, C. 2023. Phytoremediation of pollutants: Applicability and future perspective. *Plants* 12(13), 2462.

Beltrán, E.M., Pablos, M.V., Torija, C.F., Porcel, M.Á. and González-Doncel, M. 2020. Uptake of atenolol, carbamazepine and triclosan by crops irrigated with reclaimed water in a Mediterranean scenario. *Ecotoxicology and Environmental Safety* 191, 110171.

Bhagat, J., Singh, N. and Shimada, Y. 2024. Southeast Asia's environmental challenges: Emergence of new contaminants and advancements in testing methods. *Frontiers in Toxicology* 6, 1322386.

Cerar S, Mali N (2016): Assessment of presence, origin and seasonal variations of persistent organic pollutants in groundwater by means of passive sampling and multivariate statistical analysis. Journal of Geochemical Exploration 170, 78-93

Chakraborty, P., Selvaraj, S., Nakamura, M., Prithiviraj, B., Cincinelli, A. and Bang, J. 2017. PCBs and PCDD/Fs in soil from informal e-waste recycling sites and open dumpsites in India: Levels, congener profiles and health risk assessment. *Science of the Total Environment* 621, 930–938.

Chang, C.-J., Ish, J.L., Chang, V.C., Daniel, M., Jones, R.R. and White, A.J. 2024. Exposure to per- and polyfluoroalkyl substances and breast cancer risk: A systematic review and meta-analysis of epidemiologic studies. *American Journal of Epidemiology* 193(8), 1182–1196.

Chen, J., Tanguay, R.L., Tal, T.L., Gai, Z., Ma, X., Bai, C., Tilton, S.C., Jin, D., Yang, D., Huang, C. and Dong, Q. 2014. Early life perfluorooctanesulphonic acid (PFOS) exposure impairs zebrafish organogenesis. *Aquatic Toxicology* 150, 124–132.

Chen, X., Lu, X., Zhao, R., Su, G., Meng, J., Li, Q., Hua, Y. and Shi, B. 2024. Occurrence and risks of PPCPs of a typical mountainous region: Implications for sustainable urban water systems. *Science of the Total Environment* 951, 175714.

Chen, X., Qadeer, A., Liu, M., Deng, L., Zhou, P., Mwizerwa, I.T., Liu, S., Ajmal, Z., Xingru, Z. and Jiang, X. 2023. Bioaccumulation of emerging contaminants in aquatic biota: PFAS as a case study. In *Emerging Aquatic Contaminants*. Kumar, M., Mohapatra, S. and Weber, K. (eds), pp. 347–374, Elsevier.

Coronado-Apodaca, K.G., Rodríguez-De Luna, S.E., Araújo, R.G., Oyervides-Muñoz, M.A., González-Meza, G.M., Parra-Arroyo, L., Sosa-Hernandez, J.E., Iqbal, H.M.N. and Parra-Saldivar, R. 2023. Occurrence, transport, and detection techniques of emerging pollutants in groundwater. *MethodsX* 10, 102160.

Crini, N., Lichtfouse, E., Fourmentin, M., Ribeiro, A.R.L., Noutsopoulos, C., Mapelli, F., Fenyvesi, É., Vieira, M., Picos Corrales, L., Moreno-Piraján, J., Giraldo, L., Huq, M., Soltan, J., Torri, G., Magureanu, M., Bradu, C. and grégorio, C. 2022. Removal of emerging contaminants from wastewater using advanced treatments. A review. *Environmental Chemistry Letters* 20, 1333–1375.

Czupryński, P., Płotka, M., Glamowski, P., Żukowski, W. and Bajda, T. 2022. An assessment of an ion exchange resin system for the removal and recovery of Ni, Hg, and Cr from wet flue gas desulphurization wastewater: A pilot study. *RSC Advances* 12(9), 5145–5156.

Delbes, G., Blázquez, M., Fernandino, J.I., Grigorova, P., Hales, B.F., Metcalfe, C., Navarro-Martín, L., Parent, L., Robaire, B., Rwigemera, A., Van Der Kraak, G., Wade, M. and Marlatt, V. 2022. Effects of endocrine disrupting chemicals on gonad development: Mechanistic insights from fish and mammals. *Environmental Research* 204, 112040.

Demir, E., Turna, F. and Akkoyunlu, G. 2022. Ingested polystyrene microplastics as a carrier of heavy metals (cadmium or silver): Uptake, gut damage, oxidative stress, and DNA damage in Drosophila larvae. *SSRN Electronic Journal.*

DeWitt, J.C., Blossom, S.J. and Schaider, L.A. 2019. Exposure to per-fluoroalkyl and polyfluoroalkyl substances leads to immunotoxicity: Epidemiological and toxicological evidence. *Journal of Exposure Science & Environmental Epidemiology* 29(2), 148–156.

Dhangar, K. and Kumar, M. 2020. Tricks and tracks in removal of emerging contaminants from the wastewater through hybrid treatment systems: A review. *Science of the Total Environment* 738, 140320.

Feng, Y., Wang, Z., Duan, H. and Shao, B. 2024. Tris(1,3-dichloro-2-propyl) phosphate induces endoplasmic reticulum stress and mitochondrial-dependent apoptosis in mouse spermatocyte GC-2 cells. *Food and Chemical Toxicology* 185, 114506.

Fenton, S.E., Ducatman, A., Boobis, A., DeWitt, J.C., Lau, C., Ng, C., Smith, J.S. and Roberts, S.M. 2021. Per- and polyfluoroalkyl substance toxicity and human health review: Current state of knowledge and strategies for informing future research. *Environmental Toxicology and Chemistry* 40(3), 606–630.

Firdessa, R., Oelschlaeger, T. and Moll, H. 2014. Identification of multiple cellular uptake pathways of polystyrene nanoparticles and factors affecting the uptake: Relevance for drug delivery systems. *European Journal of Cell Biology* 93, 323–337.

Gallego, S., Montemurro, N., Béguet, J., Rouard, N., Philippot, L., Pérez, S. and Martin-Laurent, F. 2021. Ecotoxicological risk assessment of wastewater irrigation on soil microorganisms: Fate and impact of wastewater-borne micropollutants in lettuce-soil system. *Ecotoxicology and Environmental Safety* 223, 112595.

Gibson, G., Cundy, A., Kafwamfwa, N. and Stewart, A. 2023. "Old" and "new" contaminants and their management: Learning from the past, looking to the future. *Environmental Geochemistry and Health* 45(4), 1091–1105.

Häder, D., Banaszak, A., Villafañe, V., Narvarte, M., González, R. and Helbling, E. 2020. Anthropogenic pollution of aquatic ecosystems: Emerging problems with global implications. *Science of the Total Environment* 713, 136586.

Hassanien, A., Saadaoui, I., Schipper, K., Al-Marri, S., Dalgamouni, T., Aouida, M., Saeed, S. and Al-Jabri, H.M. 2022. Genetic engineering to enhance microalgal-based produced water treatment with emphasis on CRISPR/Cas9: A review. *Frontiers in Bioengineering and Biotechnology* 10, 1104914.

Hu, M. and Palić, D. 2020. Micro- and nano-plastics activation of oxidative and inflammatory adverse outcome pathways. *Redox Biology* 37, 101620.

Hurley, S., Goldberg, D., Park, J.-S., Petreas, M., Bernstein, L., Anton-Culver, H., Neuhausen, S., Nelson, D. and Reynolds, P. 2019. A breast cancer case-control study of polybrominated diphenyl ether (PBDE) serum levels among California women. *Environment International* 127, 412–419.

Jain, M., Kumar, D., Chaudhary, J., Kumar, S., Sharma, S. and Singh Verma, A. 2023. Review on E-waste management and its impact on the environment and society. *Waste Management Bulletin* 1(3), 34–44.

Jantzen, C.E., Annunziato, K.A., Bugel, S.M. and Cooper, K.R. 2016. PFOS, PFNA, and PFOA sub-lethal exposure to embryonic zebrafish have different toxicity profiles in terms of morphometrics, behavior and gene expression. *Aquatic Toxicology* 175, 160–170.

Jones, R.R., Fisher, J.A., Medgyesi, D.N., Buller, I.D., Liao, L.M., Gierach, G., Ward, M.H. and Silverman, D.T. 2023. Ethylene oxide emissions and incident breast cancer and non-Hodgkin lymphoma in a US cohort. *Journal of the National Cancer Institute* 115(4), 405–412.

Kadadou, D., Tizani, L., Alsafar, H. and Hasan, S.W. 2024. Analytical methods for determining environmental contaminants of concern in water and wastewater. *MethodsX* 12, 102582.

Kafle, A., Timilsina, A., Gautam, A., Adhikari, K., Bhattarai, A. and Aryal, N. 2022. Phytoremediation: Mechanisms, plant selection and enhancement by natural and synthetic agents. *Environmental Advances* 8, 100203.

Kanafin, Y.N., Satayeva, A., Abdirova, P., Inglezakis, V.J., Arkhangelsky, E. and Poulopoulos, S.G. 2023. Membrane bioreactor and advanced oxidation processes for combined treatment of synthetic wastewater containing naproxen, bisphenol A, and sulfamethoxazole. *Journal of Water Process Engineering* 55, 104250.

Kathi, S. and El Din Mahmoud, A. 2024. Trends in effective removal of emerging contaminants from wastewater: A comprehensive review. *Desalination and Water Treatment* 317, 100258.

Khan, S., Naushad, M., Muthusamy, G., Iqbal, J. and Alfadul, S. 2021. Emerging contaminants of high concern for the environment: Current trends and future research. Environmental Research 207, 112609.

Koumaki, E., Noutsopoulos, C., Mamais, D., Fragkiskatos, G. and Andreadakis, A. 2021. Fate of emerging contaminants in high-rate activated sludge systems. *International Journal of Environmental Research and Public Health* 18(2), 400.

Kumar, M., Mohapatra, S. and Weber, K. 2023. *Emerging Aquatic Contaminants: One Health Framework for Risk Assessment and Remediation in the Post COVID-19 Anthropocene*, Elsevier.

Kuppan, N., Padman, M., Mahadeva, M., Srinivasan, S. and Devarajan, R. 2024. A comprehensive review of sustainable bioremediation techniques: Eco friendly solutions for waste and pollution management. *Waste Management Bulletin* 2(3), 154–171.

Larasati, A., Fowler, G.D. and Graham, N.J.D. 2021. Insights into chemical regeneration of activated carbon for water treatment. *Journal of Environmental Chemical Engineering* 9(4), 105555.

Legler, J. 2008. New insights into the endocrine disrupting effects of brominated flame retardants. *Chemosphere* 73(2), 216–222.

Leung, S.C.E., Wanninayake, D., Chen, D., Nguyen, N.-T. and Li, Q. 2023. Physicochemical properties and interactions of perfluoroalkyl substances (PFAS): Challenges and opportunities in sensing and remediation. *Science of the Total Environment* 905, 166764.

Li, P., Karunanidhi, D., Subramani, T. and Srinivasamoorthy, K. 2021a. Sources and consequences of groundwater contamination. *Archives of Environmental Contamination and Toxicology* 80(1), 1–10.

Li, X., Gao, H., Li, P., Chen, W., Tang, S., Liu, L., Zhou, G., Xia, T., Wang, A. and Zhang, S. 2021b. Impaired sperm quantity and motility in adult rats following gestational and lactational exposure to environmentally relevant levels of PBDE-47: A potential role of thyroid hormones disruption. *Environmental Pollution* 268, 115773.

Li, X., Shen, X., Jiang, W., Xi, Y. and Li, S. 2024. Comprehensive review of emerging contaminants: Detection technologies, environmental impact, and management strategies. *Ecotoxicology and Environmental Safety* 278, 116420.

Lin, H.W., Feng, H.X., Chen, L., Yuan, X.J. and Tan, Z. 2020. Maternal exposure to environmental endocrine disruptors during pregnancy is associated with pediatric germ cell tumors. *Nagoya Journal of Medical Science* 82(2), 323–333.

Liu, S., Yang, R., Yin, N., Wang, Y.-L. and Faiola, F. 2019. Environmental and human relevant PFOS and PFOA doses alter human mesenchymal stem cell self-renewal, adipogenesis and osteogenesis. *Ecotoxicology and Environmental Safety* 169, 564–572.

Liu, S., Zhao, G., Li, J., Zhao, H., Wang, Y., Chen, J. and Zhao, H. 2017. Association of polybrominated diphenylethers (PBDEs) and hydroxylated metabolites (OH-PBDEs) serum levels with thyroid function in thyroid cancer patients. *Environmental Research* 159, 1–8.

Loganathan, P., Vigneswaran, S., Kandasamy, J., Cuprys, A.K., Maletskyi, Z. and Ratnaweera, H. 2023. Treatment trends and combined methods in removing pharmaceuticals and personal care products from wastewater: A review. *Membranes* 13(2), 158.

Lukač Reberski, J., Selak, A., Lapworth, D.J., Maurice, L.D., Terzić, J., Civil, W. and Stroj, A. 2023. Emerging organic contaminants in springs of the highly karstified Dinaric region. *Journal of Hydrology* 621, 129583.

Maddela, N.R., Ramakrishnan, B., Kakarla, D., Venkateswarlu, K. and Megharaj, M. 2022. Major contaminants of emerging concern in soils: A perspective on potential health risks. *RSC Advances* 12(20), 12396–12415.

Maddela, N.R., Venkateswarlu, K. and Megharaj, M. 2020. Tris(2-chloroethyl) phosphate, a pervasive flame retardant: Critical perspective on its emissions into the environment and human toxicity. *Environmental Science: Processes & Impacts* 22(9), 1809–1827.

Mahinroosta, R. and Senevirathna, L. 2020. A review of the emerging treatment technologies for PFAS contaminated soils. *Journal of Environmental Management* 255, 109896.

Majeed, Z.M. and Mashaan Rabee, A. 2025. The toxicity of polyethylene microplastics on molecular and biochemical parameters in albino mice. *Journal of Ecological Engineering* 26(3), 274–282.

Maji, S., Roy, C. and Sinha, S. 2023. Gas chromatography-mass spectrometry (GC-MS): A comprehensive review of synergistic combinations and their applications in the past two decades. *Journal of Analytical Sciences and Applied Biotechnology* 5, 72–85.

Marotta, V., Russo, G., Gambardella, C., Grasso, M., La Sala, D., Chiofalo, M.G., D'Anna, R., Puzziello, A., Docimo, G., Masone, S., Barbato, F., Colao, A., Faggiano, A. and Grumetto, L. 2019. Human exposure to bisphenol AF and diethylhexylphthalate increases susceptibility to develop differentiated thyroid cancer in patients with thyroid nodules. *Chemosphere* 218, 885–894.

Masinga, P., Simbanegavi, T.T., Makuvara, Z., Marumure, J., Chaukura, N. and Gwenzi, W. 2024. Emerging organic contaminants in the soil–plant-receptor continuum: Transport, fate, health risks, and removal mechanisms. *Environmental Monitoring and Assessment* 196(4), 367.

Meher, A.K. and Zarouri, A. 2025. Environmental applications of mass spectrometry for emerging contaminants. *Molecules* 30(2), 364.

Morin-Crini, N., Lichtfouse, E., Liu, G., Balaram, V., Ribeiro, A.R.L., Lu, Z., Stock, F., Carmona, E., Teixeira, M.R., Picos-Corrales, L.A., Moreno-Piraján, J.C., Giraldo, L., Li, C., Pandey, A., Hocquet, D., Torri, G. and Crini, G. 2022. Worldwide cases of water pollution by emerging contaminants: A review. *Environmental Chemistry Letters* 204(20), 2311.

Mwizerwa, I., Chen, X., Hu, B. and Ismail, M. 2020a. Enhanced landfill leachate treatment using spiral symmetry stream anaerobic bioreactor and sequential batch reactor. *Journal of Environmental Accounting and Management* 9, 159–171.

Mwizerwa, I.T., Husein, M., Houssini, K., Praise, S. and Kateita, J. 2024a. *Innovative and Hybrid Technologies for Wastewater Treatment and Recycling*, pp. 326–345, CRC Press.

Mwizerwa, I.T., Sun, Z., Huang, T., Tang, J., Hasan, I.M. and Zhao, X. 2024b. Single atom photocatalytic micropollutants degradation performance: A review. *Desalination and Water Treatment* 320, 100656.

Mwizerwa, I.T., Wang, Y. and Chen, X. 2020b. A systematic review on anaerobic textile industrial wastewater treatment: Influence of processes, microbial communities and bioreactors. *Journal of Environmental Accounting and Management* 8(1), 73–91.

Naeem, A., Mahmood, T., Ali, R., Momin, S. and Khan, A. 2022. Technologies for removal of emerging contaminants from wastewater. In *Wastewater Treatment*. Ince, M. and Kaplan Ince, O. (eds), IntechOpen. https://doi.org/10.5772/intechopen.104466

Ohajinwa, C.M., van Bodegom, P.M., Osibanjo, O., Xie, Q., Chen, J., Vijver, M.G. and Peijnenburg, W. 2019. Health risks of polybrominated diphenyl ethers (PBDEs) and metals at informal electronic waste recycling sites. *International Journal of Environmental Research and Public Health* 16(6), 906.

Olawade, D.B., Ige, A.O., Olaremu, A.G., Ijiwade, J.O. and Adeola, A.O. 2024. The synergy of artificial intelligence and nanotechnology towards advancing innovation and sustainability: A mini-review. *Nano Trends* 8, 100052.

Oppeltová, P., Vlček, V., Geršl, M., Chaloupský, P., Ulrich, O., Sedláček, J., Vavrouchová, H., Kohoutková, K., Klepárník, R. and Šimečková, J. 2024. Occurrence and path pollution of emerging organic contaminants in mineral water of Hranice hypogenic Karst. *Frontiers in Environmental Science* 12, 1339818.

Otálora, A., Lerma, T., Arrieta Urango, Y. and Palencia, M. 2021. Emerging organic pollutants in aqueous environments: Detection, monitoring, and removal techniques. *Journal of Science with Technological Applications* 10, 92–153.

Parida, V.K., Saidulu, D., Majumder, A., Srivastava, A., Gupta, B. and Gupta, A.K. 2021. Emerging contaminants in wastewater: A critical review on occurrence, existing legislations, risk assessment, and sustainable treatment alternatives. *Journal of Environmental Chemical Engineering* 9(5), 105966.

Pei, X., Heng, X. and Chu, W. 2022. Polystyrene nano/microplastics induce microbiota dysbiosis, oxidative damage, and innate immune disruption in zebrafish. *Microbial Pathogenesis* 163, 105387.

Petrie, B., Barden, R. and Kasprzyk-Hordern, B. 2015. A review on emerging contaminants in wastewaters and the environment: current knowledge, understudied areas and recommendations for future monitoring. *Water research*, 72, 3-27.

Rajput, I., Yaqoob, S., Sun, Y., Sanganyado, E. and Liu, W. 2021. Polybrominated diphenyl ethers exert genotoxic effects in pantropic spotted dolphin fibroblast cell lines. *Environmental Pollution* 271, 116131.

Renzelli, V., Gallo, M., Morviducci, L., Marino, G., Ragni, A., Tuveri, E., Faggiano, A., Mazzilli, R., Natalicchio, A., Zatelli, M.C., Montagnani, M., Fogli, S., Giuffrida, D., Argentiero, A., Danesi, R., D'Oronzo, S., Gori, S., Franchina, T., Russo, A., Monami, M., Sciacca, L., Cinieri, S., Colao, A., Avogaro, A., Di Cianni, G., Giorgino, F. and Silvestris, N. 2023. Polybrominated diphenyl ethers (PBDEs) and human health: Effects on metabolism, diabetes and cancer. *Cancers* 15(17), 4237.

Resmi, M.R. and Vaishnavi 2024. Microplastic pollution in Indian aquatic ecosystems: Insights and implications. *Water, Air, & Soil Pollution* 236(1), 62.

Roche, I.V., Ubalde-Lopez, M., Daher, C., Nieuwenhuijsen, M. and Gascon, M. 2024. The health-related and learning performance effects of air pollution and other urban-related environmental factors on school-age children and adolescents: A scoping review of systematic reviews. *Current Environmental Health Reports* 11(2), 300–316.

Rodrigues, P., Oliva-Teles, L., Guimarães, L. and Carvalho, A.P. 2022. Occurrence of pharmaceutical and pesticide transformation products in freshwater: update on environmental levels, toxicological information and future challenges. *Reviews of Environmental Contamination and Toxicology*, 260(1), 14.

Rouiller-Fabre, V., Guerquin, M.J., N'Tumba-Byn, T., Muczynski, V., Moison, D., Tourpin, S., Messiaen, S., Habert, R. and Livera, G. 2015. Nuclear receptors and endocrine disruptors in fetal and neonatal testes: A gapped landscape. *Frontiers in Endocrinology* 6, 58.

Santos, D., Xiao, Y., Chaukura, N., Hill, J.M., Selvasembian, R., Zanta, C. and Meili, L. 2022. Regeneration of dye-saturated activated carbon through advanced oxidative processes: A review. *Heliyon* 8(8), e10205.

Satyam, S. and Patra, S. 2024. Innovations and challenges in adsorption-based wastewater remediation: A comprehensive review. *Heliyon* 10(9), e29573.

Shahid, M.K., Kashif, A., Fuwad, A. and Choi, Y. 2021. Current advances in treatment technologies for removal of emerging contaminants from water: A critical review. *Coordination Chemistry Reviews* 442, 213993.

Shipp, D.W., Sinjab, F. and Notingher, I. 2017. Raman spectroscopy: Techniques and applications in the life sciences. *Advances in Optics and Photonics* 9(2), 315–428.

Shoshaa, R., Ashfaq, M.Y. and Al-Ghouti, M.A. 2023. Recent developments in ultrafiltration membrane technology for the removal of potentially toxic elements, and enhanced antifouling performance: A review. *Environmental Technology & Innovation* 31, 103162.

Singh, J., Yadav, B.K., Schneidewind, U. and Krause, S. 2024a. Microplastics pollution in inland aquatic ecosystems of India with a global perspective on sources, composition, and spatial distribution. *Journal of Hydrology: Regional Studies* 53, 101798.

Singh, N.B., Kumar, B., Usman, U.L. and Susan, M.A.B.H. 2024b. Nano revolution: Exploring the frontiers of nanomaterials in science, technology, and society. *Nano-Structures & Nano-Objects* 39, 101299.

Tang, K.H. 2025. Genotoxicity of microplastics on living organisms: Effects on chromosomes, DNA and gene expression. *Environments* 12(1), 10.

Tang, K.H.D., Li, R., Li, Z. and Wang, D. 2024. Health risk of human exposure to microplastics: A review. *Environmental Chemistry Letters* 22(3), 1155–1183.

Tesic, B., Samardzija Nenadov, D., Tomanic, T., Fa Nedeljkovic, S., Milatovic, S., Stanic, B., Pogrmic-Majkic, K. and Andric, N. 2023. DEHP decreases steroidogenesis through the cAMP and ERK1/2 signaling pathways in FSH-stimulated human granulosa cells. *Cells* 12(3), 398.

Tripathi, M., Singh, S., Pathak, S., Kasaudhan, J., Mishra, A., Bala, S., Garg, D., Singh, R., Singh, P., Singh, P.K., Shukla, A.K. and Pathak, N. 2023. Recent strategies for the remediation of textile dyes from wastewater: A systematic review. *Toxics* 11(11), 940.

Van, D.-A., Ngo, T.H., Huynh, T.H., Nakada, N., Ballesteros, F. and Tanaka, H. 2021. Distribution of pharmaceutical and personal care products (PPCPs) in aquatic environment in Hanoi and Metro Manila. *Environmental Monitoring and Assessment* 193, 1–15.

van Gerwen, M., Colicino, E., Guan, H., Dolios, G., Nadkarni, G.N., Vermeulen, R.C.H., Wolff, M.S., Arora, M., Genden, E.M. and Petrick, L.M. 2023. Per- and polyfluoroalkyl substances (PFAS) exposure and thyroid cancer risk. *EBioMedicine* 97, 104831.

Vasilachi, I.C., Asiminicesei, D.M., Fertu, D.I. and Gavrilescu, M. 2021. Occurrence and fate of emerging pollutants in water environment and options for their removal. *Water* 13(2), 181.

Vázquez-Tapia, I., Salazar-Martínez, T., Acosta-Castro, M., Meléndez-Castolo, K.A., Mahlknecht, J., Cervantes-Avilés, P., Capparelli, M.V. and Mora, A. 2022. Occurrence of emerging organic contaminants and endocrine disruptors in different water compartments in Mexico: A review. *Chemosphere* 308, 136285.

Wan Ismail, W.N. and Mohktar, S.U. 2020. Various methods for removal, treatment, and detection of emerging water contaminants. In *Emerging Contaminants*. Nuro, A. (ed), IntechOpen. https://doi.org/10.5772/intechopen.93375

Wee, S.Y. and Aris, A.Z. 2023. Environmental impacts, exposure pathways, and health effects of PFOA and PFOS. *Ecotoxicology and Environmental Safety* 267, 115663.

Wu, X., Zhang, D., Chen, Y., Shen, J., Li, X., Zheng, Q., Ma, J., Xu, J., Rao, M., Liu, X. and Lu, S. 2023. Organophosphate ester exposure among Chinese waste incinerator workers: Urinary levels, risk assessment and associations with oxidative stress. *Science of the Total Environment* 854, 158808.

Xuan, L., Ju, Z., Skonieczna, M., Zhou, P.K. and Huang, R. 2023. Nanoparticles-induced potential toxicity on human health: Applications, toxicity mechanisms, and evaluation models. *MedComm* 4(4), e327.

Yadav, D., S, R., Verma, P., Singh, P., Devi, D., Kumar, P., Hussain, C.M., Gaurav, G. and Kumar, K. 2021. Environmental and health impacts of contaminants of emerging concerns: Recent treatment challenges and approaches. *Chemosphere* 272, 129492.

Zhang, Y., Zhang, H., Zhang, A., Héroux, P., Sun, Z. and Liu, Y. 2023. Remediation of atrazine-polluted soil using dielectric barrier discharge plasma and biochar sequential batch experimental technology. *Chemical Engineering Journal* 458, 141406.

Zhu, B., Han, J., Lei, L., Hua, J., Zuo, Y. and Zhou, B. 2021. Effects of SiO_2 nanoparticles on the uptake of tetrabromobisphenol A and its impact on the thyroid endocrine system in zebrafish larvae. *Ecotoxicology and Environmental Safety* 209, 111845.

2 Environmental Behaviors of Emerging Organic Contaminants

Yajing Wu, Ming Zhang, and Jianjie Fu

2.1 INTRODUCTION

Emerging organic contaminants (EOCs) encompass a wide spectrum of both newly developed and previously utilized chemicals that have recently come under scrutiny due to their potential adverse impacts on human health and the environment. The issue of EOCs has become increasingly significant, as these substances are pervasive and may pose considerable health risks. These contaminants include a multitude of compounds, such as pharmaceuticals and personal care products (PPCPs), pesticides, hormones, surfactants, flame retardants (FRs), plasticizers, and industrial additives, along with their metabolites and various transformation products (Chen et al. 2023).

Advancements in the field of trace detection methodologies have facilitated the discovery of an increased variety of EOCs across diverse environmental matrices (Chen et al. 2023). Although these contaminants typically appear in low concentrations, their escalating application has precipitated more substantial releases into the environment, consequently enhancing their accumulation in various ecosystems. The persistent introduction and subsequent dissemination of EOCs carry significant medium- and long-term ecological and health implications. The origins of EOCs in the soil are multifaceted, encompassing both natural and human-induced contributions, which may originate from both recent developments and historical accumulations. Once these contaminants are integrated into the soil matrix, they are subject to a complex interplay of accumulation, transformation, and degradation processes, orchestrated by a range of chemical and biological factors. It is imperative to delineate the emission routes and transport mechanisms of these contaminants to effectively assess their potential impacts on human health and the broader ecological systems.

This chapter delves deeply into the predominant sources that contribute to soil contamination, thoroughly examines the movement of pollutants across interconnected ecological systems including soil, water, and air (Figure 2.1). Furthermore, it provides a detailed account of prominent pollution incidents related to EOCs. The insights and observations offered here establish a robust framework for addressing the myriad challenges associated with soil pollution and for crafting comprehensive governance strategies that aim to alleviate the risks linked to EOCs.

 DOI: 10.1201/9781003509523-2

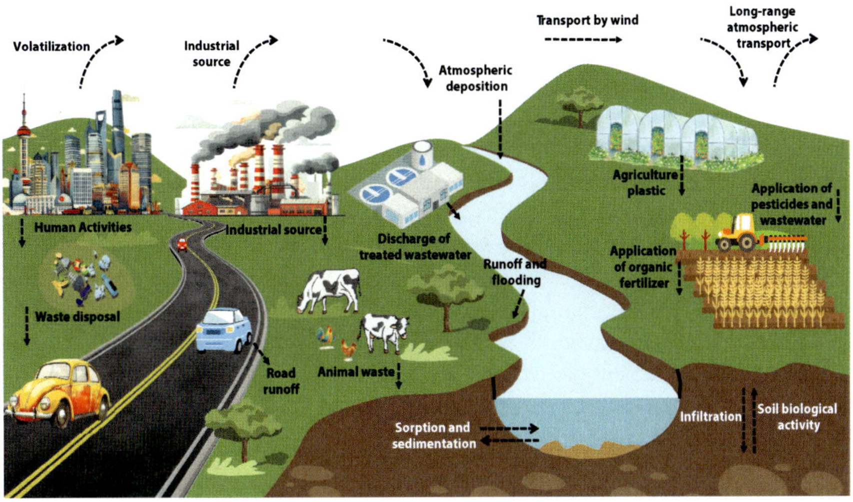

FIGURE 2.1 The pathway and fate of EOCs in soil.

2.2 SOURCES OF EOCs

Soil is widely acknowledged as a primary repository for chemical wastes. The escalating concentrations of EOCs in terrestrial environments are increasingly recognized as posing considerable threats to ecosystem integrity and human health. These EOCs are introduced into the environment through multiple channels, broadly classified into point and nonpoint (diffuse) sources. Point source pollution is specifically associated with distinct, identifiable locations of discharge that have a localized impact on the environment, with typical examples including the application of pesticides, the utilization of organic fertilizers, the discharge of wastewater, the disposal of solid waste, and emissions from industrial operations. Conversely, nonpoint source pollution originates from indeterminate sources and is dispersed across wider areas, often emanating from activities such as atmospheric deposition and surface water runoff, which complicates efforts to manage and pinpoint the sources of pollution.

2.2.1 Use of Pesticides

Since the initiation of agricultural practices, the menace of pests and crop diseases has been ever-present, compelling the development of various mitigation strategies to curb crop damage and enhance yield. To this end, the utilization of pesticides has become an integral component of modern agricultural practices, extensively employed to ensure crop protection. According to a report by the Food and Agriculture Organization, agricultural pesticide usage saw an uptick to 3.70 million tons of active ingredients in the year 2022, which not only represents a 4% increase from the previous year but also signifies a 13% rise over the past decade, and a doubling of use since 1990. Particularly in low-income regions, there has been a notable

153% increase in pesticide use over this period. Beyond agricultural fields, pesticides are also extensively used in urban and peri-urban agriculture—engaging approximately 800 million people globally—to control weeds and disease vectors. However, the widespread use of pesticides is not without its drawbacks, as it will generally remain in environmental media.

Pesticides are introduced into the soil environment through various channels including spray drift, surface runoff, treatments on seeds, and the washing off residues from plant surfaces by rainfall or irrigation. Neonicotinoid pesticides, which are widely used insecticides, serve as a notable example. Despite the prohibition of neonicotinoid insecticides in some areas due to their detrimental effects on pollinators such as honeybees, they continue to pose significant concerns as EOCs. Data from the UK Food and Environment Research Agency reveal that between 2012 and 2016, around 120 tons of neonicotinoids were utilized on approximately 2 million hectares of farmland in the UK. In the Philippines, investigations have found residues of these chemicals in 78% of the soil samples tested, with concentrations varying from 0.017 to 113.5 μg/kg based on the type of crop grown (Bonmatin et al. 2021). Research conducted in the cocoa-producing areas of Ghana detected neonicotinoids in more than half of the soil samples, with levels ranging between 4.3 and 251.4 ng/g (Dankyi et al. 2014). Urban soils have also been impacted, with these contaminants detected in parks and residential areas in Tianjin, China, at concentrations ranging from 0.23 to 230.76 ng/g (Zhou et al. 2018).

Pesticide pollution can arise from several nonintentional sources, including accidental spills, pipe leaks, breaches in containers, or through the unsuitable disposal practices at waste sites and subterranean storage facilities. Compromised storage conditions, deteriorated containers, and inadequate handling processes intensify these problems. Furthermore, a lack of awareness regarding the correct dosages and methods of pesticide application frequently contributes to their excessive use, leading to the buildup of agrochemicals in the soil. This accumulation is often exacerbated by the use of contaminated water for irrigation, the excessive application of tainted compost, and the utilization of home-produced fertilizers.

2.2.2 Organic Fertilizer

The recovery of organic agricultural waste represents a sustainable and eco-friendly method of disposal. Utilizing organic fertilizers, including animal manure, compost, septic tank sludge, sewage sludge, municipal biosolids, and residues from food processing, is crucial in the implementation of waste recycling strategies.

2.2.2.1 Feces and Animal Waste

The utilization of veterinary antibiotics in concentrated animal feeding operations poses a substantial environmental challenge, especially in countries such as China, Brazil, India, the United States, and Australia, where the practice of intensive livestock farming is widespread. PPCPs enter the bodies of both humans and animals, where they undergo only partial metabolic transformation before being excreted as either unchanged agents or metabolites through urine and feces. The management of animal manure, which can include systems such as solid, slurry, or lagoon-based setups, or a combination of these, varies depending on the content of manure solids and the methods used for collection, storage, transportation, and application in fields. Common pretreatment processes such as anaerobic digestion and composting are often utilized. Yet, an increasingly pressing issue is the accumulation of antimicrobial residues and the proliferation of antimicrobial-resistant organisms in soils that have been amended with organic fertilizers made from human or animal waste. Antimicrobials are commonly administered to livestock for disease prevention and growth promotion, with studies showing that approximately 75%–90% of these substances remain unmetabolized in the bodies of both animals and humans. As a result, these residues accumulate in various waste products such as manure, municipal sewage, wastewater sludge, and biosolids, further exacerbating environmental contamination.

2.2.2.2 Sewage Sludge and Biosolids

The composition of wastewater varies considerably, and the diverse treatment technologies employed generate a range of semisolid and solid byproducts that can be used as organic fertilizers. These byproducts include untreated fecal sludge from septic systems, along with primary and secondary wastewater sludge processed through digestion and sedimentation techniques, and biosolids that undergo advanced tertiary treatment and stabilization processes. In 2020, several European countries, including Norway, Sweden, Finland, France, Ireland, and Denmark, reported that more than 70% of their sewage sludge was applied to land. Data from the U.S. Environmental Protection Agency show that in 2023, around 4 million dry tons of treated sewage sludge were generated. Out of this, 60% was applied directly to land to enhance soil fertility, 24% was directed toward landfill disposal, and 14% was processed through incineration (EPA 2023). These sludge residues are rich in essential soil nutrients such as potassium, nitrogen, manganese, and iron, contributing significantly to soil health and productivity. Despite these uses, the input of industrial wastewater often carries lipophilic and hydrophobic pollutants, which tend to persist in the solid fraction even after extensive treatment processes.

Sewage sludge and biosolids continue to be notable sources of EOCs. For example, an analysis of 75 preserved sludge samples from 67 wastewater treatment facilities across the United States revealed the presence of 20 varieties of organophosphate-triesters at an average concentration of 1290 ng/g and 11 varieties of organophosphate-diesters at 78.4 ng/g (Wang, Kannan, et al. 2019). Similarly, the range of organophosphate esters (OPEs) in sludge samples from the Pearl River Delta in China was noted to be between 1.54 and 2650 ng/g (Lao et al. 2023). Annually, the application of sludge in China is estimated to introduce 330 kg of organophosphate-trimesters and 134 kg of diesters into the soil (Fu et al. 2017). Furthermore, due to their hydrophobic nature, plastic additives such as plasticizers and antioxidants tend to accumulate in sludge and biosolids. For example, phthalates, owing to their low solubility in wastewater, are commonly concentrated in biosolids.

Microplastics persistently accumulate in sewage sludge due to the absence of regulatory frameworks in Europe and the United States that consider plastics as contaminants requiring removal. Standard sludge treatment processes, including lime stabilization, anaerobic digestion, and thermal drying, fail to effectively eliminate these particles. Research indicates that more than 90% of microplastics that enter wastewater treatment facilities are ultimately retained in the sludge. The density of microplastic particles in sludge has been quantified, showing ranges from 6.44×10^8 to 1.67×10^{12} particles per 1000 tons (Harley-Nyang et al. 2023). Annually, the application of sludge to fields results in substantial transfers of microplastics to agricultural soils—estimated at 63,000–430,000 tons in Europe and 44,000–300,000 tons in North America, figures which notably exceed those observed for marine environments. This makes sewage sludge a primary vector for microplastic pollution in terrestrial ecosystems.

2.2.2.3 Compost

Organic fertilizers, derived from the composting and fermentation of biowaste sourced from municipal, household, and industrial activities, including green and food residues, are essential for sustainable agriculture. This composting process is crucial for augmenting soil fertility and structure. The European Composting Network recorded in 2022 reveal that the European Union's production of such compost was approximately 42 million tonnes, of which around 21 million tonnes will be applied to farmland. However, alongside nutrients, compost can also serve as a vector for EOCs. Recent scholarly work further underscores a significant issue: the inadvertent introduction of microplastic contaminants into the soil through such organic fertilizers. Suboptimal waste management and insufficient segregation practices contribute to plastic contamination in compost made from biowaste. Although procedures such as pre- and post-composting screening and manual sorting remove the majority of visible plastics, residual plastics are still detected in the compost's final composition. Poor waste management and inadequate sorting practices lead to plastic contamination in biowaste compost. Although screening and manual sorting before and after composting remove most visible plastics, microplastics that are difficult to remove still remain. A study by Okori et al. measured plastic concentrations in three different compost types from composting facilities, finding them to range from 2.38 to 180 mg/kg (Okori et al. 2024). When considering typical application rates, this equates to annual plastic additions to soils of 0.016–1.26 kg/ha with a 7 t/ha application rate and 0.08–6.3 kg/ha with a 35 t/ha application rate. These estimates, while conservative, do not encompass contributions from microplastic or nanoplastic particles. Higher concentrations of plastics have been recorded in other research; for instance, compost samples in Slovenia demonstrated an average plastic content of 1200 mg/kg (Gajst 2016). Urban-derived compost is notably a significant pathway for the transmission of plastic pollutants into soil, posing increased risks in agricultural zones with steep slopes, such as vineyards where erosion may further disseminate plastic-infused soil into adjacent environments.

The degradation of plastics results in the liberation of additives, which subsequently behave as EOCs and infiltrate soils through compost applications. An analysis conducted on 12 bio-based fertilizers uncovered the presence of microplastics, with concentrations varying from 15 to 2912 particles/g of dry weight. Phthalate levels were similarly detected, ranging from 20 to 8000 μg/kg, highlighting a significant environmental concern (Estoppey et al. 2024). Another research shown that the content of phthalic acid esters (PAEs) in fertilizers ranges from 1.17 to 2795 μg/kg

(Mo et al. 2008). Plant cultivation activities in summer increase fertilizer application compared to spring and winter, which has led to relatively high soil PAEs levels in summer in many regions. These findings corroborate previous studies indicating that the introduction of fertilizers to soil environments significantly enhances the accumulation of phthalates, with variations observed depending on the specific types and quantities of fertilizers used. Moreover, a comparative study highlighted that phthalate concentrations were substantially greater in fields dedicated to vegetable and fruit production than in those used for rice and cotton cultivation. This discrepancy is likely due to the more intensive application of fertilizers in horticultural fields. These findings collectively emphasize the role of fertilizers not only in enhancing crop yields but also as significant vectors for the transmission of phthalate contaminants into agricultural soil, raising concerns about the long-term sustainability of such agricultural practices.

2.2.3 Irrigation Wastewater

The composition and levels of pollutants found in wastewater depend significantly on the source and the employed treatment methodologies. A diverse assortment of persistent, bioactive chemical compounds, which includes analgesics, antibiotics, antiepileptic medications, beta-blockers, lipid-regulating agents, steroids, and hormone-based treatments, are consequently released into ecosystems through both wastewater effluents and sludge. Due to their hydrophilic nature, part of these EOCs often remain in aqueous environments even after undergoing water treatment processes.

The practice of using untreated wastewater for irrigation introduces substantial health hazards by incorporating various pollutants into the ecosystem. While the utilization of untreated wastewater in agricultural irrigation is officially reported by only a select few countries, the volume utilized in certain areas is considerable, with estimates of up to 1,840 hm^3 per year in the Mezquital Valley of Mexico over the past decade. In contrast, regions facing acute water shortages, like the Near East and North Africa, commonly resort to using treated wastewater for irrigation. In such areas, Israel stands out where about 40% of the agricultural produce is cultivated using treated wastewater. This method significantly reduces reliance on chemical fertilizers; however, it also presents persistent concerns about the safety of agricultural outputs due to the potential presence of residual pollutants, which may compromise food safety and public health.

Despite the implementation of secondary and tertiary treatment methods, wastewater treatment technologies still often struggle to completely eliminate EOCs, a persistent challenge noted by researchers. The effectiveness of routine surveillance and monitoring of these hazardous substances is generally insufficient, which allows a variety of EOCs to persist in treated effluents. These include a range of PPCPs, antibiotics, phthalates and their derivatives, perfluorinated compounds, OPEs, and microplastics, all identified as significant pollutants. Take PPCPs as an example, which are widely used in daily life. Residues from personal care products, including cosmetics, detergents, soaps, sunscreens, toothpaste, and perfumes, cling to the skin and enter aquatic systems through activities such as bathing and showering, thereby continuously polluting the environment. Antibiotics represent another major group of contaminants; a significant portion of these substances, including tetracyclines (75%–80%), erythromycin (50%–90%), and lincomycin (60%), are not fully metabolized by humans or animals and thus enter the environment predominantly via wastewater irrigation, posing risks to soil health and water quality (Cycoń et al. 2019). The concentrations of antibiotics in treated wastewater can fluctuate considerably, typically ranging from nanograms to micrograms per milliliter, influenced by the effectiveness of the treatment processes employed. Furthermore, wastewater from food packaging and processing sectors frequently contains pesticides and fungicides, which subsequently contaminate both water and soil. These residues, stemming from crop treatment and storage procedures, find their way into wastewater systems, exacerbating pollution levels.

2.2.4 Agricultural Plastic Waste

In recent decades, the application of plastics within the agricultural sector has seen a significant increase, a trend documented extensively in recent studies. Research indicates that plastics, such as mulch films, greenhouse coverings, irrigation pipes, and protective netting, play a pivotal role in these improvements. By 2016, the global market for agricultural plastic films was estimated to weigh around 4 million tons, with forecasts suggesting an annual increase rate of 5.6% through 2030. Furthermore, data from 2019 reveal that the European Union's agricultural sector utilized approximately 707,800 tons of nonpackaging plastic products. Out of this total, 54% was designated for livestock production, while the remaining 46% supported crop production activities.

Plastic mulch films have become essential in modern agriculture, providing substantial benefits by insulating seedlings, decreasing water evaporation, and controlling both weeds and pests, which together contribute to increased agricultural yields. Over two decades, from 1991 to 2011, the adoption of plastic mulch in China has surged, resulting in a fourfold increase in usage. During the same period, European countries also utilized plastic mulch extensively, covering about 427,000 ha (Steinmetz et al. 2016), with projections indicating further expansion in use. In addition to mulch films, the utilization of greenhouses, constructed primarily from plastics, facilitates the extension of growing seasons and enhances food production throughout the year. While affluent nations often prefer glass structures for their greenhouses, developing countries find plastic films an economically viable alternative, offering significant cost savings. Plastic is also deployed across other agricultural applications including seedling trays, pots, containers for pesticides, baling films, and mosquito nets. Despite the pivotal role these items play in enhancing agricultural efficiency, the predominance of single-use plastics in these applications has led to low recycling rates, thus exacerbating the challenge of plastic waste management.

The practice of inadequately recovering and disposing of agricultural plastics has led to significant contamination of soils across various regions. Fragmentation of plastic waste from agricultural practices into microplastics and nanoplastics occurs through various mechanisms such as UV exposure, weathering, and mechanical stress. Research has demonstrated that these fragments, once deposited on farmland, continue to degrade due to environmental conditions such as freeze-thaw cycles and interaction with soil biota. Investigations have revealed notable quantities of polyethylene residues, such as an average of 0.11 ± 0.07 m^2 plastic/m^2 of soil in horticultural soils as identified by Ramos et al. (2015). Piehl et al. (2018) detected 206 macro-plastics/ha and 0.34 ± 0.36 microplastics/kg soil. Huang et al. (2020) further confirmed that the extensive use of plastic mulch contributes significantly to microplastic contamination in agricultural soils.

Recent studies have highlighted the presence of carcinogenic and endocrine-disrupting compounds within plastic residues found in agricultural soils (Rillig et al. 2021). Among these additives, phthalates, which are only loosely bonded to the plastic polymers, are particularly notorious for their capacity to migrate from the plastics into the surrounding soil and water bodies. This migration is alarmingly evident in agricultural soils where plastic mulches are used, as studies have consistently shown a substantial increase in soil phthalate levels in such areas (Xu et al. 2020). Kong et al. (2012) further documented that soils under plastic film in agricultural settings exhibit phthalate concentrations that are 74% higher than those in cropland soils and 208% higher than in vegetable garden soils without plastic film. The tendency of these chemicals to volatilize from the soil also contributes significantly to atmospheric pollution. The degree of phthalate release is influenced by several factors, including the color, thickness of the plastic film, and the prevailing environmental conditions such as temperature, with darker and thicker films typically facilitating greater pollutant discharge at increased temperatures. Moreover, the chemical breakdown of these additives through oxidation processes can result in the formation of new contaminants, which may be even more harmful than their precursors, thus exacerbating the pollution of soils.

2.2.5 Human Activities

As urban areas expand and more individuals relocate from rural to urban settings, over half of the world's population now inhabits cities. Human activities in densely populated areas have become a primary driver of soil contamination, necessitating enhanced monitoring and management. Domestic sources play a significant role in this process, introducing a wide array of organic and inorganic pollutants into urban soils.

These contaminants originate from household cleaning agents, personal care products, paints, solvents, FRs, indoor pesticides, and activities related to home gardening. For instance, brominated flame retardants (BFRs), which are integral to the manufacturing processes of electronic circuitry, construction materials, and various textiles, play a substantial role in the contamination of urban soils. Their presence is especially significant around areas associated with waste disposal and industrial activities. Such locations often result in the contamination of dairy and poultry products, such as milk and eggs. Another typical group of urban contaminants is phthalates, widely used as plasticizers in industrial applications, food packaging materials, and PPCPs, and are recognized as predominant pollutants in urban soils. Their widespread occurrence in these environments is primarily due to their notable volatility. Detailed analyses of urban soils in Paris have revealed that levels of diethyl phthalate, di-*n*-butyl phthalate, and diisononyl phthalate are substantially elevated, being 12, 10, and 7 times higher, respectively, compared with their concentrations in rural agricultural soils (Tran et al. 2015). Moreover, investigations in residential areas of Xi'an, China, have identified significantly increased concentrations of low molecular weight phthalates, which are predominantly linked to the usage of PPCPs (Wang et al. 2018). Studies conducted on soils in Moscow further corroborate that urban emissions, especially proximate to landfill sites, serve as a principal source for phthalate contamination (Brodskiy et al. 2019).

Moreover, vehicular traffic intensifies soil contamination as it produces considerable quantities of microplastics and rubber particles from tire wear. These particles, mainly composed of synthetic polymers, are dispersed into the soils adjacent to roads through both atmospheric deposition and water runoff, thus presenting risks to the soil biota. Notably, the annual emissions of tire dust are substantial, with figures reported at approximately 10,000 tons in Sweden and reaching up to 110,000 tons in Germany (Kumar et al. 2025). The concentration of polybrominated diphenyl ethers (PBDEs) in road dust was 23 700 ng/g, which was significantly higher than that in surrounding indoor dust, indicating that traffic may be an important source of PBDEs in road dust (Cao et al. 2014). Additionally, in highway-proximate zones, the concentrations of OPEs ranged from 130 to 430 ng/kg, whereas in less populated areas, the detected levels of tris(2-chloroisopropyl) phosphate (TCPP), tris(1,3-dichloro-2-propyl) phosphate (TDCPP), and tributyl phosphate (TBP) were 68.0, 29.0, and 19.0 ng/kg, respectively (Marklund et al. 2005). This is because, during the use and aging of vehicles, OPEs can enter the atmosphere, dust and soil through volatilization, dust and wear. Urban soils are an important pathway for exposure of humans and other organisms to contaminants, which poses serious health risks, especially for children who are vulnerable to direct contamination through frequent contact during play.

2.2.6 Waste Disposal

Waste disposal practices, including landfilling, open burning, informal recycling, and illegal dumping, are major sources of EOCs in soil environments. These practices facilitate the release and accumulation of hazardous substances such as phthalates, bisphenol A, FRs, per- and polyfluoroalkyl substances (PFASs), and microplastics.

Leachate from landfills serves as a significant pathway for EOCs to permeate both soil and groundwater. A comprehensive analysis conducted on organic pollutants in landfill leachate worldwide highlighted the ubiquitous presence of contaminants such as phthalates, bisphenol A, and analgesics, antimicrobials. Moreover, landfills are crucial disposal sites for wastes containing perfluorooctanesulfonic acid (PFOS), perfluorooctanoic acid (PFOA), and other PFASs, which are consistently detected in landfill leachate, as corroborated by numerous studies.

Open burning and incineration of municipal and medical waste are still widely used, particularly in low- and middle-income regions, to reduce waste volumes before or instead of landfill disposal. While incineration is increasingly adopted in middle- and high-income nations, accounting for about 11% of the world's waste management strategies, this process releases various organic pollutants, including pesticides and FRs, into the atmosphere during combustion. Soils located near areas where waste is openly burned or incinerated typically exhibit higher concentrations of harmful contaminants, such as BFRs and plasticizers. For instance, PFASs levels in soil downwind of the Norlite hazardous waste incinerator in Cohoes, New York, were twice as high as those upwind. Although some PFASs compounds are destroyed during the incineration process, the ash produced still contains harmful concentrations of these substances, posing a risk of leaching into surrounding environments.

Informal waste recycling and collection, critical economic activities in many low-income communities, are also sources of significant environmental contamination. In locations like Mumbai, for instance, it is estimated that 150,000 individuals rely on collecting and recycling waste, primarily plastic, as a means of livelihood. These activities are often carried out under suboptimal conditions that do not adhere to safety standards, resulting in the accumulation of hazardous pollutants in areas where these workers operate, which frequently coincide with their living spaces. The exposure to dangerous chemicals, including phthalates and BFRs, is a common consequence for the recyclers and their families, particularly affecting children who are more susceptible to toxic influences.

The escalation of e-waste generation in the digital age has introduced additional challenges. This type of waste is laden with enduring and hazardous substances, including microplastics, PFASs, and BFRs, all of which pose substantial environmental threats. Inadequate disposal practices, such as co-disposal with municipal waste or informal burning, enhance their mobility and environmental impact. Moreover, chemicals like PBDEs, which are used as FRs, may leach into the environment through the incomplete combustion of e-waste or its natural degradation in landfills. The swift turnover of electronic devices has led to substantial quantities of toxic materials, such as phthalates and BFRs, being deposited into landfill sites. Notably, PAEs, recognized for their volatility, have been extensively detected across various soil environments, highlighting the pervasive nature of these contaminants.

Littering and illegal dumping of solid waste, particularly plastic, are further sources of soil contamination. From the period of 1950 to 2015, the world produced approximately 6.3 billion tons of plastic waste, out of which nearly 4.97 billion tons have either been deposited in landfills or accumulated in the natural environment, significantly impacting ecosystems. Microplastic particles frequently migrate into soil ecosystems via mechanisms such as runoff, wind transportation, or flooding events. Research conducted by Kim et al. demonstrated that the majority of debris washed away from highways during storm events is composed of nonbiodegradable materials, primarily plastics, with observed waste loads varying from 0.85 to 6.6 kg/ha (Kim et al. 2006). Despite the common presence of microplastics in soils near roads and illegal dumping sites, comprehensive studies quantifying the extent to which these microplastics permeate the soil through such activities are lacking, necessitating more focused research to inform effective pollution control strategies.

2.2.7 Industrial Sources

Contaminants from industrial manufacturing processes often result in the deposition of EOCs in soils near the production sites, influenced by the nature of the products manufactured and the specific waste treatment or production methods employed.

These pollutants typically originate from both intentional releases and unintentional leaks into the environment, adversely affecting the soil, adjacent water bodies, and the atmosphere. For example, soils in proximity to pesticide manufacturing plants and areas where these chemicals are heavily applied often exhibit detectable levels of pesticide residues. In Sydney, Australia, analyses of soil samples from the vicinity of a former chlorinated plastics manufacturing site revealed the presence of approximately 6.7% chlorine-rich microplastics (Fuller and Gautam 2016). Similarly, research conducted in a PFASs production region in China identified soil concentrations ranging from 0.10 to 2.34 μg/kg for perfluoroalkyl sulfonic acids (PFSAs) and 1.89 to 32.6 μg/kg for perfluoroalkyl carboxylic acids (PFCAs) (Jin et al. 2015). Research conducted across seven countries spanning four continents showed that the mean concentrations of ionizable PFASs (i-PFAS) and fluorotelomer alcohols (FTOH) in the atmosphere were around 10^2 pg/m^3, whereas the concentration of i-PFAS reached 10^4 pg/m^3 near a fluorochemical manufacturing site in China (Wang, Munir, et al. 2023). Sites of abandoned industrial facilities, such as old pesticide factories, also present significant risks of residual contamination. For instance, studies have shown that construction and demolition debris from decommissioned pesticide production facilities contain elevated concentrations of pesticides, highlighting the lasting impact of historical industrial activities on environmental health.

Aside from the manufacturing sector, the construction industry also plays a major role in contributing to environmental contamination through the release of EOCs. Construction materials are categorized into inert components such as soil, bricks, and concrete, and noninert components such as plastics, sealants, wires, insulation, and treated wood. These reactive materials frequently encompass dangerous compounds, including brominated and chlorinated flame retardants, various plasticizers, and preservatives. A significant number of these substances are classified as persistent organic pollutants (POPs). In developed countries, waste from construction and demolition activities accounts for about 30% of total waste production. Consequently, the mishandling of construction waste significantly endangers environmental health.

2.2.8 Atmospheric Deposition

Atmospheric deposition serves as the primary pathway through which EOCs are introduced into soil environments, including both dry and wet deposition modalities. These processes vary considerably among different EOCs, such as PFASs, OPEs, and microplastics, which demonstrate distinctive patterns of movement and deposition in soil matrices. The deposition rates and concentrations of these substances in soils are significantly influenced by several environmental factors. These include the proximity to industrial emission sources, the directions and speeds of local wind currents, and the specific sizes and types of particles carrying the contaminants. These elements collectively determine the geographic distribution and concentration levels of EOCs within different soil types.

EOCs enter the terrestrial environment through dry deposition after rapid evaporation and entry into the atmosphere. During this process, EOCs such as OPEs often adhere to atmospheric particulates, descending to the surface where they may be integrated into plant systems, soil substrates, and aquatic environments.

The dimensions of these particulates are critical in determining their dispersion in the environment. In the case of OPEs, frequently detected in indoor settings such as residential, commercial, and vehicular interiors, concentrations have been measured ranging from 11.4 to 1406 ng/m^3 (Wang, Zhu, et al. 2020; Dong et al. 2025). Ambient outdoor air samples from varied environments such as urban centers, rural areas, e-waste dismantling sites, and airports have shown OPE concentrations fluctuating between 1.32 and 202 ng/m^3 (Li et al. 2019; Olivero-Verbel et al. 2021). These compounds may escape indoor environments via air exchange and dust expulsion systems, reaching outdoor environments where they are found in varying concentrations in dust. Research conducted in the subtropical region of South China recorded an average OPE dry deposition rate of 2100 ± 980 ng/m^2/day onto paddy fields (Zhang et al. 2020). Moreover, wind-mediated transport mechanisms also influence the distribution of microplastics, typically originating from poorly managed waste facilities and urban infrastructures. A study around Paris noted microplastic deposition rates of 29–280 items/m^2/day, predominantly fibers, half of which were longer than 1000 μm (Dris et al. 2015). Further studies in suburban Paris found lower deposition rates, suggesting potential long-distance transportation to isolated terrestrial and aquatic habitats. Despite limited research on the atmospheric deposition of microplastics, available data indicate that this process is a notable contributor to microplastic contamination in urban and suburban soil layers.

Precipitation, encompassing both rain and snow, serves as a crucial vector for the transmission of EOCs from the atmosphere to the soil. This phenomenon is thoroughly examined in various studies across continents such as North America and Asia, highlighting its global relevance. In regions such as North Carolina, evidence suggests that the volume of EOCs deposited through wet deposition is remarkably greater, estimated at 20 times the rate observed for dry deposition. Snow samples from isolated Arctic locations as well as highways have shown high concentrations of OPEs. Specifically, Σ *n*-PFAS concentrations have been recorded at 10^5 pg/m^3 in the Arctic Ny-Ålesund, slightly higher than the concentrations found in ambient air at 10^1 pg/m^3 (Xie et al. 2015). In agricultural zones such as the rice fields in Nanning, China, wet deposition accounted for a significant 53% of the total OPE deposition (Zhang et al. 2020). Research also indicates a correlation between the distribution of OPEs in drinking water, rainwater, and groundwater, suggesting the potential for these contaminants to permeate groundwater systems and consequently integrate into soil substrates. These findings underscore the likelihood of such contaminants being deposited onto surface soils predominantly through precipitation processes.

The process of long-range atmospheric transport (LRAT), coupled with the dynamics of wet deposition, plays a pivotal role in the distribution of PFASs across geographically remote areas such as the Arctic and Antarctic. This highlights the extensive capacity for global circulation of these chemical contaminants. Similarly, POPs and other classes of organochlorine compounds are also transported over vast distances through the atmosphere before they are eventually deposited on terrestrial surfaces. These movements of pollutants are not confined by national borders as they navigate through various atmospheric pathways in both longitudinal and latitudinal directions.

2.2.9 Runoff and Flooding

Erosive runoff and flooding events are instrumental in the widespread dispersion of pollutants, channeling them from urban territories and various land uses into adjacent soil environments. Atmospheric pollutants captured within watersheds are subsequently carried and concentrated via runoff and floodwaters. A detailed investigation conducted by Bai et al. (2018) in Denver, Colorado, examined the presence of emerging contaminants in surface water sources influenced by both seasonal snowmelt and effluent outputs from several wastewater treatment facilities. Their research revealed that out of 144 EOCs monitored, 109 were detected in the water at concentrations reaching several hundred nanograms per liter despite the dilution effects of natural water flows (Bai et al. 2018). Petrie et al. (2015) gathered and analyzed data regarding the levels of pharmaceuticals surface waters across the UK, reporting approximately 70 pharmaceuticals, with the highest concentrations observed in river waters for the analgesic tramadol, up to 7731 ng/L. Furthermore, organic contaminants generally exhibit strong binding affinities for soil-derived organic matter and clay minerals, which facilitates their enhanced transport and dispersion throughout aquatic environments. Over time, these contaminants are known to accumulate in the soil, leading to enduring environmental pollution. The dynamics of contaminant movement and accumulation in soils have been further evidenced in the studies by Dias-Ferreira et al. (2016) and Paule-Mercado et al. (2016). Both studies documented heightened levels of pollutants, highlighting the downstream migration and buildup of these substances in soil ecosystems. These findings collectively emphasize the critical role of hydrological processes in the transportation and deposition of pollutants, which results in prolonged contamination of soils in affected areas.

2.3 MIGRATION OF EOCs IN WATER-SOIL-AIR

The dispersal of EOCs in soil environments follows both horizontal and vertical trajectories. Horizontally, EOCs migrate primarily through surface runoff and lateral displacement, processes where contaminants traverse the top layers of the soil. The efficiency and extent of this horizontal migration are substantially influenced by the soil's physical structure, its compositional elements, and the inherent properties of the EOCs themselves, such as solubility and sorption capacity. Vertically, EOCs penetrate deeper into the soil profile, propelled by infiltration and leaching facilitated by the soil's porosity and the activity of various soil organisms, which modify the soil matrix. Horizontal movement is crucial in defining the scope and boundary of EOC contamination across landscapes, potentially influencing adjacent land and water systems through widespread dispersion. Conversely, vertical migration impacts the depth of pollution penetration, posing significant challenges for remediation as contaminants reach groundwater or deeper soil layers less accessible for treatment. Acknowledging both the lateral and vertical pathways of EOC migration is crucial for a comprehensive assessment of the environmental risks posed by these contaminants.

2.3.1 Transport of EOCs in the Atmosphere

Once released into the atmosphere, these volatile and semi-volatile organic contaminants, such as OPEs, spread far and wide traveling on the wind. To a significant extent, their spread is determined by the physical and chemical characteristics specific to their particular type. Pollutants with higher volatility are more likely to remain airborne and so travel farther distances. However, particulate-type pollutants settle relatively fast to the ground or water bodies by the forces of gravity. Key factors influencing their atmospheric behavior include volatility, polarity, molecular weight, meteorological conditions (e.g., wind speed, temperature, precipitation), and topography. For instance, volatile organic compounds (VOCs) are highly volatile substances that will take the gas phase and can be transported by air currents tens to hundreds of kilometers. But semi-volatile organic compounds (SVOCs), in contrast, bind much more readily to atmospheric particulates, limiting their range of travel. Additionally, some types of plastic particles (those with a low density) can become airborne, polluting in a diffused fashion across different environmental contexts (Chen et al. 2020). EOCs, such as long-chain PFASs compounds, are also capable of binding to microplastic particles, which may facilitate their dispersion through airborne transport under specific environmental conditions (Borthakur et al. 2022).

The LRAT of EOCs involves a series of complex interactions dictated by both physical and chemical dynamics within the atmosphere. These contaminants are subjected to continuous processes of diffusion, deposition, and subsequent resuspension. The journey of these pollutants is heavily influenced by meteorological variables such as wind speed, wind direction and temperature, which are critical determinants of the pollutants' travel and dispersion patterns. Strong wind currents have the capacity to propel these contaminants several hundreds or even thousands of kilometers downwind, markedly expanding their area of influence. At cooler temperatures, there is a greater propensity for EOCs to adsorb onto airborne particles. These particulates facilitate the transport of EOCs to terrestrial and aquatic environments far from their initial emission points through atmospheric deposition, thereby promoting a global spread of these pollutants. The phenomena of "global distillation" or the "condensation effect" are particularly pronounced in colder, polar regions. Here, the low temperatures foster the condensation and accumulation of volatile pollutants, which migrate longitudinally from tropical and temperate zones to polar regions due to this cooling effect. POPs, such as PFASs, exhibit long atmospheric lifetimes, allowing them to reach remote areas like the Arctic and Antarctica (Wang, Wang, et al. 2019). The detection of these compounds in such remote locations underscores the vast reach and enduring presence of EOCs in the global atmospheric system, reflecting their profound impact on environmental health worldwide.

2.3.2 EOCs in the Air Settles Directly into the Soil

The transfer of EOCs from the atmosphere to soil environments primarily occurs via dry settling, wet settling, and particle-bound pathways. Gas-phase mechanisms of volatile or semi-volatile EOCs is modulated by the contaminants' vapor pressure, the prevailing temperature, and the soil's sorption capacity for these compounds. Typically, compounds that are more volatile tend to remain in their gaseous state and deposit closer to their source. whereas SVOCs are more likely to bind to soil particles, facilitating their transport over longer distances. The levels of long-chain PFASs, specifically 8:2 FTOH and 6:2 FTOH, experienced substantial reductions of up to 98% and 91%, respectively, within a 5-km perimeter surrounding the emission source (Chen et al. 2018). In the vicinity of a PFASs manufacturing plant, atmospheric concentrations of hexafluoropropylene oxide-dimer acid (Gen-X) decreased markedly from 24.6 pg/m^3 to 1.9 pg/m^3 at a distance of 5 km (D'Ambro et al. 2021). Notably, concentrations of PFOA in soil, 8 km downwind from the source, matched those observed near the source and were considerably higher than in other regions, although PFOS levels were more stable (Schroeder et al. 2021).

EOCs are transported from the atmosphere to the soil via wet deposition processes, e.g., rainfall, snowfall, and fog. This efficiently scrubs EOCs from the atmosphere, terminating in deposition to terrestrial surfaces and into soil profiles. Because the water solubilities of the EOCs used and precipitation pHs can vary greatly, the effectiveness of wet deposition to transport these contaminants is largely dictated. For example, EOCs that are highly hydrophilic, like OPEs and many agricultural pesticides, are typically transported to the soil via wet deposition when present in regions characterized by heavy precipitation. After deposition, these chemicals, because of both their water solubility and polarity, are likely to leach through the soil and actively create conditions likely to enable contamination of ground water systems expanding the area of environmental contamination. Additionally, a study in Japan found that perfluorobutanoic acid (PFBA) and perfluoropentanoic acid (PFPeA) fluxes were different than other PFASs fluxes, and the fluxes of the PFASs fit a pattern of transcontinental transport in atmospheric currents and cloud processes at sites located as far as 1000 km from the Pacific Coastline (Taniyasu et al. 2013). PFSAs has significantly lower solubility than PFCAs and the sulfonic acid groups are more hydrophobic than the carboxylic acid groups, so PFSAs interacts less with rainfall and prefers to bind to larger air particles with a diameter greater than 2.5 μm.

The environmental behavior and impact of EOCs can be significantly altered while deposition through chemical transformations. For example, precursors to PFCAs, such as FTOH, are known to experience photochemical reactions upon deposition onto snow or ice and yield PFCAs via interaction with ·OH. Volatile neutral PFAS compounds can degrade into more polar ionized forms (i-PFAS) as volatiles travel long range within the atmosphere. Due to these transformations, wet deposition processes will render them more likely to be captured in precipitation and deposited in soil profiles (Wang, Munir, et al. 2023). Transformations such as these are important, since the forms, persistence, and toxicity of these compounds may be affected to produce compounds with different environmental and health implications. Air-soil exchange is an important way to evaluate the entry of EOC from the atmosphere into terrestrial ecosystems. TCPP exhibited a high diffuse flux (from −171 to 813 ng/m^2/day), indicating predominant deposition behavior, whereas tris(2-chloroethyl) phosphate (TCEP) showed a large range of volatilization flux ranging from −15.2 to 1100 ng/m^2/day with a median of 14.1 ng/m^2/day, suggesting a high re-emission from soil to atmosphere (Wang et al. 2020). The fugacity fraction for OPEs had values near or below 0.3 near Albany Airport in New York, and close to zero in Arctic climates, implying high deposition in such colder regions (Li et al. 2019), while a value close to 1 observed in Nepal indicated net volatilization from industrial soils to the atmosphere, thus classifying this region as a significant "source" of EOCs emissions (Yadav et al. 2018). The potential for secondary pollution occurs when soil EOC concentrations are elevated and the potential for these compounds to re-volatile into the atmosphere is increased by changes in source dynamics or climatic conditions. This dual role enables soils to act both as a "sink" and a "source" for EOCs, adding layers of complexity to their environmental management.

EOCs can also be brought down to the ground through the influence of gravity, which is also a particle deposition mechanism, where they are bound to aerosol particles or dust in the atmosphere. Particulate matter is a suitable medium for EOC capture because this attachment occurs when particulate matter provides a large surface area and numerous adsorption sites. Particularly, EOCs with large molecular structures tend to stick to suspended airborne particles, transporting these EOCs by particle deposition to collect in soils removed from their sources' original emissions. Particle deposition strongly depends on meteorological factors and affects the efficiency and area. Factors such as wind speed, the intensity and frequency of precipitation as well as geographical topography govern them, the speed with which particles settle, and the pattern of deposition in space. Once settled, they can remain in the soil matrix for extended periods. While these factors can allow them to be re-entrained into the atmosphere or washed into aquatic systems through such processes as wind-driven erosion or water runoff, the pollutants also create a cyclical dispersal and migration pattern.

2.3.3 Aerial EOCs Sinks into Water

Air–water interactions greatly influence the exchange of EOCs between the atmosphere and aquatic environments, particularly for compounds that exhibit both high volatility and hydrophilicity. Transfer efficiency for these pollutants is controlled by the Henry's law constant for each pollutant, as well as; environmental variables (temperature and turbulence of the water body). Particularly, VOCs with high water solubility, such as small-molecule OPEs, are accidentally soluble in water as soon as they get released in the air. Subsequently, these pollutants can either be absorbed by aquatic organisms providing biomass or break down, becoming incorporated in the food chain and interfering with ecological balances. Air–water exchange is a continuous and dynamic process, which repeatedly draws EOCs across the air–water boundary, confounding efforts to control and mitigate their environmental impact. For example, we estimated that German North Sea dry deposition arrives with an OPE load of 710 ± 580 kg annually (Möller et al. 2011). Atmospheric transport further potentiates the levels of OPEs arriving in the Mediterranean and Black Seas, where deposition rates were measured at 21.8–404 and 93.1–470 ng/m^2/day respectively. Applied to annual terms, these figures correspond to an estimated deposition load of 13.0–260 tons/year into each of the Mediterranean and Black Seas (Castro-Jiménez et al. 2014). The significance depicted by this evidence shows a massive scale of OPE deposition on the earth's surface from the atmosphere that has a parallel to such processes in the oceanic environment.

2.3.4 Migration of EOCs in Water

Diffusion and convection constitute the main means of transport of EOCs in aquatic environments. Diffusion is primarily characterized by the autonomous migration of EOC molecules according to a concentration gradient in the water. These molecules diffuse at a rate that is strongly dependent on their molecular architecture, the temperature of the surrounding water, and the fluid dynamics of the aquatic environment. For example, EOCs made up of smaller, water-soluble molecules such as bisphenol A can easily diffuse across the water column. Under such elevated temperatures and minimal water flow, these substances diffuse more rapidly into the atmosphere, increasing the range over which they are distributed and amplifying the environmental effects of their presence.

Advective transport utilizes the bulk movement of the water itself to move the EOCs through water bodies. Factors such as the current of natural currents prevailing, rainfall pattern, and those that trigger disturbance of natural currents at sea surface with wind, all play a huge role in this process. Such dynamics accelerate the spread of pollutants. For example, in riverine systems, contamination can be extended from the area in or around the initial point of entry by the transportation of EOCs downstream. PFASs have been the subject of intense research with elevated levels ranging from 10^2 ng/L in surface and groundwater near industrial sites and wastewater treatment facilities down to 10^1 ng/L in more remote environments (Wang, Ji, et al. 2023). These results highlight wastewater from industry and residential sources as the main delivery vector for PFASs and depict wastewater as a major source of PFASs leaching into surface and groundwater systems. In addition, PFASs have the potential to seep into groundwater flow systems and potentially return to surface waters depending on the geological and hydrological properties of aquifers.

2.3.5 Diffusion of EOCs from Water to Soil

EOCs infiltrate into soils primarily through sorption and sedimentation, as they are already present in water bodies receiving wastewater irrigation or surface run off. Among these pathways, sorption plays a particularly critical role. Due to the high lipophilicity and low water solubility of some EOCs, they tend to bind tightly to organic and mineral components in the soil matrix. The efficiency and extent of this sorption is controlled by several key factors such as the content of organic matter in soil, the ambient pH, the ambient temperature as well as the intrinsic chemical properties of the EOCs. It is particularly concerned with pollutants showing a pronounced hydrophobic character and the likely retention of such pollutants in soils containing high levels of organic matter, extending their environmental residence time and ecological impact.

Extensive research has shown that many environmental conditions (e.g., soil carbon levels, organic material content, and clay mineral composition) are crucial in governing the sorption mechanisms of various EOCs. Adsorption of dissociation-capable molecules, such as some pharmaceutical agents, is more complex depending on the prevailing pH conditions. For example, Schaffer et al., 2011, studied the degradation process of cationic β-blockers in the environment with competitive ions with Ca^{2+} concentrations of 40–400 mg/L. The study found that reduced cation concentrations decrease atenolol mobility likely as a result of less competition for adsorption sites on the anionic surface. Conversely, pharmaceuticals such as diclofenac and ibuprofen, which are negatively charged under typical environmental conditions, showed no significant or little adsorption to neutral or negatively charged surfaces within chalk aquifers. Al-Khazrajy and Boxall (2016) also observed low partition coefficients for some pharmaceuticals in aquifer sediment matrices. These findings are critical for understanding how ionic EOC behavior and eventual fate are modified by ion chemistry shifts such as episodic groundwater recharge, saltwater intrusion, etc. Moreover, the interactions that EOCs have with metal ions such as those observed with tetracycline antibiotics can modify their transport and spatial distribution in soil.

The sedimentation process is instrumental in facilitating the movement of EOCs from aquatic environments to terrestrial soil systems, especially in scenarios where water currents are directed upstream. Typically, contaminants are carried by particulate matter that remains suspended within the water column. As these particles gradually lose buoyancy, they descend and are ultimately deposited onto the soil surface. The overall effectiveness and magnitude of this deposition are determined by several factors, including the behavior and velocity of water flow as well as the inherent characteristics of the particles involved. Key particle attributes such as their size, density, and surface properties play a crucial role in influencing how efficiently pollutants are transferred from water bodies to soil through the sedimentation mechanism. For instance, contaminants such as pesticides and microplastics, which are commonly present in industrial discharge and agricultural runoff, typically exist as part of suspended particulate matter within water bodies. Through the process of sedimentation, these particles accumulate progressively in the sediment layers of rivers, lakes, and other freshwater systems. Over extended periods, this accumulation can result in the gradual release of EOCs from the sediment into the overlying soil matrix. This prolonged release process has the potential to continuously degrade soil quality and compromise environmental health by introducing persistent contaminants into the terrestrial ecosystem.

2.3.6 Migration of EOCs in Soil

The movement of EOCs through soil matrices is an important mechanism controlling their environmental distribution in various compartments. Different types of EOCs exhibit varying degrees of mobility depending on their physicochemical properties. Some categories of PPCPs, e.g., ibuprofen, are certainly mobile. This high mobility permits the infiltration of these compounds into soil layers to accumulate not only within the soil but also in underlying sedimentary deposits (Zi et al. 2023). In contrast, other PPCPs such as bisphenol A exhibit a high affinity to soil particles, resulting in intimate sorption and a very long residence time of the compounds in the soil environment (Zi et al. 2023). Similarly, antibiotics such as tetracycline and ciprofloxacin—due to their positive charges—interact electrostatically and hydrophobically with soil constituents, which restricts their movement and causes them to accumulate in the upper soil layers.

EOCs migrate also based on the inherent properties of the soil itself. Soils with fine-textured pores, high adsorptive material content, and high microbial activity facilitate the vertical movement and dispersion of certain contaminants like carbamazepine and ibuprofen. In contrast, PAEs and BFRs (e.g., PBDE) frequently display a decreasing concentration gradient with depth in the soil. However, this trend is due to their strong sorption to soil particles and their rapid degradation rate, which limit their penetration down into the soil profile (Tran et al. 2022). By contrast, high-density plastics have a greater chance of staying within the soil matrix. Instead, these plastics progressively migrate into soil layers deeper in the soil profile, thereby decreasing their potential for horizontal movement across the soil surface. The movement downward of microplastics is facilitated by practices such as mechanical tillage in agricultural settings. The microplastics are held back in the deeper strata of the soil where they accumulate thanks to this agricultural intervention, thus affecting their general distribution and potential ecological influence.

The EOCs movement in soil is controlled by soil organic matter, moisture, minerals, pH, and ionic strength. Studies have empirically shown that soil organic matter interacts with EOCs. For example, soils with high organic material contain a higher ability to bind these contaminants and limit their mobility in the soil matrix. In contrast, the hydrophilicity of EOCs strongly depends on the moisture content in the soil. Adsorption of contaminants with strong affinity is increased when the soil moisture is lower. Beyond that, electrostatic forces and hydrophobic interactions play a central role in deciding the mobility of EOCs through soil environments (Li et al. 2024). For example, migration behaviors of PFASs in soil are very much governed by their chain length on carbon and their total charge. For instance, PFAS compounds with longer carbon chains will encounter more hydrophobic interactions and therefore tend to stay within the upper layers of the soil, while PFAS compounds with shorter carbon chains will be more attracted to the deeper profile owing to the stronger attraction based on electrostatic attraction (Wang, Munir, et al. 2023). The vertical spread and migration pathways of PFASs across various soil layers are defined by these particular interactions. In addition, other contaminants within the soil environment may affect the migration behavior of the EOCs. Co-occurring contaminants, like heavy metals and humic substances, can alter the surface charge of particles such as microplastics, further affecting the transport and distribution of EOCs. Leaching studies have shown that these interactions can significantly alter migration patterns of both organic and inorganic pollutants.

Biological elements play an important part in the movement of EOCs in soil environments, along with physical and chemical factors. Developments in the plant root systems form a network of channels and voids that allow the vertical transport of EOCs. As an example, microplastics and other microscopic particles can pass along these root-induced pathways and ascend upward into the soil surface or migrate downward into the lower soil layers (Li et al. 2021). It is also important that different types of soil microorganisms, such as earthworms, help redistribute EOCs (Heinze et al. 2021). For instance, earthworms build burrow systems, which make room for plastic particles with a deeper penetration into the soil strata. Other soil fauna (mites, e.g., *Hypoaspis aculeifer*; springtails, e.g., *Folsomia candida*, *Proisotoma minuta*) contribute to the movement of microplastics as mechanical interfering due to scratching and chewing. These behaviors allow the physical displacement of pollutants and redistribute them along the soil profile. These studies collectively show that biological processes can profoundly affect the migration patterns of EOCs and that soil organisms are principal mediators of both vertical and horizontal dispersal and thus the environmental spread of the contaminants.

2.3.7 Redispersion of EOCs into the Air

Volatility and re-volatilization are the principal mechanisms facilitating the migration of EOCs from soil environments into the atmosphere. Dynamics will be a function of factors such as substance volatility, temperature, and soil properties. For example, EOCs with high volatility, such as short-chain OPEs, desorb more strongly from soil particle surfaces into the air than less volatile EOCs, and are thereby at greater risk from weathering under warmer conditions, and may therefore be a significant source of local air pollution. In addition, lower soil moisture can also promote the volatilization of EOCs because it reduces the water capacity to adsorb EOCs and allows it to be more easily volatilized into the gas phase. Another important mechanism is re-volatilization, which means that EOCs that have previously settled into the soil are again released into the atmosphere. This mechanism is especially important at longer time scales and is often observed for POPs such as PFASs. As seasonal temperatures change, or in periods of prolonged heat, such as summer, these compounds may be released back into the atmosphere. The effect of this is that pollutant levels in the air are revived, and may even worsen air quality and put everyone and the ecosystems around us at risk if not viewed with caution.

In addition, biological processes were shown to play a significant role in the transfer of EOCs between soil and the atmosphere. These pollutants are absorbed and partially broken down through various soil microorganisms and by plant root systems. These contaminants are not fully decomposed, and some segments may subsequently volatilize and escape into the atmosphere. Dynamic exchange with the soil–plant system occurs with certain substances such as antibiotics or pharmaceutical residues, which may be emitted into the atmosphere via plant transpiration processes. Moreover, floral growth cycles and soil microbial activity are also major factors regulating the rate of release of the pollutants. Biological factors are therefore important not only for determining the duration of retention of EOCs by the soil but also for helping control the rate and way of EOC migration to the atmosphere through the synergistic action of plants and microbial organisms.

2.4 CASE STUDIES OF EOCs CONTAMINATION INCIDENTS

The problem of soil contamination by EOCs has intensified into a pressing environmental crisis that necessitates urgent intervention. As previously discussed, the sources of EOCs in the soil are diverse, stemming from various industrial and agricultural practices. This section overview examines multiple exemplary instances of contamination, highlighting that the processes by which contaminants entering the soil are both intricate and diverse. This comprehensive review highlights the critical need for strengthening the management of pollution sources and intensifying efforts toward the remediation of contaminated soils. Such actions are essential for reducing the environmental and health impacts of soil pollution and safeguarding public health and ecological systems.

2.4.1 PFAS Leakage at US Military Base in Japan

In August 2024, an incident involving a PFAS leak was reported at the Yokota Base of the US Military in Japan, according to Japan's Asahi Shimbun (Yajima and Tajima 2024). The source of the leak was identified as an overflow of water containing organic fluorides from the storage reservoirs within the base, leading to potential external spillage. This release of PFASs is not only a significant environmental concern but also raises alarms about the possible enduring health implications for the surrounding community. PFAS compounds are known to be linked with several health issues, such as liver damage, compromised immune system functionality, and disruptions in reproductive development. The incident at Yokota Base illustrates a classic case of soil pollution resulting from anthropogenic activities and accidental discharges, posing serious risks to both local ecosystems and the health of nearby residents.

2.4.2 The Neonicotinoid Ban and the Fight to Protect Pollinators

Pest problems in farmlands cause many farmers considerable frustration. Neonicotinoid pesticides, known for their high efficiency and broad-spectrum insecticidal properties, quickly gained popularity as powerful tools for pest control in agricultural production. Farmers widely adopted these pesticides, hoping to boost crop yields. However, with the increasing use of these chemicals, scientists began to notice some unusual phenomena—notably, a dramatic decline in bee populations. More and more research revealed that neonicotinoids affected the insects' nervous systems, causing bees and other pollinators to lose their sense of direction, unable to return to their hives, ultimately leading to their deaths (Amor et al. 2024). This phenomenon sparked global concern, with environmental organizations and the public questioning the widespread use of these pesticides. Although the decline in bee populations is primarily attributed to neonicotinoid residues in plants, soil contamination has also emerged as a significant contributing factor. This is due to the fact that neonicotinoids can persist and accumulate in agricultural soils, from which they are taken up by crops and subsequently transferred to pollinators via pollen and nectar.

The European Union took the lead, starting in 2013, by banning three types of neonicotinoids for use on outdoor crops to protect vital pollinators such as bees to respond to this crisis. This ban spurred the agricultural industry to seek safer alternatives, accelerating the development of eco-friendly pesticides while promoting integrated pest management and other environmentally friendly farming practices. Additionally, the ban led several countries to strengthen their pesticide use regulations and assessment mechanisms, ensuring that the impact of pesticides on the environment and human health was minimized. This event not only changed the global pesticide usage patterns in agriculture but also raised societal awareness of agricultural sustainability and environmental protection.

2.4.3 The Hidden Threat: How 6-PPD-Quinone Endangered Aquatic Life

On a clear autumn day, scientists in the Pacific Northwest of the United States discovered a series of unusual events—large numbers of Pacific salmon had died in a short period. After a thorough investigation, they found that the culprit behind this tragedy was a chemical additive used in tires—6-PPD. Widely used in tire manufacturing to prevent rubber aging, 6-PPD transforms into a toxic substance called 6-PPD-quinone when washed away by rain, contaminating rivers and causing catastrophic damage to aquatic ecosystems (Tian et al. 2021). This discovery revealed an invisible threat with far-reaching consequences. Although this incident is linked to EOCs in water bodies, road runoff, and surface runoff are important sources of soil EOCs. With time, the increasing impact of 6-PPD-quinone in soil is becoming a growing concern.

In response to this environmental crisis, swift action was taken. Researchers intensified studies on the toxicity of 6-PPD degradation products, while environmental agencies began requiring tire manufacturers to adjust their formulations or use alternatives to reduce the chemical's environmental impact. Additionally, policymakers pushed for new regulatory measures, mandating stricter environmental impact assessments for industrial chemicals. As global attention on ecological protection grew, this crisis also spurred the development and promotion of more environmentally friendly technologies.

REFERENCES

Al-Khazrajy, O. S. A., and A. B. A. Boxall. 2016. Impacts of compound properties and sediment characteristics on the sorption behaviour of pharmaceuticals in aquatic systems. *J. Hazard. Mater.* 317:198–209.

Amor, I. B., H. Hemmami, S. Zeghoud, et al. 2024. Neonicotinoids: History, impacts, sustainable use, and application scenario. In *Neonicotinoids in the Environment: Emerging Concerns to the Human Health and Biodiversity*, eds. R. Singh, V. K. Singh, A. Kumar, S. Tripathi and R. Bhadouria, 3–14. Cham, Switzerland: Springer Nature Switzerland AG.

Bai, X., A. Lutz, R. Carroll, et al. 2018. Occurrence, distribution, and seasonality of emerging contaminants in urban watersheds. *Chemosphere* 200:133–42.

Bonmatin, J.-M., E. A. D. Mitchell, G. Glauser, et al. 2021. Residues of neonicotinoids in soil, water and people's hair: A case study from three agricultural regions of the Philippines. *Sci. Total Environ.* 757:143822.

Borthakur, A., J. Leonard, V. S. Koutnik, S. Ravi, and S. K. Mohanty. 2022. Inhalation risks of wind-blown dust from biosolid-applied agricultural lands: Are they enriched with microplastics and PFAS? *Curr. Opin. Environ. Sci. Health* 25:100309.

Brodskiy, E. S., A. A. Shelepchikov, G. I. Agapkina, M. O. Tikhonova, T. A. Paramonova, and D. N. Lipatov. 2019. Phthalate esters' content in soils of Moscow. *Moscow University Soil Sci. Bull.* 74:88–92.

Cao, Z., F. Xu, A. Covaci, et al. 2014. Distribution patterns of brominated, chlorinated, and phosphorus flame retardants with particle size in indoor and outdoor dust and implications for human exposure. *Environ. Sci. Technol.* 48:8839–46.

Castro-Jiménez, J., N. Berrojalbiz, M. Pizarro, and J. Dachs. 2014. Organophosphate ester (OPE) flame retardants and plasticizers in the open Mediterranean and Black Seas atmosphere. *Environ. Sci. Technol.* 48:3203–9.

Chen, G., Q. Feng, and J. Wang. 2020. Mini-review of microplastics in the atmosphere and their risks to humans. *Sci. Total Environ.* 703:135504.

Chen, H., Y. Yao, Z. Zhao, et al. 2018. Multimedia distribution and transfer of per- and polyfluoroalkyl substances (PFASs) surrounding two fluorochemical manufacturing facilities in Fuxin, China. *Environ. Sci. Technol.* 52:8263–71.

Chen, L., X. Yan, X. Zhou, P. Peng, Q. Sun, and F. Zhao. 2023. Advances in the on-line solid-phase extraction-liquid chromatography-mass spectrometry analysis of emerging organic contaminants. *Trends Analyt. Chem.* 160:116976.

Cycoń, M., A. Mrozik, and Z. Piotrowska-Seget. 2019. Antibiotics in the soil environment-degradation and their impact on microbial activity and diversity. *Front. Microbiol.* 10:338.

D'Ambro, E. L., H. O. T. Pye, J. O. Bash, et al. 2021. Characterizing the air emissions, transport, and deposition of per- and polyfluoroalkyl substances from a fluoropolymer manufacturing facility. *Environ. Sci. Technol.* 55:862–70.

Dankyi, E., C. Gordon, D. Carboo, and I. S. Fomsgaard. 2014. Quantification of neonicotinoid insecticide residues in soils from cocoa plantations using a QuEChERS extraction procedure and LC-MS/MS. *Sci. Total Environ.* 499:276–83.

Dias-Ferreira, C., R. L. Pato, J. B. Varejão, A. O. Tavares, and A. J. D. Ferreira. 2016. Heavy metal and PCB spatial distribution pattern in sediments within an urban catchment—contribution of historical pollution sources. *J. Soils Sediments* 16:2594–605.

Dong, X., J. Pei, and M. Qu. 2025. Organophosphate esters (OPEs) in the air and dust of new vehicle cabins: Concentrations, sources and contributions. *Build Environ.* 267:112188.

Dris, R., J. Gasperi, V. Rocher, M. Saad, N. Renault, and B. Tassin. 2015. Microplastic contamination in an urban area: A case study in Greater Paris. *Environ. Chem.* 12:592–9.

EPA. 2023. *Basic Information About Biosolids*. Web Page, United States Environmental Protection Agency. https://www.epa.gov/biosolids/basic-information-about-biosolids

Estoppey, N., G. Castro, G. A. Slinde, et al. 2024. Exposure assessment of plastics, phthalate plasticizers and their transformation products in diverse bio-based fertilizers. *Sci. Total Environ.* 918:170501.

Fu, L., B. Du, F. Wang, J. C. W. Lam, L. Zeng, and E. Y. Zeng. 2017. Organophosphate triesters and diester degradation products in municipal sludge from wastewater treatment plants in China: Spatial patterns and ecological implications. *Environ. Sci. Technol.* 51:13614–23.

Fuller, S., and A. Gautam. 2016. A procedure for measuring microplastics using pressurized fluid extraction. *Environ. Sci. Technol.* 50:5774–80.

Gajst, T. 2016. Analysis of plastic residues in commercial compost. BS diss., Faculty of Environmental Sciences, Univ. of Nova Gorica.

Harley-Nyang, D., F. A. Memon, A. O. Baquero, and T. Galloway. 2023. Variation in microplastic concentration, characteristics and distribution in sewage sludge & biosolids around the world. *Sci. Total Environ.* 891:164068.

Heinze, W. M., D. M. Mitrano, E. Lahive, J. Koestel, and G. Cornelis. 2021. Nanoplastic transport in soil via bioturbation by Lumbricus terrestris. *Environ. Sci. Technol.* 55:16423–33.

Huang, Y., Q. Liu, W. Jia, C. Yan, and J. Wang. 2020. Agricultural plastic mulching as a source of microplastics in the terrestrial environment. *Environ. Pollut.* 260:114096.

Jin, H., Y. Zhang, L. Zhu, and J. W. Martin. 2015. Isomer profiles of perfluoroalkyl substances in water and soil surrounding a chinese fluorochemical manufacturing park. *Environ. Sci. Technol.* 49:4946–54.

Kim, L. H., J. Kang, M. Kayhanian, K. I. Gil, M. K. Stenstrom, and K. D. Zoh. 2006. Characteristics of litter waste in highway storm runoff. *Water Sci. Technol.* 53:225–34.

Kong, S., Y. Ji, L. Liu, et al. 2012. Diversities of phthalate esters in suburban agricultural soils and wasteland soil appeared with urbanization in China. *Environ. Pollut.* 170:161–8.

Kumar M., K. Mrunalini, Z. Zhou, et al. 2025. Sources, distribution and migration of micro-nano-plastics from terrestrial to aquatic environments and its environmental and health impacts: a systematic review. *J Soils Sediments* 25: 1767–1801.

Lao, Z., H. Li, Z. Liao, et al. 2023. Spatiotemporal transitions of organophosphate esters (OPEs) and brominated flame retardants (BFRs) in sediments from the Pearl River Delta, China. *Sci. Total Environ.* 855:158807.

Li, H., X. Lu, S. Wang, B. Zheng, and Y. Xu. 2021. Vertical migration of microplastics along soil profile under different crop root systems. *Environ. Pollut.* 278:116833.

Li, W., Y. Wang, and K. Kannan. 2019. Occurrence, distribution and human exposure to 20 organophosphate esters in air, soil, pine needles, river water, and dust samples collected around an airport in New York state, United States. *Environ. Int.* 131:105054.

Li, Y., C. Liu, H. Yang, et al. 2024. Leaching of chemicals from microplastics: A review of chemical types, leaching mechanisms and influencing factors. *Sci. Total Environ.* 906:167666.

Marklund, A., B. Andersson, and P. Haglund. 2005. Traffic as a source of organophosphorus flame retardantsand plasticizers in snow. *Environ. Sci. Technol.* 39:3555–62.

Mo, C.-H., Q.-Y. Cai, Y.-H. Li, and Q.-Y. Zeng. 2008. Occurrence of priority organic pollutants in the fertilizers, China. *J. Hazard. Mater.* 152:1208–13.

Möller, A., Z. Xie, A. Caba, R. Sturm, and R. Ebinghaus. 2011. Organophosphorus flame retardants and plasticizers in the atmosphere of the North Sea. *Environ. Pollut.* 159:3660–5.

Okori, F., J. Lederer, A. J. Komakech, T. Schwarzböck, and J. Fellner. 2024. Plastics and other extraneous matter in municipal solid waste compost: A systematic review of sources, occurrence, implications, and fate in amended soils. *Environ. Adv.* 15:100494.

Olivero-Verbel, R., T. Moreno, J. Fernández-Arribas, et al. 2021. Organophosphate esters in airborne particles from subway stations. *Sci. Total Environ.* 769:145105.

Paule-Mercado, M. A., J. S. Ventura, S. A. Memon, D. Jahng, J.-H. Kang, and C.-H. Lee. 2016. Monitoring and predicting the fecal indicator bacteria concentrations from agricultural, mixed land use and urban stormwater runoff. *Sci. Total Environ.* 550:1171–81.

Petrie, B., R. Barden, and B. Kasprzyk-Hordern. 2015. A review on emerging contaminants in wastewaters and the environment: Current knowledge, understudied areas and recommendations for future monitoring. *Water Res.* 72:3–27.

Piehl, S., A. Leibner, M. G. J. Löder, R. Dris, C. Bogner, and C. Laforsch. 2018. Identification and quantification of macro- and microplastics on an agricultural farmland. *Sci. Rep.* 8:17950.

Ramos, L., G. Berenstein, E. A. Hughes, A. Zalts, and J. M. Montserrat. 2015. Polyethylene film incorporation into the horticultural soil of small periurban production units in Argentina. *Sci. Total Environ.* 523:74–81.

Rillig, M. C., S. W. Kim, T.-Y. Kim, and W. R. Waldman. 2021. The global plastic toxicity debt. *Environ. Sci. Technol.* 55:2717–9.

Schaffer, M., H. Börnick, T. Licha, K. Nödler, and E. Worch. 2011. Column experiments to simulate the transport of the basic beta-blocker atenolol in sandy aquifers under competitive conditions. *Geophys. Res. Abstr.* 13:EGU2011–3018.

Schroeder, T., D. Bond, and J. Foley. 2021. PFAS soil and groundwater contamination via industrial airborne emission and land deposition in SW Vermont and Eastern New York State, USA. *Environ. Sci.: Process. Impacts* 23:291–301.

Steinmetz, Z., C. Wollmann, M. Schaefer, et al. 2016. Plastic mulching in agriculture. Trading short-term agronomic benefits for long-term soil degradation? *Sci. Total Environ.* 550: 690–705.

Taniyasu, S., N. Yamashita, H.-B. Moon, et al. 2013. Does wet precipitation represent local and regional atmospheric transportation by perfluorinated alkyl substances? *Environ. Int.* 55:25–32.

Tian, Z., H. Zhao, K. T. Peter, et al. 2021. A ubiquitous tire rubber–derived chemical induces acute mortality in Coho Salmon. *Science* 371:185–9.

Tran, B. C., M.-J. Teil, M. Blanchard, F. Alliot, and M. Chevreuil. 2015. Fate of phthalates and BPA in agricultural and non-agricultural soils of the Paris area (France). *Environ. Sci. Pollut. R.* 22:11118–26.

Tran, H. T., C. Lin, X.-T. Bui, et al. 2022. Phthalates in the environment: Characteristics, fate and transport, and advanced wastewater treatment technologies. *Bioresour. Technol.* 344:126249.

Wang, L., M. Liu, W. Tao, et al. 2018. Pollution characteristics and health risk assessment of phthalate esters in urban soil in the typical semi-arid city of Xi'an, Northwest China. *Chemosphere* 191:467–76.

Wang, X., C. Wang, T. Zhu, P. Gong, J. Fu, and Z. Cong. 2019. Persistent organic pollutants in the polar regions and the Tibetan Plateau: A review of current knowledge and future prospects. *Environ. Pollut.* 248:191–208.

Wang X., Q. Zhu, X. Yan, Y. Wang, C. Liao, and G. Jiang. 2020. A review of organophosphate flame retardants and plasticizers in the environment: Analysis, occurrence and risk assessment. *Sci. Total Environ.* 731: 139071.

Wang Y., Y. Ji, V. Tishchenko, and Q. Huang. 2023. Removing per- and polyfluoroalkyl substances (PFAS) in water by foam fractionation. *Chemosphere* 311: 137004.

Wang, Y., P. Kannan, R. U. Halden, and K. Kannan. 2019. A nationwide survey of 31 organophosphate esters in sewage sludge from the United States. *Sci. Total Environ.* 655:446–53.

Wang, Y., Z. Li, F. Tan, Y. Xu, H. Zhao, and J. Chen. 2020. Occurrence and air-soil exchange of organophosphate flame retardants in the air and soil of Dalian, China. *Environ. Pollut.* 265:114850.

Wang, Y., U. Munir, and Q. Huang. 2023. Occurrence of per- and polyfluoroalkyl substances (PFAS) in soil: Sources, fate, and remediation. *Soil Environ. Health* 1:100004.

Xie, Z, Z. Wang, W. Mi, A. Möller, H. Wolschke, and R. Ebinghaus. 2015. Neutral Poly-/perfluoroalkyl Substances in Air and Snow from the Arctic. *Sci. Rep.* 5: 8912.

Xu, H., Y. Song, L. Cang, and D. Zhou. 2020. Ion exchange membranes enhance the electrokinetic in situ chemical oxidation of PAH-contaminated soil. *J. Hazard. Mater.* 382:121042.

Yadav, I. C., N. L. Devi, J. Li, and G. Zhang. 2018. Organophosphate ester flame retardants in Nepalese soil: Spatial distribution, source apportionment and air-soil exchange assessment. *Chemosphere* 190:114–23.

Yajima, D., and N. Tajima. Inspection starts at Yokota Air Base over leak of water with PFAS. National Report. *The Asahi Shimbun*. December 20, 2024. https://www.asahi.com/ajw/articles/15558303.

Zhang, Z., G. Lin, T. Lin, R. Zhang, L. Jin, and Y. Di. 2020. Occurrence, behavior, and fate of organophosphate esters (OPEs) in subtropical paddy field environment: A case study in Nanning City of South China. *Environ. Pollut.* 267:115675.

Zhou, Y., X. Lu, X. Fu, et al. 2018. Development of a fast and sensitive method for measuring multiple neonicotinoid insecticide residues in soil and the application in parks and residential areas. *Anal. Chim. Acta* 1016:19–28.

Zi, S., J. Xu, Y. Zhang, D. Wu, and J. Liu. 2023. Transport of bisphenol A, bisphenol S, and three bisphenol F isomers in saturated soils. *Environ. Sci. Pollut. R.* 30:116091–104.

3 Occurrence of Emerging Organic Contaminants in Soil

Areej Tahir, Anam Khalid, Zarish Imtiaz, Faiqa Faryal, Mehmood Aslam, Maosen Zhao, and Mujtaba Baqar

3.1 INTRODUCTION TO EOCs

Since industrialization, numerous organic chemicals have been intentionally invented or discovered as byproducts of processes that rapidly garnered recognition due to their wide range of applications. However, their unregulated use has led to their dispersion into the environment, raising scientific concerns about their unknown behavior and toxicological impacts. Broadly, such chemicals are termed emerging organic contaminants (EOCs) that might originate from natural or human activities, potentially affecting ecological integrity and human health. EOCs encompass various groups of chemicals such as pesticides, industrial chemicals, pharmaceuticals and personal care products (PPCPs), nanomaterials, food additives, illicit drugs, and several other chemicals that vary with respect to nature, effects, and type of applications (Calvo-Flores et al., 2018). Nevertheless, among EOCs, polychlorinated biphenyls (PCBs), organochlorine pesticides (OCPs), perfluoroalkyl and polyfluoroalkyl substances (PFAS), plasticizers (phthalates), PPCPs, short- and medium-chain chlorinated paraffins (SCCPs and MCCPs), and organophosphate esters (OPEs) have emerged due to increasing concerns associated with their environmental concentrations in the environment and potential health effects (Calvo-Flores et al., 2018). Technological advancements now allow the detection of EOCs even at very low levels in the environment, thereby enabling regulatory authorities to establish future monitoring programs (Raghav et al., 2013).

After originating from various sources, EOCs are likely to disperse into soils, urban runoff, and aquatic water bodies. A number of assessment reports on EOCs in aquatic environments are available from different regions worldwide, including Asia, Australia, America, Africa, and Europe (Kroon et al., 2020). In contrast, there is a notable lack of information regarding the fate and transport of EOCs in terrestrial environments, particularly within soil matrices (Pullagurala et al., 2018). Studies suggest that soil properties, such as the higher affinity of soil particles for exogenous compounds such as EOCs, are responsible for their higher contaminant

DOI: 10.1201/9781003509523-3

loads, ranging from picogram (pg) to microgram per kilogram (μg/kg) of soil (Biel-Maeso et al., 2021). For instance, the anti-inflammatory drug (Naproxen) exhibits higher adsorption rates to solid particles as compared with dissolving in liquid media (Teijón et al., 2014). Furthermore, the adsorption capacity of EOCs to soil particles impacts the mobility of other EOCs as seen in the case of triclosan, which has a higher adsorption rate to polyethylene (29.3 mg/μg/h) than polystyrene (0.27 mg/μg/h) in soil (Chen et al., 2021). Similarly, equilibrium capacities of EOCs also vary among themselves and in relation to soil. It is noteworthy that human activities and environmental phenomena also play a vital role in influencing the interactions between EOCs and soil properties, leading to their bioaccumulation in organisms (Mammo et al., 2020).

Primarily, humans and animals are exposed to EOCs in soil through the ingestion of contaminated food or soil. Inhalation and dermal contact serve as additional exposure routes for EOCs. Plants and crops have been reported to accumulate EOCs most significantly in their roots and least in their grains or fruits (Pullagurala et al., 2018). Additionally, EOCs are likely to disrupt plants through altering their genetic makeup, causing oxidation stress, and affecting plant biomass (Chen et al., 2014). However, further research is needed to understand the toxic effects of EOCs on human and ecological integrity through soil exposure. This chapter covers the occurrence of EOCs, such as PCBs, OCPs, PFAS, SCCPs, MCCPs, OPEs, phthalates, and PPCPs in soil, as well as the regulatory measures currently in place.

3.2 SOURCE OF EOCs IN SOIL

The exploitation of soil has drastically increased since the industrial era due to the invention of new products, often with an unregulated management. Soil being a natural resource provides many services for various life forms, including recycling natural elements, acting as a landfill for waste produced from anthropogenic activities, and serving as a sink for air and water-transported chemicals/EOCs. The physicochemical properties of soil also determine its affinity toward pollutants; EOCs, for example, might behave differently under varying pH levels, moisture content, temperature, soil composition, and microbial influences (Biswas et al., 2018). EOCs tend to sequester more in organic matter with rubbery or glassy phases. Similarly, the desorption of EOCs from soil depends on their chemical nature and their reactions toward physical, chemical, or biological treatments (Fang et al., 2018). Soil behavior toward EOCs varies under aerobic and anaerobic conditions; however, the fate and transport of EOCs in anaerobic soils are yet to be thoroughly addressed (Fang et al., 2020). Temperature and rainfall fluctuations due to changing climate patterns might increase the migration of EOCs and their metabolites from the soil. This could lead to a higher release of particle-bound EOCs and turn soil into a source of EOC transport to other environmental matrices (Maddela et al., 2022). Therefore, soil health is a significantly important concern within the scientific community, as EOCs discharging from industrial and domestic sources can impact both human health and ecosystem services.

3.3 OCCURRENCE OF EOCs IN SOIL

3.3.1 Phthalates

A number of human activities contribute to the environmental release of phthalate esters with plastics, construction, and textile industries being major contributors alongside waste dumping, vehicles, and personal care products (Abdel Daiem et al., 2012). As seen in Table 3.1, most studies on the occurrence of phthalates in soil have been conducted in China. Variations in phthalate sources are observed with different land use types, ranging from point to nonpoint sources. For instance in agricultural soils, mulch, wastewater, sludge treatments, fertilizers, and atmospheric deposition are common sources for the occurrence of phthalate esters (Niu et al., 2014). In China, two studies identified di-*n*-butyl phthalate (DnBP) and di-2-ethylhexyl phthalate (DEHP) as prevalent phthalate additives in orchards and farmlands, with phthalate levels ranging from 0.05 to 10.4 mg/kg (Kong et al., 2012; Wang et al.,

TABLE 3.1
Comparative Concentrations of Phthalates (ng/g), Number of Detected Phthalates Across Different Land Use Types Globally

Country	Land Use Type	Number of Detected Phthalates (*n*)	Concentration	References
China	Agricultural areas	15	75–6369	Niu et al. (2014)
	Agricultural areas	6	163–947	Wang et al. (2017)
	Agricultural areas	6	0.20–0.50	Zhang et al. (2015)
	Urban	6	510–8000	Li et al. (2006)
	Urban	5	2.00–3100	Xia et al. (2011)
	Surface	5	2300–24,700	Zhang et al. (2014)
	Vegetable	15	100–2100	Li et al. (2016)
	Agricultural	6	50.0–10,400	Kong et al. (2012)
	Surface	12	530–5500	Zhu et al. (2010)
	Nonindustrialized area	6	2200–158,000	Wu et al. (2015)
	Electronics manufacturing area	6	8630–172,000	
	E-waste recycling sites	5	12,600–46,700	Lü et al. (2018)
Czech Republic	Agricultural areas	2	500–2800	Daňková et al. (2013)
France	Rural area	9	ND–0.20	Tran et al. (2015)
Scotland	Agricultural areas	1	25.0–1600	Rhind et al. (2013)
Denmark	Agricultural	6	14.0–2500	Vikelsøe et al. (2002)
Serbia	Surface	6	190–2200	Škrbić et al. (2016)

ND, not detected.

2015a). Additionally, agricultural soils in the Shandong province and the Yangtze River delta of China recorded phthalate additives with concentrations of 756–1590 μg/kg and 5.42–1580 ng/g, respectively (Wei et al., 2020; Zhou et al., 2020). In some areas, croplands treated with sludge contain considerable levels of phthalates (0.004–97.4 mg/kg) (Meng et al., 2014). Such treated soil would not only result in higher phthalates uptake by plants but will also lead to significant intake by humans through food and soil ingestion. In the Czech Republic and France, phthalates have also been detected in agricultural soils, with concentrations ranging from 0.46 to 2.80 mg/kg and Not detected to 242 ng/g, respectively (Daňková et al., 2013; Tran et al., 2015).

In connection with urban soils, traffic, PPCPs, and waste disposal activities contribute to phthalates contamination. Moreover, in China, park soil was recorded with 0.08 mg/kg of DnBP, while roadside soil had relatively higher concentrations of six phthalates, ranging from 2.8 to 51.1 mg/kg (Kong et al., 2012; Wu et al., 2015). At an e-waste recycling site in China, six phthalates were detected at concentrations of 2.84–51.05 mg/kg, with DEHP and DnBP as dominant phthalates (Liu et al., 2020). Similarly, the soil surrounding electronic manufacturing plants in China was estimated to contain 8.63–171.64 mg/kg of phthalates (Wu et al., 2015). In Scotland, soil from industrial areas recorded phthalate concentrations ranging from 25 to 1600 μg/kg (Rhind et al., 2013). Phthalate presence in soil has been positively correlated with organic matter content but negatively related to soil pH, indicating different behaviors of phthalates toward soil properties (Li et al., 2020a). Nevertheless, further studies are required to understand phthalates distribution in diverse soils and the fate of these chemicals.

3.3.2 Pharmaceuticals and Personal Care Products

PPCPs encompass chemicals used in a variety of cosmetic products such as face creams, hair dyes, shampoos, perfumes, and medicines for humans and animals. Agricultural activities, solid and wastewater discharge from municipal or industrial sectors, and livestock are considered primary sources of PPCPs (Bartrons & Peñuelas, 2017). Even minor amounts of PPCPs can induce adverse impacts on the health of organisms, as these chemicals become active in their systems. In the soil matrix, pharmaceutical products, such as analgesics, antibiotics, anti-inflammatory products, hormones, and antimicrobial agents have been detected, along with cosmetic products such as preservatives, fragrances, insect repellents, sunscreens, and 4-methyl-benzilidene-camphor (Liu & Wong, 2013). PPCPs such as ciprofloxacin in the soil can also affect plant growth and morphology (Guruge et al., 2019). As shown in Table 3.2, different PPCPs have been identified in soils across North and South America, Asia, and Europe, with antibiotics being the dominant group.

Likewise, Chen et al. (2013a) also reported approximately 33 PPCPs, including salicylic acid, cotinine, sulfamethazine, caffeine, paraxanthine, trimethoprim, cholesterol, fluoxetine, and progesterone, up to a concentration of 13.1 ng/g in wastewater-irrigated soil in North America. On the other hand, biosolids were also identified as a major source of PPCPs such as azole biocides, climbazole, clotrimazole, and miconazole (Chen et al., 2013b). Similarly, sludge is another source of PPCPs with

TABLE 3.2
Classification and Compounds of PPCPs, and Mean Concentration (ng/g) of PPCPs Present in Soil Globally

Classification	Compounds	Location	Mean Concentration	References
Lipid-regulating drugs	Clofibric acid	Germany	ND	Ternes et al. (2007)
		USA	4.27	Chen et al. (2013a)
		USA	ND	Xu et al. (2009)
	Fenofibric acid	Germany	ND	Ternes et al. (2007)
	Bezafibrate	Germany	ND	
	Gemfibrozil	USA	0	Kinney et al. (2006)
Analgesics/anti-inflammatory drugs	Ibuprofen	Germany	319	Karnjanapiboonwong et al. (2011)
			7.88	Chen et al. (2013a)
			ND	Xu et al. (2009)
			ND	Ternes et al. (2007)
	Naproxen	USA	23.8	Chen et al. (2013a)
			ND	Xu et al. (2009)
	Ketoprofen		ND	
	Diclofenac	Germany	ND	Ternes et al. (2007)
		USA	ND	Xu et al. (2009)
	Diclofenac sodium		6.82	Chen et al. (2013a)
	Acetaminophen		1640	Kinney et al. (2006)
	Salicylic acid	China	4.50	
			9.10	
			10.7	
Antiseptics	Triclosan	USA	8.16	Karnjanapiboonwong et al. (2011)
			138	Chen et al. (2013a)
			ND	Xu et al. (2009)
		China	<1.58	Chen et al. (2011)
			1.80	
			<1.58	
Antibiotics	Ciprofloxacin	USA	0.68	Karnjanapiboonwong et al. (2011)
	Trimethoprim	Germany	ND	Ternes et al. (2007)
		USA	60.0	Kinney et al. (2006)
		China	<2.15	Chen et al. (2011)
			2.60	
			<2.15	
	Sulfamethoxazole	Germany	ND	Ternes et al. (2007)
		USA	332	Kinney et al. (2006)
	Erythromycin	Germany	ND	Ternes et al. (2007)
		USA	6270	Kinney et al. (2006)

(Continued)

TABLE 3.2 (*Continued*)
Classification and Compounds of PPCPs, and Mean Concentration (ng/g) of PPCPs Present in Soil Globally

Classification	Compounds	Location	Mean Concentration	References
	Oxytetracycline	China	6.20	Chen et al. (2011)
			7.50	
			212	
	Tetracycline		<2.67	
			6.90	
			19.9	
Antiepileptic	Carbamazepine	Germany	ND	Ternes et al. (2007)
		USA	549	Kinney et al. (2006)
	Primidone	China	<1.11	Chen et al. (2011)
			3.30	
			ND	
Endocrine-disrupting chemicals	Bisphenol A	USA	31.0	Chen et al. (2013a)
			ND	Xu et al. (2009)
		China	3.33	Chen et al. (2011)
			<3.40	
			2.70	
		Spain	4.60	Sanchez-Brunete et al. (2009)
	4-Nonylphenol	China	36.9	Chen et al. (2011)
			25.8	
			55.4	
		USA	136	Chen et al. (2013a)
			ND	Xu et al. (2009)
Steroid hormones	Estrone (E1)	USA	136	Chen et al. (2013a)
			ND	Xu et al. (2009)
	17Î²-Estradiol (E2)		3.33	Karnjanapiboonwong et al. (2011)
			ND	Chen et al. (2013a)
	17Î±-Ethynylestradiol (EE2)		2.70	Karnjanapiboonwong et al. (2011)
			ND	Xu et al. (2009)
	Estriol (E3)		7.73	Karnjanapiboonwong et al. (2011)
Psychomotor stimulants	Caffeine	USA	ND	
		Germany	ND	Ternes et al. (2007)
		USA	317	Kinney et al. (2006)
Antidepressants	Fluoxetine		376	
Beta-blockers	Warfarin		2770	
Synthetic musk fragrance	Galaxolide		1.98	Chase et al. (2012)
		Germany	ND	Ternes et al. (2007)
	Tonalide	USA	5.46	Chase et al. (2012)
		Germany	ND	Ternes et al. (2007)

(Continued)

TABLE 3.2 (*Continued*)
Classification and Compounds of PPCPs, and Mean Concentration (ng/g) of PPCPs Present in Soil Globally

Classification	Compounds	Location	Mean Concentration	References
Flame retardants	Tetrabromobisphenol A	Spain	0.30	Sanchez-Brunete et al. (2009)
	Tetrachlorobisphenol A		ND	
X-ray contrast media	Diatrizoate	Germany	ND	Ternes et al. (2007)
	Iopamidol		ND	

ND, not detected.

antibiotics norfloxacin and triclosan recorded at high concentrations of up to 20 μg/g in China (Newbold et al., 2009). Moreover, the metabolic activities of microorganisms, such as respiration, were greatly affected by the presence of PPCPs such as sulfamethoxazole and ciprofloxacin (Qin et al., 2015). Apart from the concentrations of PPCPs, soil partitioning behavior with other matrices also plays a role in the accumulation of PPCPs. For instance, doxycycline, an antibiotic, has a high partitioning coefficient (3.2 L/kg) between solid and liquid, leading to high sorption to solids (Liu et al., 2015). Additionally, soil pH, carbon content, and mineral composition affect the transport and sorption of PPCPs in the soil through chemical bonding, ion exchange, and the formation of complex chemicals with Ca^{2+}, Mg^{2+}, Fe^{3+}, or Al^{3+} ions (Díaz-Cruz et al., 2009). Overall, the presence of PPCPs in the soil is a major concern, as their intake via the food chain might induce cumulative effects on the biological systems of organisms, potentially resulting in carcinogenic and noncarcinogenic diseases.

3.3.3 Polychlorinated Biphenyls

PCBs are industrial chemicals commonly used in electrical equipment for lubrication, as coolants, and as additives. However, their usage has led to widespread dispersion into various environmental matrices and biota globally. Previous studies have documented the occurrence of PCBs in soils from various countries, with notable differences in both the number of congeners analyzed and concentration ranges. In Ethiopia, 16 PCB congeners were detected at concentrations ranging from 0.40 to 18.5 μg/kg (Prasse et al., 2012). In India, analysis of 28 congeners revealed concentrations between 3.33 and 34.8 μg/kg (Kumar et al., 2013). Similarly, in Portugal, 21 congeners were reported at levels ranging from 0.18 to 34.0 μg/kg (Cachada et al., 2012). The highest concentrations were observed in Azerbaijan, where 7 congeners ranged from 0.02 to 147 μg/kg (Ukalska-Jaruga et al., 2020a). As seen in Table 3.3, several studies in China have examined the occurrence of PCBs across different land-use types. In agricultural soils, PCBs ranged from 0.01–0.60 μg/kg (Nakata et al., 2005), 0.06–4.36 μg/kg for 209 congeners (Mao et al., 2021), <0.10–130 μg/kg for 18 congeners (Sun et al., 2016), and 0.05–1.83 μg/kg for 29 congeners (Xu et al.,

TABLE 3.3
Comparative Concentrations (ng/g) and Number of Detected PCB Congeners Across Different Land Use Types Globally

Country	Land Use Type	Number of Detected Congeners	Groups Detected	ΣPCBs	References
Algeria	Urban land	7	Tri- to hepta- and deca-PCB	ND–19.3	Halfadji et al. (2019)
Azerbaijan	Agricultural soils	7	Tri- to hepta-PCB	0.02–147	Ukalska-Jaruga et al. (2020a)
China	Mountain	25	Tri to hepta- and deca-PCB	47.1–423	Wang et al. (2009)
	Agricultural soil	7	Tri to hepta-PCB	0.01–0.62	Nakata et al. (2005)
	Agricultural soils	209	Mono- to deca-PCB	0.06–4.36	Mao et al. (2021)
	Agricultural soil	18	Tri to hepta-PCB	<0.10–130	Sun et al. (2016)
	Agricultural area	29	Di to nona-PCB	0.05–1.83	Xu et al. (2019a)
	E-waste dismantling area	19	Tri to hexa- and deca-PCB	9.93–2610	Liu et al. (2020)
Italy	Agricultural area	80	Mono to deca-chlorinated	100–30,000	Di Guardo et al. (2020)
Nepal	Urban soils	6	Tri-, penta to hepta-PCB	0.01–9.70	Pokhrel et al. (2018)
Poland	Agricultural land	7	Tri to hepta-PCB	0.41–4.77	Maliszewska-Kordybach et al. (2013)
	Dumpsite area	7	Tri to hepta-PCB	2.50–12.0	Melnyk et al. (2015)
Sweden	Nonagricultural land	13	Tri to octa-PCB	0.50–55.0	Armitage et al. (2006)
Tunisia	Wastewater irrigated soil	18	Tri to hepta-PCB	11.3–21.9	Haddaoui et al. (2016)
Turkey	Agricultural land	15	Tri to octa-PCB	ND–0.30	Can-Güven et al. (2022)
	Mountain area	12	Tri to hepta-PCB	0.08–0.29	Turgut et al. (2012)

ND, not detected.

2019a). High concentrations, ranging from 47.1 to 423 μg/kg for 25 congeners of PCBs, were found in mountain areas of China, with mono- to tetra-PCBs accounting for 90% of the total PCBs detected (Wang et al., 2009).

In contrast, a mountain area in Turkey reported relatively lower concentrations, ranging from 0.08 to 0.29 μg/kg for 12 congeners, with indicator PCBs constituting 76%–79% (Turgut et al., 2012). Di Guardo et al. (2020) studied the relationship between PCBs and soil depth as well as soil type. They found that PCB concentrations decrease with depth due to the hydrophobic nature of PCBs that prevent their vertical movement. The top layer consisted of hydrophobic PCBs, including penta-, hexa- and deca-PCBs congeners, while lesser hydrophobic congeners, such as mono-, di- and tri-PCBs congeners, were present below 36 cm soil depth. In Tunisia, 18 congeners of PCBs were studied, with hexa-PCB congeners being predominant, while penta- and hepta-PCBs were present at slightly lower levels (Haddaoui et al., 2016). In urban areas of Algeria, seven PCB congeners were observed, with concentrations ranging from below the detection limit to 19.34 μg/kg (Halfadji et al., 2019). In urban soils of Nepal, Pokhrel et al. (2018) studied six congeners, ranging from 0.01 to 9.7 μg/kg. An e-waste dismantling area in China exhibited high concentrations of PCBs in soil, ranging from 9.93 to 2610 μg/kg for 19 congeners, with tri-PCBs (PCB-28) accounting for 628 μg/kg, followed by tetra-PCBs (PCB-52) at 431 μg/kg (Liu et al., 2020). Overall, it appears that anthropogenic sources continue to contribute PCBs to the surrounding environment, particularly soil, necessitating effective monitoring and treatment technologies to control sources and rehabilitate contaminated areas.

3.3.4 Organochlorine Pesticides

OCPs are among the most effective and widely used pesticides, primarily used to control agricultural pests and disease vectors. However, due to their persistent nature, these agrochemicals remain present in the environment long after their use has ceased. Research conducted in China revealed that soil samples collected from eight districts of Wuhan city contained significant amounts of OCPs, with concentrations varying across different land-use types. Agricultural soils showed high levels of ΣHCHs (10.7 ng/g) and ΣDDTs (987 ng/g) (Tadesse, 2021). The presence of OCPs varies significantly depending on the land-use type, as shown in Table 3.4. For instance, a study conducted in Changchun, Northeast China, indicated that land-use type significantly impacted the residues of certain OCPs in soil. Urban soils had higher levels of ΣHCHs (16.3 ng/g), while industrial and agricultural soils had higher levels of ΣDDTs (11.8 ng/g) (Gao et al., 2013; Zhao et al., 2023). In agricultural lands, the highest concentrations of ΣOCPs were found in Italy (1914 ng/g) (Thiombane et al., 2018), followed by Poland (1038 ng/g) (Maliszewska-Kordybach et al., 2013). Some lands in China (Gao et al., 2013), Iran (Jorfi et al., 2019), Italy (Thiombane et al., 2018), Mexico (Núñez-Gastélum et al., 2019), Pakistan (Syed et al., 2014), Spain (Moreno et al., 2008), and Turkey (Can-Güven et al., 2022) showed concentrations below detection limits. In most studies, ΣDDTs were found in greater concentrations compared with ΣHCHs, except in Nigeria (Tesi et al., 2020), Poland (Ukalska-Jaruga et al., 2020b), and Spain (Moreno et al., 2008), where ΣHCHs were higher. The lowest range of ΣOCPs was detected in fallow lands of Argentina (0.18–0.71 ng/g) (Lupi et al., 2016), agricultural lands of Kazakhstan (0.139–0.714 ng/g), and Poland (0.118–1.39 ng/g) (Łozowicka et al., 2016). The highest reported concentrations of ΣOCPs were found in dumping sites of Kazakhstan (121,052 ng/g) (Mamirova et al., 2021) and Pakistan (1469 ng/g) (Syed et al., 2014).

Comparing data from different studies, it's evident that there are significant differences in OCP concentrations. These differences could be due to variations in agricultural practices, industrial activities, and environmental regulations. According to

TABLE 3.4
Comparative Concentrations (ng/g) and Number of Detected Isomers of OCPs Across Different Land Use Types Globally

Country	Land Use Type	Number of Isomers Detected	DDTs	HCBs	HCHs	ΣOCPs	References
Argentina	Agricultural land	10	0.02–0.25	—	0.01–0.13	0.18–0.71	Lupi et al. (2016)
China	Surface soil	8	ND–184	ND–6.35	ND–42.6	ND–219	Gao et al. (2013)
	Agricultural soil	13	0.70–84.0	0.08–3.2	0.03–2.40	—	Nakata et al. (2005)
	Agricultural soils	8	7.20–2910	—	2.30–760	—	Shi et al. (2013)
	Agricultural soil	8	2.50–987	—	0.19–135	5.44–994	Tadesse (2021)
	Urban land	18	4.40–24.0	ND–8.67	2.62–60.1	15.6–92.8	Zhao et al. (2023)
Iran	Agricultural land	20	ND–2.08	—	ND	ND–5.16	Jorfi et al. (2019)
Italy	Agricultural land	24	ND–633	ND–13.4	ND–47.3	ND–1914	Thiombane et al. (2018)
Kazakhstan	Agricultural land	6	0.13–0.71	—	0.006–0.007	0.14–0.71	Łozowicka et al. (2016)
	Dumping site	24	777–14,072	4.70	25.5–488	121,052	Mamirova et al. (2021)
Mexico	Agricultural land	5	ND–172	—	—	ND–335	Núñez-Gastélum et al. (2019)
Nigeria	Agricultural land	20	ND–63.5	—	ND–156	5.82–397	Tesi et al. (2020)
Pakistan	Dumping site	13	ND–821	—	ND–122	ND–1469	Syed et al. (2014)
Poland	Agricultural land	6	1.29–163	—	0.04–0.48	3.38–386	Maliszewska-Kordybach et al. (2013)
	Agricultural soil	6	0.61–485	—	0.98–1009	4.03–1038	Ukalska-Jaruga et al. (2020b)
	Agricultural land	6	0.11–1.39	—	0.004–0.006	0.12–1.39	Łozowicka et al. (2016)
Spain	Agricultural soil	8	ND–16.4	—	ND–40.6		Moreno et al. (2008)
Turkey	Agricultural land	22	ND–44.4	—	ND–0.113	ND–28.1	Can-Güven et al. (2022)
	Mountainous land	28	3.22–24.6	0.04–1.28	0.14–1.51	4.44–25.8	Turgut et al. (2012)

ND, not detected.

Tadesse (2021), the level of OCPs is also dependent on soil properties such as moisture content and total organic carbon (TOC). High organic matter in the soil can bind OCPs due to their hydrophobic nature, which also facilitates microbial degradation of OCPs. However, Tadesse (2021) reported a negative correlation between moisture content and all OCPs except o, p′-DDT. Furthermore, the presence of OCPs was not influenced by the soil pH. Similarly, Shi et al. (2013) found a strong negative correlation between the abundance of protozoa and residual levels of OCPs in agricultural soil. Conclusively, despite the ban on OCPs in most countries, they are still present in derivative forms or are being released from obsolete sources, affecting life in the soil and impacting life forms dependent on it.

3.3.5 Perfluoroalkyl and Polyfluoroalkyl Substances

PFAS are released into the environment through both direct and indirect sources. Direct sources include the direct release of PFAS compounds and their derivatives into the environment during the usage, production, and disposal of PFAS products (Buck et al., 2011). Indirect sources include the release of PFAS through the abiotic and biotic degradation of PFAS precursors. The primary source of PFAS contamination is industrial emission, either through direct discharge of industrial wastewater into water bodies or through air emissions. These compounds ultimately seep into the soil or become part of surface water. Another major source is the use of aqueous film-forming foam (AFFF) application and waste disposal sites (Qiao et al., 2023).

Soil acts as a major sink for PFAS contamination, accumulating these compounds for a long period, and serving as a secondary source by reintroducing them into other matrices through percolation, plant uptake, runoff, and volatilization (Ma et al., 2022). Johnson et al. (2022) reported significant concentrations of PFAS in soil than any other matrices such as air, surface water, and groundwater. PFAS originating from primary or secondary sources in the soil matrix demonstrates the major contribution of wastewater irrigation (Dalahmeh et al., 2018; Shigei et al., 2020), pesticides (Nascimento et al., 2018) and biosolids applications (Johnson et al., 2022), fluorochemical manufacturing plants (Zhang et al., 2020), wastewater treatment plants (Kim et al., 2019), atmospheric deposition (Sim et al., 2021), and other industrial discharges (Ma et al., 2022). PFAS enter into the ecological food web through rhizospheric uptake by plants, and subsequent transfer to higher trophic levels (Bolan et al., 2021; Zhang et al., 2020). A study reported the bioaccumulation of PFAS for more than 100 years in AFFF-contaminated soils (Høisæter & Breedveld, 2022). As soil pH decreases, an increase in cation concentrations facilitates PFAS adsorption in soil (Wang et al., 2015b). Qi et al. (2014) reported that soil texture plays a vital role in the adsorption of PFAS in soil. PFAS was also found in deep soil cores in France after the usage of PFAS fire-extinguishing material after three decades (Dauchy et al., 2019).

The occurrence of PFAS in different land-use types is given in Table 3.5, which includes data from around the world, including Asia, Europe, North America, and South America. The highest concentration of $\sum$PFAS = 10,477 ng/g was observed at a firefighting training facility in France, while the lowest was found to be in the soil near FMPs in Shandong, China ($\sum$PFAS = 1.62 ng/g) (Dauchy et al., 2019). In

TABLE 3.5
Comparative Concentrations (ng/g) and Number of Detected PFAS Across Different Land Use Types Globally

Country	Land Use Type	Number of Detected Compounds	ΣPFAS (ng/g)	References
Pakistan	Agricultural land	44	Range: 14.4–466	Baqar et al. (2024)
France	Three Decades old - firefighting training	29	10,477	Dauchy et al. (2019)
China	Fluorochemical production	12	1.62	Zhang et al. (2020)
Canada	Firefighting training (Near Airport)	256	668	Liu et al. (2021)
Finland	Firefighting training	54	1286	Reinikainen et al. (2022)
China	Firefighting training (Fire stations)	18	57.9	Zhu et al. (2022)
China	Firefighting training (Near Airport)	18	6.14	Zhu et al. (2022)
China	Chemical	18	8.01	Zhu et al. (2022)
China	Electroplating	18	8.46	Zhu et al. (2022)
China	Waste disposal	18	11.4	Zhu et al. (2022)
Brazil	Pesticide application	19	1.51	Nascimento et al. (2018)
Uganda	Wastewater contaminated irrigation	26	5.03	Dalahmeh et al. (2018)
France	Three decades old - nonsource site	29	37.0	Dauchy et al. (2019)
South Korea	Industrial complexes and a wastewater treatment plant	17	1.92	Kim et al. (2019)
Jordan	Wastewater contaminated irrigation	20	0.65	Shigei et al. (2020)

ND, not detected.

AFFF-contaminated soil, ∑PFOS concentrations were found near the firefighting training areas in Finland (380 ng/g) (Reinikainen et al., 2022), Canada (285 ng/g) (Liu et al., 2021), France (574 ng/g) (Dauchy et al., 2019), the United States (183 ng/g) (Nickerson et al., 2020), and China (37.9 ng/g) (Zhu et al., 2022). It is noted that the short-chain PFAS usually accumulate in the surface soil during dry seasons and then migrate to deeper soil layers during precipitation infiltration (Röhler et al., 2021). A research study estimated global background levels of certain PFAS, such as PFHpA (0.06 ng/g), perfluorodecanoic acid (PFDA) (0.01 ng/g), PFOA (0.05ng/g), PFHxA

(0.06 ng/g), and PFOS (0.06 ng/g) in soil (Washington et al., 2019). From all these above studies, it can be inferred that there is a need for regulatory bodies to monitor PFAS inputs to soil from different sources.

3.3.6 Short- and Medium-Chain Chlorinated Paraffins

Considering the diverse range of physicochemical properties, chlorinated paraffins (CPs) have found their way into several products and processes. CPs are used as plasticizers and flame retardants in polyvinyl chloride (PVC), as lubricants and coolants in metalworking, as additives in various plastics, leather, rubber, and paint products, as well as in sealants and adhesives (Lassen et al., 2014). The primary sources of CPs, especially SCCPs and MCCPs in soil, are CP manufacturing plants and the use of CP-containing sludge. Secondary sources include air and water-transported CPs, and anthropogenic activities (Wei et al., 2016). Contaminated soil may also release SCCPs and MCCPs to the surrounding environment through re-evaporation, diffusion, and runoff (Zhou et al., 2023). Moreover, environmental factors such as humidity, wind speed, and temperature influence the re-evaporation rate of SCCPs and MCCPs from surface soils (Semeena & Lammel, 2005). Regarding the occurrence of CPs, soil properties play a vital role along with the physicochemical properties of CPs. Since CPs are known to be hydrophobic, their affinity for soil is greater, as depicted by the partitioning coefficient (log K_{oc}: 4.5–10) of organic carbon (Environment Canada, 2008). The accumulation of CPs in the soil might further impact the health and bodily functions of soil organisms if the concentrations of SCCPs and MCCPs exceed 5.28 μg/g and 28 μg/g, respectively (Wang et al., 2019a). Moreover, plants and crops would also be affected by CP-contaminated soil, subsequently entering the food chain (Song et al., 2011).

In Table 3.6, the presence of SCCPs and MCCPs in soil, globally, has been reported in Western Europe, Africa, and Asia. In the case of soil, the main factors governing the levels of CPs in soil include the type of land use, distance from industrial areas, soil profile, and proximity to the coast. For instance, increased levels of SCCPs and MCCPs have been recorded at an e-waste disassembling sites in China (Halse et al., 2015; Yuan et al., 2017). The main reason behind this is the use of outdated techniques for the treatment of electronic waste, which is an important source of CPs (van Mourik et al., 2016). Previous studies have revealed that SCCPs are prevalent in soils (Chen et al., 2014), surface particulates (Zeng et al., 2016), paddy seeds, and snails (Yuan et al., 2017) in e-waste dismantling areas in Taizhou. Soil samples in the vicinity of CP production plants in China had concentrations up to 1.84 μg/g for SCCPs and 2.07 μg/g for MCCPs (Xu et al., 2016). Meanwhile, in agricultural soils, SCCPs and MCCPs concentrations ranged from 39 to 609 and 127 to 1969 ng/g, respectively (Aamir et al., 2019), followed by 50.1–266 ng/g for SCCPs in coastal soils (Zhao et al., 2019). Moreover, surface soils in China from industrial areas and e-waste recycling centers (Xu et al., 2019b) reported concentrations of up to 220 μg/g for SCCPs and 4400 μg/g for MCCPs. Similarly, Nipen et al. (2022) and Moeckel et al. (2020) recorded several folds higher CP concentrations in the soil near e-waste recycling centers in East and West Africa compared with European soil levels (not detected—417 ng/g) (Halse et al., 2015), and Pakistani soil levels (not detected—

TABLE 3.6
Comparative Concentrations (ng/g) of SCCPs and MCCPs across Different Land Use Types Globally

Country	Sampling Area	CPs Studied	Dominant Congeners	Median (Range) Concentrations	References
China	E-waste dismantling area	SCCPs	$C_{12,13}Cl_{7,8}$	2320 (10.0–220,000)	Xu et al. (2019b)
		MCCPs	$C_{14}Cl_{7,8}$	27,000 (1050–4,400,000)	
		MCCPs		121 (15.0–740)	
	Various land use types	SCCPs	$C_{11,13}Cl_{7,8}$	43.0 (<LOD[a]–14,285)	Li et al. (2021)
		MCCPs	$C_{14}Cl_{7,8}$	38.0 (<LOD–6760)	
	Urban	SCCPs	$C_{11,13}Cl_{5,7}$	<LOD–620	Wang et al. (2017)
		MCCPs	$C_{14,15}Cl_{7,8}$	2.00–190	
	Industrial	SCCPs	$C_{10,12}Cl_{6,7}$	2.00–240	Wang et al. (2013a)
		MCCPs	$C_{14}Cl_{6–8}$	2.00–1500	
	Suburban	SCCPs	$C_{10,11}Cl_{6,7}$	160–1500	Zeng et al. (2011)
	Roadside	SCCPs	$C_{11,13}Cl_{7–8}$	0.40–420	Wang et al. (2013b)
	Wood land	SCCPs	$C_{10}Cl_{6–7}$	18.0–116	Chen et al. (2013b)
	Control area	SCCPs	$C_{11,13}Cl_{8}$	<LOD–795	Xia (2012)
Tanzania	Rural-urban areas	SCCPs	$C_{13}Cl_{7,8}$ $C_{14}Cl_{7,8}$	51.0 (<LOD–5300)	Nipen et al. (2022)
		MCCPs		170 (<LOD–5100)	
Ghana/Sierra Leone	E-waste	SCCPs	—	69.0–28,000	Moeckel et al. (2020)
		MCCPs		<LOD–1400	
UK	Suburban	SCCPs	—	Mean:31.0 (<LOD–317)	Halse et al. (2017)
Norway	Suburban	SCCPs		Mean:18.0 (<LOD–417)	
Switzerland	Various land use types	SCCPs	$C_{10–12}Cl_{5–7}$	3.00–35.0	Bogdal et al. (2017)
		MCCPs	$C_{14}Cl_{7,8}$	5.10–160	
Pakistan	Industrial, residential and commercial areas	SCCPs	$C_{10}Cl_{7,8}$ $C_{14,16}Cl_{7,8}$	54.0 (<LOD–130)	Tahir et al. (2024)
		MCCPs		33.0 (<LOD–74.0)	

[a] LOD, limit of detection.

130 ng/g) (Tahir et al., 2024). In rural and urban areas of Tanzania (Nipen et al., 2022) and China (Li et al., 2021), SCCPs and MCCPs were detected at levels up to 14.0 and 6.80 µg/g, respectively, which were higher than in surface soils of suburban areas in the United Kingdom and Norway (Not Detected—417 ng/g) (Halse et al., 2017). However, studies have suggested that CPs adsorb to soil particles without the influence of soil properties such as organic carbon (Wang et al., 2013a). Contrary to this, a significant association was observed between the concentration of SCCPs and organic carbon in the vertical profile of soil (Wei et al., 2016).

In terms of congeners, the abundance of higher chlorine congeners (Cl_{7-9}) in the soil is consistent among global studies mentioned in Table 3.6. For instance, industrial areas of China (Xu et al., 2019b; Li et al., 2021), urban areas of Tanzania (Nipen et al., 2022), and rural/urban areas of Switzerland (Bogdal et al., 2017) were reported to have Cl_{6-9} as dominant congeners. However, variation was reported in carbon chain lengths among different studies, with C_{10} to C_{14} as dominating congeners of SCCPs and MCCPs, respectively. It was also reported that lighter congeners tend to transport more easily along the vertical direction of soil as compared with the heavier congeners of CPs (Wei et al., 2016). It is evident from these studies that long-range atmospheric transport and subsequent settling seem to be the dominant factors in contaminating soils with CPs (Wei et al., 2016; Wang et al., 2013b). Due to the wastewater application in agricultural fields and/or sludge application, soils may also be considered a potential major reservoir of CPs (Zeng et al., 2011). From the studies reported, it can be assessed that air and water are the main drivers for the transportation of CPs over long distances, especially when the area within proximity of the source is suspected to be highly contaminated.

3.3.7 Organophosphate Esters

OPEs have been widely used as flame retardants and plasticizers in various products related to textiles, furniture, and plastics (Greaves & Letcher, 2017). These products, along with anthropogenic activities, are responsible for the distribution of OPEs into the environment and soil matrix. Owing to the hydrophobic nature of OPEs, such as tris (1-chloro-2-propyl)phosphate (TCIPP), triphenyl phosphate (TPHP), and trimethylolpropane phosphate (TMPP), tend to adsorb to minerals and organic matter in the soil (Cui et al., 2017), affecting plants, the ecological community of the soil, and the associated food chain (Zhang et al., 2020). To date, several studies have documented the occurrence of OPEs in soil across Asia, Europe, and the Arctic regions, as mentioned in Table 3.7.

In the USA, ΣOPEs were observed in the range of 0.10–10,000 ng/g, with TCIPP detected as the dominant OPE compound (Li et al., 2019), alongside TPHP, TMPP, and 2-ethylhexyldiphenyl phosphate. The presence of TCIPP in soil is mainly attributed to its wide application of OPE (Ge et al., 2020). According to land use types, OPEs have not only been reported from urban areas, including industrial sites and waste processing sites, but also from agricultural areas (Ge et al., 2020; Wang et al., 2018; Yadav et al., 2018a). For instance, in Vietnam, surface soils near open e-waste recycling areas were found with TPHP as the prominent compound in the range

TABLE 3.7
Number of Congeners and Median Concentrations (ng/g) of OPEs Detected in Soil Worldwide from Different Land Use Types

Country	Sample Type	TCEP	TCIPP	TDCIPP	TPHP	TMPP	EHDPP	TEP	TNBP	TIBP	TBOEP	TEHP	∑OPEs (*n*)	References
China[a]	Contaminated soil	92.0	21.0	—	26.0	119	11.0	—	22.0	47.0	200	—	398 (8)	Wan et al. (2016)
China	Contaminated soil	6.39	29.1	2.32	7.02	1.84	0.18	3.47	1.09	4.89	47.3	2.87	696 (12)	Wang et al. (2018)
China	Contaminated soil	<MDL	899	146	10700	—	—	—	—	—	—	17.8	12,000 (13)	Ge et al. (2020)
Vietnam	Contaminated soil	2.10	—	2.50	51.0	97.0	69.0	—	—	—	—	—	N/A (12)	Matsukami et al. (2015)
America	Airport soil	<MDL–0.10	<MDL–52.4	<MDL–4.53	<MDL–3.11	0.64–10.8	<MDL–8.14	<MDL–2.41	<MDL–0.88	<MDL	<MDL–3.41	<MDL–3.43	1.16–73.1 (20)	Li et al. (2019)
China	Industrial soil	32.4	111	37.7	80.9		13.0	17.2	25.5	20.3	73.5	21.7	433 (10)	Sun and Zhu (2021)
Nepal	Industrial city soil	12.7	31.7	19.1	12.9	41.3	25.6	—	16.5	—	—	26.1	186 (8)	Yadav et al. (2018b)
China	Urban soil	4.00	1.00	5.00	2.00	12.0	4.00	5.00	10.0	—	54.0	1.00	75.0 (11)	Cui et al. (2017)
China[a]	Urban soil	1.70	1.04	0.97	10.6		3.23	0.17	3.86	0.97	18.8	0.78	42.1 (10)	Sun and Zhu (2021)
Germany	Urban soil	13.0–18.2	2.57–8.33	—	—	—	—	—	—	—	2.30–13.0	—	(4)	Fries and Mihajlović (2011)

(Continued)

TABLE 3.7 (*Continued*)
Number of Congeners and Median Concentrations (ng/g) of OPEs Detected in Soil Worldwide from Different Land Use Types

Country	Sample Type	TCEP	TCIPP	TDCIPP	TPHP	TMPP	EHDPP	TEP	TNBP	TIBP	TBOEP	TEHP	∑OPEs (*n*)	References
Nepal	Urban soil	25.3	165	29.3	49.4	230	51.2	—	16.5	—	—	94.8	662	Yadav et al. (2018a)
China	Farmland soil	—	1.12	0.04	0.19	0.13	0.04	0.04	0.11	0.13	0.66	0.14	(8)	Ji et al. (2019)
China[a]	Farmland soil	2.50	0.20	0.21	2.22	—	0.36	0.07	0.70	0.18	1.24	0.21	3.91	Sun and Zhu (2021)
China[a]	Paddy field soil	0.13	1.04	0.06	0.13	0.11	—	—	0.10	0.06	0.10	0.31	(12)	Zhang et al. (2020)
Japan	Greenhouse soil	—	—	—	—	350	—	—	—	—	—	—	7.89	Cho et al. (1996)
China[a]	Urban soil and rural soil	3.07	14.5	1.36	0.21	0.29	0.18	0.09	2.56	0.28	0.36	0.47	2.10	Wang et al. (2019b)
Arctic[a]		1.02	2.05	0.30	1.20	—	—	—	3.89	—	1.01	0.30	(11)	Han et al. (2020)
China[a]	Forest soil	0.11	0.58	0.07	0.03	0.003	—	—	0.03	0.02	0.01	0.03	—	Zhang et al. (2020)
Germany[a]	Campus soil	4.96	1.23	0.07	3.61	—	—	—	9.00	—	—	—	17.5	Mihajlović et al. (2011)
Germany	Grasslands	<LOQ–18.2	0.59–8.33	—	—	—	—	—	—	—	—	—	(13)	Fries and Mihajlović (2011)

[a] Indicates studies stating mean OPEs concentrations.

of 11.0–3300 ng/g (Matsukami et al., 2015). Similarly, in China, multiple recycling units detected TCIPP as dominant compounds, including e-waste treatment areas (TCIPP:19.1–2270 ng/g) (Ge et al., 2020), plastic treatment areas (TCIPP: 4.0–52.0 ng/g) (Wan et al., 2016), and multi-waste recycling areas (TCIPP: 34.9–1370 ng/g) (Wang et al., 2018). Moreover, agricultural soils near these recycling sites were also observed to be contaminated, likely due to soil–air exchange (Wan et al., 2016; Wang et al., 2018).

In the industrial areas of China and Turkey, surface soils contained 38.0–468 ng/g of OPEs, indicating that the number of industrial units is directly associated with the levels of OPEs (Sun & Zhu, 2021). Studies have also found TMPP in air-bases and commercial airports due to its application in the hydraulic oil of aircraft. For instance, a U.S. Air Force base and a commercial airport were detected with 0.02–130 and 9.18–691 ng/g of TMPP, respectively (Li et al., 2019). Additionally, anthropogenic activities, such as traffic and plastic use, contribute to OPE contamination in the environment. Traffic in Guangzhou, China, was found responsible for 41.0–1370 ng/g of OPEs in the soil (Cui et al., 2017). Similarly, soils in urban areas of China and Nepal were detected with OPEs levels of 10.1–315 and 65.0–27,500 ng/g, respectively (Yadav et al., 2018b). It has been suggested that transportation and PVC plastics are the main sources of OPEs in urban soil, elevating OPEs levels compared with rural soils (Li et al., 2020b). Nevertheless, agricultural soils were also found to be contaminated with OPEs. Cho et al. (1996) reported up to 350 ng/g TMPP in greenhouse soils due to its use in film plastics, which adsorbed to surface soil. In contrast, OPEs (6.72 ng/g) were detected in farm soils without mulch films (Ji et al., 2019). Zhang et al. (2020) suggested that the similar profiles of OPEs in urban and rural soils were attributed to urban sources of emissions affecting paddy fields. Sun and Zhu (2021) indicated that atmospheric deposition, wastewater irrigation, and sludge usage in farmlands contribute to OPEs contamination in the paddy fields. In addition to point sources, atmospheric deposition and long-distance transport of OPEs are significant factors contaminating remote areas. For instance, OPEs such as tris(2-carboxyethyl)phosphine (TCEP), TCIPP, and TPHP were found up to 4.96 ng/g in German grassland soil, with rainwater attributed as the main source (Mihajlović et al., 2011). Fries and Mihajlović (2011) suggested atmospheric deposition as the main source for remote regions in Germany. In the case of Arctic soil, OPEs were found in the range of 1.33–17.5 ng/g, with atmospheric OPEs ascribed as the source (Han et al., 2020). This implies the need for effective policies to control the sources and effects of OPEs on soil to mitigate potential health risks, including cancer.

3.4 REGULATORY MEASURES FOR EOCs IN SOIL

Compared with water and air media, the need to enforce control measures for soil protection has historically been less considered. According to scientific and regulatory experts, only a few EOCs have been regulated. To address this concern, a list of substances has been developed based on their effects on human health and soil ecology. In 2002, an environmental action program was established to highlight the importance of soil and outline necessary measures to prevent soil contamination

(Montanarella, 2003). Additionally, various networks and campaigns in Europe, such as SNOWMAN, CLARINET, NICOLE2, EURODEMO+, and EUGRIS5, were initiated to create awareness regarding soil prevention, rehabilitation, and treatment of contaminated soils (Robinson et al., 2012). However, several European countries, including Italy, Netherlands, Germany, France, Belgium, and Denmark have devised their own policies and regulations to control soil contamination and degradation (Rodrigues et al., 2009). With respect to the EOCs discussed in this chapter, worldwide regulations are in place to avoid or reduce soil contamination. For instance, PPCPs are managed in many countries in Europe, the United States, the United Kingdom, Australia, and Canada by restricting their use (Maddela et al., 2022). Furthermore, the unlawful use of POPs is still observed in many developing nations, most likely as a result of their affordability, efficiency, and inadequate laws (Sohail et al., 2018). The Stockholm Convention was established to put bans and limitations on the manufacturing and use of POPs and other hazardous compounds, which was administered worldwide. Despite the fact that the Stockholm Convention has been in place since 2001, the manufacturing and use of POPs have been prohibited; yet, they are still documented in various matrices (Ullah et al., 2019). In addition to this, screening levels for some POPs are recommended by some regulatory authorities such as the USEPA, which has defined indicator values for residential and industrial soils. Similarly, Canada and the Netherlands have also developed standards for some POPs such as PCBs and OCPs as mentioned in Table 3.8.

For instance, the USEPA suggested screening levels of 0.23 and 0.94 mg/kg for PCBs in residential and industrial soils, respectively, with a high risk of posing cancer among humans (USEPA, 2024). Among OCPs, DDT and its derivatives posed cancerous risks to individuals if the residential and industrial soils reported concentrations of 1.9–2.3 and 8.5–9.6 mg/kg, respectively (USEPA, 2024). In the Netherlands, the recommended intervention values are 1.7–34 mg/kg for DDT, DDE, and DDD in soil on any type of land (Ministry of Infrastructure and Water Management, 2013). Similarly, Canada provided soil quality guidelines for agricultural and residential land as 0.7 μg/g of ΣDDTs, while European countries such as Finland and Poland suggest soil having 1 and 2.01 mg/kg of ΣDDTs as unacceptable (CCME, 2020; European Environment Agency, 2023).

Moreover, among PPCPs, bisphenol A exceeding concentrations of 320 and 4100 mg/kg in residential and industrial soil, respectively, would cause endocrinal disruptions among individuals (USEPA, 2024). Similarly, phthalates with values of 36 mg/kg DnBP, 60 mg/kg DEHP (Ministry of Infrastructure and Water Management, 2013), 630 mg/kg DnBP for residential soil, 8200 mg/kg DnBP for industrial soil, 39 mg/kg DEHP for residential soil, and 160 mg/kg DEHP for industrial soil (USEPA, 2024) are recommended as intervention values for soil. In the case of SCCPs, Japan has set a screening value of 10 mg/kg for the soil of all types of land use (Tsunemi, 2009). Additionally, the screening data for soil are not yet available for most PPCPs, OPEs, and MCCPs, which might be due to fewer studies available on their toxicological effects and soil profile. However, few international standards and transboundary conventions are available for other environmental matrices such as wastewater, biosolids, and air, which might contribute EOCs to soil. The regulatory bodies in action include federal laws and regulations, state laws and regulations, and

TABLE 3.8
Regional Screening Levels for PCBs, OCPs, Phthalates, PPCPs OPEs, and PFAS in Soil

Country	∑OCPs[c] (mg/kg)	∑Phthalate[d] (mg/kg)	∑PCBs (mg/kg)	PPCPs (mg/kg)	∑PFAS (mg/kg)	References
USA[a]	0.04–1300	39.0–6300	0.0004–0.13[e]	3200[h], 1.90[i]	1.90×10^{-5} to 2500	USEPA (2024)
USA[b]	0.17–16,000	160–82,000	0.0002–0.52[e]	41,000[h], 25.0[i]	7.80×10^{-5} to 33,000	
Netherlands*	0.32–34.0	36.0–60.0	1[f]			Ministry of Infrastructure and Water Management (2013)
Canada**	0.01–0.70	—	—	—	—	CCME (2020)
Canada***	0.70–2.00	—	—	—	—	
Belgium[b]	—	—	10.4[g]	—	—	European Environment Agency (2023)
Finland[b]	—	—	5.00[g]	—	—	
Italy[b]	—	—	5.00[g]	—	—	
Poland[b]	—	—	2.75[g]	—	—	
Spain[b]	—	—	0.80[g]	—	—	
Finland[a]	1.00	—	0.50	—	—	
Netherlands[a]	4.00	—	1.00	—	—	
Poland[a]	2.01	—	—	—	—	
Austria[a]	—	—	1.00	—	—	
Belgium[a]	—	—	0.90	—	—	
Czechia[a]	—	—	5.00	—	—	
Italy[a]	—	—	0.00	—	—	
Lithuania[a]	—	—	0.10	—	—	
Poland[a]	—	—	0.55	—	—	
Spain[a]	—	—	0.08	—	—	

[a] Regional screening levels for residential soil.
[b] Regional screening levels for industrial soil.
[c] ∑OCPs = sum of Aldrin, Endosulfan, Endrin, Heptachlor, Acetochlor, Chlordane, DDT, DDE, DDD, Mirex, HCH, HCB, and Lindane.
[d] ∑Phthalate = sum of DnBP and DEHP.
[e] Range of 12 PCBs including PCB 126, 169, 81, 77, 123, 118, 105, 157, 156, 114, 167, 189.
[f] Sum of 7 PCBs.
[g] Sum of 7 PCBs (28, 52, 101, 118, 138, 153, 180).
[h] For Bisphenol A.
[i] For Warfarin.
*Standards for all type land use.
**Standards for agricultural soil.
***Standards for residential soil.

international laws and regulations, including the Basel Convention (Maddela et al., 2022). Therefore, regulatory policies are required to be developed and enforced to avoid the aggravation of soil quality and associated life forms.

3.5 CONCLUSION AND FUTURE PROSPECTS

Conclusively, soil acts not only as a sink for EOCs but also as a secondary source for these compounds. Other environmental matrices, such as air and water upon interacting with contaminated soil can become contaminated. Moreover, the presence of EOCs in soil poses significant risks to the health of nontarget biota, including soil organisms, humans, and plants. This is particularly concerning due to the persistent nature of EOCs, their longer degradation periods, as well as the harmful byproducts of EOCs produced. However, international reports have linked synthetic chemicals to fatalities and various diseases, highlighting the increasing concern over environmental pollution. Despite these concerns, the lack of effective laws and communication between regulatory bodies and researchers allows industries to continue discharging chemicals into the surrounding environment. To address this, You et al. (2022) suggest using the susceptible-infectious-susceptible (SIS) model to raise public awareness and prompt governmental actions to evaluate risks and control the use of EOCs.

The aforementioned information on the occurrence of EOCs also highlights several knowledge gaps that warrant future research. For instance, there is a lack of information regarding the effect of soil properties on the degradation and leaching rates of phthalates. Additionally, the relative leaching rates of different plasticizer classes across various soil types remain largely unknown. Further studies are needed to understand the source apportionment of plasticizers in soils used for different land use types, particularly urban areas. Future research should also focus on soil core studies to establish relationships between polymers and plasticizers. While DEHP has been detected in cropland soils mainly associated with the sewage sludge application to croplands, its persistence and long-term effects, along with those of other plasticizers and phthalates, require further investigations. Globally, limited studies have been available on the occurrence of phthalates in soil, with China being an exception. Despite their widespread use, particularly in urbanizing cities of Asia and Africa where waste management is often ineffective, our standing for phthalates and other plasticizers remains insufficient. Understanding the degradation rates and the factors influencing them is crucial for both phthalates and other plasticizers in soil. Therefore, the effects of phthalates and other plasticizers on the soil community must be investigated to comprehend their bioaccumulation potential across the food chain.

In the case of OPEs, toxicological data for soil organisms are lacking, which would help in investigating the effects of OPEs and their byproducts on soil communities. Moreover, associations among OPEs products, soil properties, degradation, and microbial response are yet to be studied in actual soil. The structural similarity of OPEs to OCPs is concerning as they both might pose similar effects on biota and environmental integrity. Given the high levels of OPEs in the environment, their potential effects on the ecological communities are significant. Similarly, for OCPs and PCBs, regular monitoring is necessary to check soil quality and assess the contamination load. Stringent policies and the development of standards would be required to prevent further contamination of soil and other environmental matrices. Additionally, a few studies are available on the occurrence of MCCPs in soil globally compared with SCCPs. Understanding these levels will help to estimate their usage and source apportionment in urban and remote areas. All of the abovementioned EOCs, along with PPCPs, require strict monitoring to avoid unforeseen outcomes. Developing reliable and effective remediation techniques to control EOCs at their source is essential to prevent environmental pollution.

REFERENCES

Aamir, M., Yin, S., Zhou, Y., Xu, C., Liu, K., & Liu, W. (2019). Congener-specific C(10)C(13) and C(14)C(17) chlorinated paraffins in Chinese agricultural soils: Spatio-vertical distribution, homologue pattern and environmental behavior. *Environmental Pollution, 245*, 789–798. https://doi.org/10.1016/j.envpol.2018.10.132

Abdel daiem, M. M., Rivera-Utrilla, J., Ocampo-Pérez, R., Méndez-Díaz, J. D., & Sánchez-Polo, M. (2012). Environmental impact of phthalic acid esters and their removal from water and sediments by different technologies: A review. *Journal of Environmental Management, 109*, 164–178. https://doi.org/10.1016/j.jenvman.2012.05.014

Armitage, J. M., Hanson, M., Axelman, J., & Cousins, I. T. (2006). Levels and vertical distribution of PCBs in agricultural and natural soils from Sweden. *Science of the Total Environment, 371*(1), 344–352. https://doi.org/10.1016/j.scitotenv.2006.07.028

Baqar, M., Saleem, R., Zhao, M., Zhao, L., Cheng, Z., Chen, H., Yao, Y., & Sun, H. (2024). Combustion of high-calorific industrial waste in conventional brick kilns: An emerging source of PFAS emissions to agricultural soils. *Science of the Total Environment, 906*, 167612. https://doi.org/10.1016/j.scitotenv.2023.167612

Bartrons, M., & Peñuelas, J. (2017). Pharmaceuticals and personal-care products in plants. *Trends in Plant Science, 22*(3), 194–203. https://doi.org/10.1016/j.tplants.2016.12.010

Biel-Maeso, M., Burke, V., Greskowiak, J., Massmann, G., Lara-Martín, P. A., & Corada-Fernández, C. (2021). Mobility of contaminants of emerging concern in soil column experiments. *Science of the Total Environment, 762*, 144102.

Biswas, B., Qi, F., Biswas, J. K., Wijayawardena, A., Khan, M. A., & Naidu, R. (2018). The fate of chemical pollutants with soil properties and processes in the climate change paradigm: A review. *Soil Systems, 2*(3), 51. https://doi.org/10.3390/soilsystems2030051

Bolan, N., Sarkar, B., Vithanage, M., Singh, G., Tsang, D. C. W., Mukhopadhyay, R., Ramadass, K., Vinu, A., Sun, Y., Ramanayaka, S., Hoang, S. A., Yan, Y., Li, Y., Rinklebe, J., Li, H., & Kirkham, M. B. (2021). Distribution, behaviour, bioavailability and remediation of poly- and per-fluoroalkyl substances (PFAS) in solid biowastes and biowaste-treated soil. *Environment International,* 155, 106600.

Bogdal, C., Niggeler, N., Glüge, J., Diefenbacher, P. S., Wächter, D., & Hungerbühler, K. (2017). Temporal trends of chlorinated paraffins and polychlorinated biphenyls in Swiss soils. *Environmental Pollution, 220*(Pt B), 891–899. https://doi.org/10.1016/j.envpol.2016.10.073

Buck, R. C., Franklin, J., Buck, R. C., Franklin, J., Berger, U., Conder, J. M., Cousins, I. T., De Voogt, P., Jensen, A. A., Kannan, K., Mabury, S. A., & van Leeuwen, S. P. (2011). Perfluoroalkyl and polyfluoroalkyl substances in the environment: Terminology, classification, and origins. *Integrated Environmental Assessment and Management, 7*(4), 513–541.

Cachada, A., Pato, P., Rocha-Santos, T., da Silva, E. F., & Duarte, A. C. (2012). Levels, sources and potential human health risks of organic pollutants in urban soils. *Science of the Total Environment, 430*, 184–192. https://doi.org/10.1016/j.scitotenv.2012.04.075

Calvo-Flores, F. G., Isac-García, J., & Dobado, J. A. (2018). *Emerging Pollutants: Origin, Structure, and Properties*. Wiley-VCH, Weinheim, Germany.

Can-Güven, E., Gedik, K., & Kurt-Karakuş, P. B. (2022). Organochlorine pesticides and polychlorinated biphenyls from a greenhouse area on the Mediterranean coast of Turkey: Distribution, air-soil exchange, enantiomeric signature, and source implications. *Atmospheric Pollution Research*, *13*(1). https://doi.org/10.1016/j.apr.2021.101263

Canadian Council of Ministers of the Environment, Winnipeg. (CCME) (2020). *Soil Quality Guidelines for the Protection of Environmental and Human Health*.

Chase, D. A., Karnjanapiboonwong, A., Fang, Y., Cobb, G. P., Morse, A. N., & Anderson, T. A. (2012). Occurrence of synthetic musk fragrances in effluent and non-effluent impacted environments. *Science of the Total Environment*, *416*, 253–260.

Chen, F., Ying, G.-G., Kong, L.-X., Wang, L., Zhao, J.-L., Zhou, L.-J., & Zhang, L.-J. (2011). Distribution and accumulation of endocrine-disrupting chemicals and pharmaceuticals in wastewater irrigated soils in Hebei, China. *Environmental Pollution*, *159*(6), 1490–1498. https://doi.org/10.1016/j.envpol.2011.03.016

Chen, L., Huang, Y., Han, S., Feng, Y., Jiang, G., Tang, C., Ye, Z., Zhan, W., Liu, M., & Zhang, S. (2013b). Sample pretreatment optimization for the analysis of short chain chlorinated paraffins in soil with gas chromatography-electron capture negative ion-mass spectrometry. *Journal of Chromatography. A*, 1274, 36–43. https://doi.org/10.1016/j.chroma.2012.12.010

Chen, R., Wang, Y., Wang, P., & Jiang, G. (2014). Spatial distribution of short chain chlorinated paraffins in soils from Taizhou, an e-waste dismantling area. *Environmental Chemistry*, *33*, 873–879.

Chen, W., Xu, J., Lu, S., Jiao, W., Wu, L., & Chang, A. C. (2013a). Fates and transport of PPCPs in soil receiving reclaimed water irrigation. *Chemosphere*, *93*(10), 2621–2630. https://doi.org/10.1016/j.chemosphere.2013.09.088

Chen, X., Gu, X., Bao, L., Ma, S., & Mu, Y. (2021). Comparison of adsorption and desorption of triclosan between microplastics and soil particles. *Chemosphere*, *263*, 127947.

Cho, K. J., Hirakawa, T., Mukai, T., Takimoto, K., & Okada, M. (1996). Origin and stormwater runoff of TCP (tricresyl phosphate) isomers. *Water Research*, *30*(6), 1431–1438.

Cui, K., Wen, J., Zeng, F., Li, S., Zhou, X., & Zeng, Z. (2017). Occurrence and distribution of organophosphate esters in urban soils of the subtropical city, Guangzhou, China. *Chemosphere*, *175*, 514–520. https://doi.org/10.1016/j.chemosphere.2017.02.070

Dalahmeh, S., Tirgani, S., Komakech, A. J., Niwagaba, C. B., & Ahrens, L. (2018). Per-and polyfluoroalkyl substances (PFASs) in water, soil and plants in wetlands and agricultural areas in Kampala, Uganda. *Science of the Total Environment*, *631*, 660–667.

Daňková, R., Jarošová, A., & Poláková, Š. (2013). Monitoring of phthalates in Moravian agricultural soil in 2011 and 2012. *Mendelnet*, 44, 563–567.

Dauchy, X., Boiteux, V., Colin, A., Hémard, J., Bach, C., Rosin, C., & Munoz, J.-F. (2019). Deep seepage of per-and polyfluoroalkyl substances through the soil of a firefighter training site and subsequent groundwater contamination. *Chemosphere*, *214*, 729–737.

Di Guardo, A., Raspa, G., Terzaghi, E., Vergani, L., Mapelli, F., Borin, S., Zanardini, E., Morosini, C., Anelli, S., Nastasio, P., Sale, V. M., & Armiraglio, S. (2020). PCB vertical and horizontal movement in agricultural soils of a highly contaminated site: Role of soil properties, cultivation history and PCB physico-chemical parameters. *Science of the Total Environment*, *747*, 141477. https://doi.org/10.1016/j.scitotenv.2020.141477

Díaz-Cruz, M. S., García-Galán, M. J., Guerra, P., Jelic, A., Postigo, C., Eljarrat, E., Farré, M., López de Alda, M. J., Petrovic, M., Barceló, D., Petrovic, M., & Barceló, D. (2009). Analysis of selected emerging contaminants in sewage sludge. *TrAC Trends in Analytical Chemistry*, *28*(11), 1263–1275. https://doi.org/10.1016/j.trac.2009.09.003

Environment Canada. (2008). *Chlorinated Paraffins. Follow-up Report on a PSL1 Assessment for Which Data Were Insufficient to Conclude Whether the Substances Were "Toxic" to the Environment and to the Human Health*. PMN P-14-0683, P-14-0684.

European Environment Agency. (2023). *Soil Monitoring in Europe – Indicators and Thresholds for Soil Health Assessments* (pp. 87–88). EEA Report 08/2022. https://www.eea.europa.eu/publications/soil-monitoring-in-europe

Fang, Y., Kim, E., & Strathmann, T. J. (2018). Mineral- and base-catalyzed hydrolysis of organophosphate flame retardants: Potential major fate-controlling sink in soil and aquatic environments. *Environmental Science & Technology*, *52*(4), 1997–2006. https://doi.org/10.1021/acs.est.7b05911

Fang, Y., Vanzin, G., Cupples, A. M., & Strathmann, T. J. (2020). Influence of terminal electron-accepting conditions on the soil microbial community and degradation of organic contaminants of emerging concern. *Science of the Total Environment*, *706*, 135327. https://doi.org/10.1016/j.scitotenv.2019.135327

Fries, E., & Mihajlović, I. (2011). Pollution of soils with organophosphorus flame retardants and plasticizers. *Journal of Environmental Monitoring: JEM*, *13*(10), 2692–2694. https://doi.org/10.1039/c1em10538h

Gao, J., Zhou, H., Pan, G., Wang, J., & Chen, B. (2013). Factors influencing the persistence of organochlorine pesticides in surface soil from the region around the Hongze Lake, China. *Science of the Total Environment*, *443*, 7–13. https://doi.org/10.1016/j.scitotenv.2012.10.086

Ge, X., Ma, S., Zhang, X., Yang, Y., Li, G., & Yu, Y. (2020). Halogenated and organophosphorous flame retardants in surface soils from an e-waste dismantling park and its surrounding area: Distributions, sources, and human health risks. *Environment International*, *139*, 105741. https://doi.org/10.1016/j.envint.2020.105741

Greaves, A. K., & Letcher, R. J. (2017). A review of organophosphate esters in the environment from biological effects to distribution and fate. *Bulletin of Environmental Contamination and Toxicology*, *98*(1), 2–7. https://doi.org/10.1007/s00128-016-1898-0

Guruge, K. S., Goswami, P., Tanoue, R., Nomiyama, K., Wijesekara, R. G. S., & Dharmaratne, T. S. (2019). First nationwide investigation and environmental risk assessment of 72 pharmaceuticals and personal care products from Sri Lankan surface waterways. *Science of the Total Environment*, *690*, 683–695. https://doi.org/10.1016/j.scitotenv.2019.07.042

Haddaoui, I., Mahjoub, O., Mahjoub, B., Boujelben, A., & Di Bella, G. (2016). Occurrence and distribution of PAHs, PCBs, and chlorinated pesticides in Tunisian soil irrigated with treated wastewater. *Chemosphere*, *146*, 195–205. https://doi.org/10.1016/j.chemosphere.2015.12.007

Halfadji, A., Touabet, A., Portet-Koltalo, F., Le Derf, F., & Merlet-Machour, N. (2019). Concentrations and source identification of polycyclic aromatic hydrocarbons (PAHs) and polychlorinated biphenyls (PCBs) in agricultural, urban/residential, and industrial soils, East of Oran (Northwest Algeria). *Polycyclic Aromatic Compounds*, *39*(4), 299–310. https://doi.org/10.1080/10406638.2017.1326947

Halse, A. K., Schlabach, M., Schuster, J. K., Jones, K. C., Steinnes, E., & Breivik, K. (2015). Endosulfan, pentachlorobenzene and short-chain chlorinated paraffins in background soils from Western Europe. *Environmental Pollution*, *196*, 21–28.

Halse, A. K., Schlabach, M., Schuster, J. K., Jones, K. C., Steinnes, E., & Breivik, K. (2017). Corrigendum to "Endosulfan, pentachlorobenzene and short-chain chlorinated paraffins in background soils from Western Europe". *Environtal Pollution*, 196, 21–28. *Environmental Pollution*, *220*(Pt B), 1477–1479. https://doi.org/10.1016/j.envpol.2016.10.084

Han, X., Hao, Y., Li, Y., Yang, R., Wang, P., Zhang, G., Zhang, Q., & Jiang, G. (2020). Occurrence and distribution of organophosphate esters in the air and soils of Ny-Ålesund and London Island, Svalbard, Arctic. *Environmental Pollution*, *263*(Pt A), 114495. https://doi.org/10.1016/j.envpol.2020.114495

Høisæter, Å., & Breedveld, G. D. (2022). Leaching potential of per-and polyfluoroalkyl substances from source zones with historic contamination of aqueous film forming foam-a surfactant mixture problem. *Environmental Advances*, *8*, 100222.

Ji, Y., Wang, Y., Yao, Y., Ren, C., Lan, Z., Fang, X., Zhang, K., Sun, W., Alder, A. C., & Sun, H. (2019). Occurrence of organophosphate flame retardants in farmland soils from Northern China: Primary source analysis and risk assessment. *Environmental Pollution*, *247*, 832–838. https://doi.org/10.1016/j.envpol.2019.01.036

Johnson, G. R., Brusseau, M. L., Carroll, K. C., Tick, G. R., & Duncan, C. M. (2022). Global distributions, source-type dependencies, and concentration ranges of per- and polyfluoroalkyl substances in groundwater. *Science of the Total Environment*, *841*, 156602.

Jorfi, S., Atashi, Z., Akhbarizadeh, R., Khorasgani, Z. N., & Ahmadi, M. (2019). Distribution and health risk assessment of organochlorine pesticides in agricultural soils of the Aghili plain, Southwest Iran. *Environmental Earth Sciences*, *78*(20), 1–9. https://doi.org/10.1007/s12665-019-8605-5

Karnjanapiboonwong, A., Suski, J. G., Shah, A. A., Cai, Q., Morse, A. N., & Anderson, T. A. (2011). Occurrence of PPCPs at a wastewater treatment plant and in soil and groundwater at a land application site. *Water, Air, & Soil Pollution*, *216*, 257–273.

Kim, H., Ekpe, O. D., Lee, J.-H., Kim, D.-H., & Oh, J.-E. (2019). Field-scale evaluation of the uptake of Perfluoroalkyl substances from soil by rice in paddy fields in South Korea. *Science of the Total Environment*, *671*, 714–721.

Kinney, C. A., Furlong, E. T., Werner, S. L., & Cahill, J. D. (2006). Presence and distribution of wastewater-derived pharmaceuticals in soil irrigated with reclaimed water. *Environmental Toxicology and Chemistry: An International Journal*, *25*(2), 317–326.

Kong, S., Ji, Y., Liu, L., Chen, L., Zhao, X., Wang, J., Bai, Z., & Sun, Z. (2012). Diversities of phthalate esters in suburban agricultural soils and wasteland soil appeared with urbanization in China. *Environmental Pollution*, *170*, 161–168.

Kroon, F. J., Berry, K. L. E., Brinkman, D. L., Kookana, R., Leusch, F. D. L., Melvin, S. D., Neale, P. A., Negri, A. P., Puotinen, M., Tsang, J. J., van de Merwe, J. P., & Williams, M. (2020). Sources, presence and potential effects of contaminants of emerging concern in the marine environments of the Great Barrier Reef and Torres Strait, Australia. *Science of the Total Environment*, *719*, 135140.

Kumar, B., Verma, V. K., Kumar, S., & Sharma, C. S. (2013). Probabilistic health risk assessment of polycyclic aromatic hydrocarbons and polychlorinated biphenyls in urban soils from a tropical city of India. *Journal of Environmental Science and Health, Part A*, 48(10), 1253–1263. https://doi.org/10.1080/10934529.2013.776894

Lassen, C., Sørensen, G., Crookes, M., Christensen, F., Jeppesen, C. N., Warming, M., Mikkelsen, S. H., & Nielsen, J. M. (2014). Survey of short-chain and medium-chain. *Aquatic*, *85535*, 84–88.

Li, C., Chen, J., Wang, J., Han, P., Luan, Y., Ma, X., & Lu, A. (2016). Phthalate esters in soil, plastic film, and vegetable from greenhouse vegetable production bases in Beijing, China: Concentrations, sources, and risk assessment. *Science of the Total Environment*, *568*, 1037–1043. https://doi.org/10.1016/j.scitotenv.2016.06.077

Li, F., Shi, R., Wang, Y., He, A., Han, Z., Zheng, X., Li, C., Gao, W., Wang, Y., & Jiang, G. (2021). The effect of anthropogenic activities on the environmental fate of chlorinated paraffins in surface soil in an urbanized zone of northern China. *Environmental Pollution*, *288*, 117766. https://doi.org/10.1016/j.envpol.2021.117766

Li, W., Wang, Y., & Kannan, K. (2019). Occurrence, distribution and human exposure to 20 organophosphate esters in air, soil, pine needles, river water, and dust samples collected around an airport in New York state, United States. *Environment International, 131*, 105054. https://doi.org/10.1016/j.envint.2019.105054

Li, X., Liu, W., Zhang, C., Song, P., & Wang, J. (2020a). Fate of phthalic acid esters (PAEs) in typical greenhouse soils of different cultivation ages. *Bulletin of Environmental Contamination and Toxicology, 104*(2), 301–306. https://doi.org/10.1007/s00128-019-02756-1

Li, X., Ma, J., Fang, D., Shi, T., & Gong, Y. (2020b). Organophosphate flame retardants in soils of Zhejiang Province, China: Levels, distribution, sources, and exposure risks. *Archives of Environmental Contamination and Toxicology, 78*(2), 206–215. https://doi.org/10.1007/s00244-019-00675-0

Li, X. H., Ma, L. L., Liu, X. F., Fu, S., Cheng, H. X., & Xu, X. B. (2006). Phthalate ester pollution in urban soil of Beijing, People's Republic of China. *Bulletin of Environmental Contamination & Toxicology, 77*(2), 252–259.

Liu, C., Wei, B. K., Bao, J. S., Wang, Y., Hu, J. C., Tang, Y. E., Chen, T., & Jin, J. (2020). Polychlorinated biphenyls in the soil–crop–atmosphere system in e-waste dismantling areas in Taizhou: Concentrations, congener profiles, uptake, and translocation. *Environmental Pollution, 257*, 113622. https://doi.org/10.1016/j.envpol.2019.113622

Liu, J., Lu, G., Xie, Z., Zhang, Z., Li, S., & Yan, Z. (2015). Occurrence, bioaccumulation and risk assessment of lipophilic pharmaceutically active compounds in the downstream rivers of sewage treatment plants. *Science of the Total Environment, 511*, 54–62. https://doi.org/10.1016/j.scitotenv.2014.12.033

Liu, J.-L., & Wong, M.-H. (2013). Pharmaceuticals and personal care products (PPCPs): A review on environmental contamination in China. *Environment International, 59*, 208–224. https://doi.org/10.1016/j.envint.2013.06.012

Liu, M., Munoz, G., Vo Duy, S., Sauvé, S., & Liu, J. (2021). Per-and polyfluoroalkyl substances in contaminated soil and groundwater at airports: A Canadian case study. *Environmental Science & Technology, 56*(2), 885–895.

Łozowicka, B., Kaczyński, P., Wolejko, E., Piekutin, J., Sagitov, A., Toleubayev, K., Isenova, G., & Abzeitova, E. (2016). Evaluation of organochlorine pesticide residues in soil and plants from east Europe and central Asia. *Desalination and Water Treatment, 57*(3), 1310–1321. https://doi.org/10.1080/19443994.2014.996008

Lü, H., Mo, C.-H., Zhao, H.-M., Xiang, L., Katsoyiannis, A., Li, Y.-W., Cai, Q.-Y., & Wong, M.-H. (2018). Soil contamination and sources of phthalates and its health risk in China: A review. *Environmental Research, 164*, 417–429.

Lupi, L., Bedmar, F., Wunderlin, D. A., & Miglioranza, K. S. B. (2016). Organochlorine pesticides in agricultural soils and associated biota. *Environmental Earth Sciences, 75*(6). https://doi.org/10.1007/s12665-015-5140-x

Ma, D., Zhong, H., Lv, J., Wang, Y., & Jiang, G. (2022). Levels, distributions, and sources of legacy and novel per-and perfluoroalkyl substances (PFAS) in the topsoil of Tianjin, China. *Journal of Environmental Sciences, 112*, 71–81.

Maddela, N. R., Ramakrishnan, B., Kakarla, D., Venkateswarlu, K., & Megharaj, M. (2022). Major contaminants of emerging concern in soils: A perspective on potential health risks. *RSC Advances, 12*(20), 12396–12415. https://doi.org/10.1039/d1ra09072k

Maliszewska-Kordybach, B., Smreczak, B., & Klimkowicz-Pawlas, A. (2013). The levels and composition of persistent organic pollutants in alluvial agriculture soils affected by flooding. *Environmental Monitoring and Assessment, 185*(12), 9935–9948. https://doi.org/10.1007/s10661-013-3303-3

Mamirova, A., Pidlisnyuk, V., Amirbekov, A., Ševců, A., & Nurzhanova, A. (2021). Phytoremediation potential of *Miscanthus sinensis* And in organochlorine pesticides contaminated soil amended by Tween 20 and activated carbon. *Environmental Science and Pollution Research, 28*(13), 16092–16106. https://doi.org/10.1007/s11356-020-11609-y

Mammo, F. K., Amoah, I. D., Gani, K. M., Pillay, L., Ratha, S. K., Bux, F., & Kumari, S. (2020). Microplastics in the environment: Interactions with microbes and chemical contaminants. *Science of the Total Environment*, *743*, 140518. https://doi.org/10.1016/j.scitotenv.2020.140518

Mao, S., Liu, S., Zhou, Y., An, Q., Zhou, X., Mao, Z., Wu, Y., & Liu, W. (2021). The occurrence and sources of polychlorinated biphenyls (PCBs) in agricultural soils across China with an emphasis on unintentionally produced PCBs. *Environmental Pollution*, *271*, 116171. https://doi.org/10.1016/j.envpol.2020.116171

Matsukami, H., Tue, N. M., Suzuki, G., Someya, M., Tuyen, L. H., Viet, P. H., Takahashi, S., Tanabe, S., & Takigami, H. (2015). Flame retardant emission from e-waste recycling operation in northern Vietnam: Environmental occurrence of emerging organophosphorus esters used as alternatives for PBDEs. *Science of the Total Environment*, *514*, 492–499. https://doi.org/10.1016/j.scitotenv.2015.02.008

Melnyk, A., Dettlaff, A., Kuklińska, K., Namieśnik, J., & Wolska, L. (2015). Concentration and sources of polycyclic aromatic hydrocarbons (PAHs) and polychlorinated biphenyls (PCBs) in surface soil near a municipal solid waste (MSW) landfill. *Science of the Total Environment*, *530–531*, 18–27. https://doi.org/10.1016/j.scitotenv.2015.05.092

Meng, X.-Z., Wang, Y., Xiang, N., Chen, L., Liu, Z., Wu, B., Dai, X., Zhang, Y.-H., Xie, Z., & Ebinghaus, R. (2014). Flow of sewage sludge-borne phthalate esters (PAEs) from human release to human intake: Implication for risk assessment of sludge applied to soil. *Science of the Total Environment*, *476–477*, 242–249. https://doi.org/10.1016/j.scitotenv.2014.01.007

Mihajlović, I., Miloradov, M. V., & Fries, E. (2011). Application of Twisselmann extraction, SPME, and GC-MS to assess input sources for organophosphate esters into soil. *Environmental Science & Technology*, *45*(6), 2264–2269. https://doi.org/10.1021/es103870f

Ministry of Infrastructure and Water Management. (2013). *Soil Remediation Circular*. https://rwsenvironment.eu/subjects/soil/legislation-and/#:~:text=The%20Soil%20Remediation%20Circular%202013,and%20applies%20to%20dry%20land. (Accessed on: 20th July, 2024)

Moeckel, C., Breivik, K., Nøst, T. H., Sankoh, A., Jones, K. C., & Sweetman, A. (2020). Soil pollution at a major West African E-waste recycling site: Contamination pathways and implications for potential mitigation strategies. *Environment International*, *137*, 105563. https://doi.org/10.1016/j.envint.2020.105563

Montanarella, L. (2003). The EU thematic strategy on soil protection. In R. J. A. Jones & L. Montanarella (eds.), *Land Degradation in Central and Eastern Europe,* (pp. 15–29). European Soil Bureau Research Report No.10, EUR 20688 EN, (2003), 324 pp. Office for Official Publications of the European Communities, Luxembourg.

Moreno, D. V., Ferrera, Z. S., Rodríguez, J. J. S., Aixalà, E. P., & Ballarín, F. B. (2008). Determination of organochlorine pesticides from agricultural soils using pressurized liquid extraction. *Soil and Sediment Contamination*, *17*(1), 1–11. https://doi.org/10.1080/15320380701741263

Nakata, H., Hirakawa, Y., Kawazoe, M., Nakabo, T., Arizono, K., Abe, S. I., Kitano, T., Shimada, H., Watanabe, I., Li, W., & Ding, X. (2005). Concentrations and compositions of organochlorine contaminants in sediments, soils, crustaceans, fishes and birds collected from Lake Tai, Hangzhou Bay and Shanghai city region, China. *Environmental Pollution*, *133*(3), 415–429. https://doi.org/10.1016/j.envpol.2004.07.003

Nascimento, R. A., Nunoo, D. B. O., Bizkarguenaga, E., Schultes, L., Zabaleta, I., Benskin, J. P., Spanó, S., & Leonel, J. (2018). Sulfluramid use in Brazilian agriculture: A source of per-and polyfluoroalkyl substances (PFASs) to the environment. *Environmental Pollution*, *242*, 1436–1443.

Newbold, R. R., Jefferson, W. N., & Padilla-Banks, E. (2009). Prenatal exposure to Bisphenol A at environmentally relevant doses adversely affects the murine female reproductive tract later in life. *Environmental Health Perspectives*, *117*(6), 879–885. https://doi.org/10.1289/ehp.0800045

Nickerson, A., Rodowa, A. E., Adamson, D. T., Field, J. A., Kulkarni, P. R., Kornuc, J. J., & Higgins, C. P. (2020). Spatial trends of anionic, zwitterionic, and cationic PFASs at an AFFF-impacted site. *Environmental Science & Technology*, *55*(1), 313–323.

Nipen, M., Vogt, R. D., Bohlin-Nizzetto, P., Borgå, K., Mwakalapa, E. B., Borgen, A. R., Jørgensen, S. J., Ntapanta, S. M., Mmochi, A. J., Schlabach, M., & Breivik, K. (2022). Spatial trends of chlorinated paraffins and dechloranes in air and soil in a tropical urban, suburban, and rural environment. *Environmental Pollution*, *292*(Pt A), 118298. https://doi.org/10.1016/j.envpol.2021.118298

Niu, L., Xu, Y., Xu, C., Yun, L., & Liu, W. (2014). Status of phthalate esters contamination in agricultural soils across China and associated health risks. *Environmental Pollution*, *195*, 16–23.

Núñez-Gastélum, J. A., Hernández-Carreón, S., Delgado-Ríos, M., Flores-Marguez, J. P., Meza-Montenegro, M. M., Osorio-Rosas, C., Cota-Ruiz, K., & Gardea-Torresdey, J. L. (2019). Study of organochlorine pesticides and heavy metals in soils of the Juarez valley: An important agricultural region between Mexico and the USA. *Environmental Science and Pollution Research*, *26*(36), 36401–36409. https://doi.org/10.1007/s11356-019-06724-4

Pokhrel, B., Gong, P., Wang, X., Chen, M., Wang, C., & Gao, S. (2018). Distribution, sources, and air–soil exchange of OCPs, PCBs and PAHs in urban soils of Nepal. *Chemosphere*, *200*, 532–541. https://doi.org/10.1016/j.chemosphere.2018.01.119

Prasse, C., Zech, W., Itanna, F., & Glaser, B. (2012). Contamination and source assessment of metals, polychlorinated biphenyls, and polycyclic aromatic hydrocarbons in urban soils from Addis Ababa, Ethiopia. *Toxicological & Environmental Chemistry*, *94*(10), 1954–1979. https://doi.org/10.1080/02772248.2012.737794

Pullagurala, V. L. R., Rawat, S., Adisa, I. O., Hernandez-Viezcas, J. A., Peralta-Videa, J. R., & Gardea-Torresdey, J. L. (2018). Plant uptake and translocation of contaminants of emerging concern in soil. *Science of the Total Environment*, *636*, 1585–1596.

Qi, Y., Zhang, T. C., & Ren, Y. (2014). Testosterone sorption and desorption: Effects of soil particle size. *Journal of Hazardous Materials*, *279*, 493–501.

Qiao, B., Song, D., Fang, B., Yu, H., Li, X., Zhao, L., Yao, Y., Zhu, L., Chen, H., & Sun, H. (2023). Nontarget screening and fate of emerging per-and polyfluoroalkyl substances in wastewater treatment plants in Tianjin, China. *Environmental Science & Technology*, *57*(48), 20127–20137.

Raghav, M., Eden, S., Mitchell, K., & Witte, B. (2013). Contaminants of emerging concern in water and wastewater. In *Contaminants of Emerging Concern in Water and Wastewater* (pp. 1–12). https://doi.org/10.1016/c2016-0-05074-x

Reinikainen, J., Perkola, N., Äystö, L., & Sorvari, J. (2022). The occurrence, distribution, and risks of PFAS at AFFF-impacted sites in Finland. *Science of the Total Environment*, *829*, 154237.

Rhind, S. M., Kyle, C. E., Kerr, C., Osprey, M., Zhang, Z. L., Duff, E. I., Lilly, A., Nolan, A., Hudson, G., & Towers, W. (2013). Concentrations and geographic distribution of selected organic pollutants in Scottish surface soils. *Environmental Pollution*, *182*, 15–27.

Robinson, D.A., Hockley, N., Dominati, E., Lebron, I., Scow, K.M., Reynolds, B., Emmett, B.A., Keith, A.M., de Jonge, L.W., Schjønning, P., & Moldrup, P. (2012). Natural capital, ecosystem services, and soil change: Why soil science must embrace an ecosystems approach. *Vadose Zone Journal*, *11*(1), 2011–0051.

Rodrigues, S. M., Pereira, M. E., da Silva, E. F., Hursthouse, A. S., & Duarte, A. C. (2009). A review of regulatory decisions for environmental protection: Part I: Challenges in the implementation of national soil policies. *Environment International*, *35*(1), 202–213. https://doi.org/10.1016/j.envint.2008.08.007

Röhler, K., Haluska, A. A., Susset, B., Liu, B., & Grathwohl, P. (2021). Long-term behavior of PFAS in contaminated agricultural soils in Germany. *Journal of Contaminant Hydrology*, *241*, 103812.

SánchezBrunete, C., Miguel, E., & Tadeo, J. L. (2009). Determination of tetrabromobisphenolA, tetrachlorobisphenolA and bisphenolA in soil by ultrasonicassisted extraction and gas chromatography–mass spectrometry. *Journal of Chromatography A, 1216*(29), 5497–5503. https://doi.org/10.1016/j.chroma.2009.05.065

Semeena, V. S., & Lammel, G. (2005). The significance of the grasshopper effect on the atmospheric distribution of persistent organic substances. *Geophysical Research Letters*, *32*(7). https://doi.org/10.1029/2004GL022229

Shi, Y., Lu, Y., Meng, F., Guo, F., & Zheng, X. (2013). Occurrence of organic chlorinated pesticides and their ecological effects on soil protozoa in the agricultural soils of North Western Beijing, China. *Ecotoxicology and Environmental Safety*, *92*, 123–128. https://doi.org/10.1016/j.ecoenv.2013.03.006

Shigei, M., Ahrens, L., Hazaymeh, A., & Dalahmeh, S. S. (2020). Per-and polyfluoroalkyl substances in water and soil in wastewater-irrigated farmland in Jordan. *Science of the Total Environment*, *716*, 137057.

Sim, W., Park, H., Yoon, J.-K., Kim, J.-I., & Oh, J.-E. (2021). Characteristic distribution patterns of perfluoroalkyl substances in soils according to land-use types. *Chemosphere*, *276*, 130167.

Škrbić, B. D., Ji, Y., Đurišić-Mladenović, N., & Zhao, J. (2016). Occurrence of the phthalate esters in soil and street dust samples from the Novi Sad city area, Serbia, and the influence on the children's and adults' exposure. *Journal of Hazardous Materials, 312*, 272–279. https://doi.org/10.1016/j.jhazmat.2016.03.045

Sohail, M., Eqani, S. A. M. A. S., Podgorski, J., Bhowmik, A. K., Mahmood, A., Ali, N., Sabo-Attwood, T., Bokhari, H., & Shen, H. (2018). Persistent organic pollutant emission via dust deposition throughout Pakistan: Spatial patterns, regional cycling and their implication for human health risks. *Science of the Total Environment*, *618*, 829–837. https://doi.org/10.1016/j.scitotenv.2017.08.224

Song, Y., Wu, N., Han, J., Shen, H., Tan, Y., Ding, G., Xiang, J., Tao, H., & Jin, S. (2011). Levels of PCDD/Fs and DL-PCBs in selected foods and estimated dietary intake for the local residents of Luqiao and Yuhang in Zhejiang, China. *Chemosphere*, *85*(3), 329–334.

Sun, J., Pan, L., Tsang, D. C. W., Zhan, Y., Liu, W., Wang, X., Zhu, L., & Li, X. (2016). Polychlorinated biphenyls in agricultural soils from the Yangtze River Delta of China: Regional contamination characteristics, combined ecological effects and human health risks. *Chemosphere*, *163*, 422–428. https://doi.org/10.1016/j.chemosphere.2016.08.038

Sun, Y., & Zhu, H. (2021). A pilot study of organophosphate esters in surface soils collected from Jinan City, China: Implications for risk assessments. *Environmental Science and Pollution Research International*, *28*(3), 3344–3353. https://doi.org/10.1007/s11356-020-10730-2

Syed, J. H., Malik, R. N., & Muhammad, A. (2014). Organochlorine pesticides in surface soils and sediments from obsolete pesticides dumping site near Lahore city, Pakistan: Contamination status and their distribution. *Chemistry and Ecology*, *30*(1), 87–96. https://doi.org/10.1080/02757540.2013.829051

Tadesse, A. W. (2021). Occurrences, potential sources and health impacts of organochlorine pesticides in soil from Wuhan, Central China. *Bulletin of Environmental Contamination and Toxicology*, *107*(2), 296–311. https://doi.org/10.1007/s00128-021-03245-0

Tahir, A., Abbasi, N. A., He, C., & Ahmad, S. R. (2024). Spatial distribution and air-soil exchange of short and medium chain chlorinated paraffins in Lahore, Pakistan. *Science of the Total Environment*, 953, 176054.

Teijón, G., Candela, L., Šimůnek, J., Tamoh, K., & Valdes-Abellán, J. (2014). Fate and transport of naproxen in a sandy aquifer material: Saturated column studies and model evaluation. *Soil and Sediment Contamination: An International Journal*, *23*(7), 736–750.

Ternes, T. A., Bonerz, M., Herrmann, N., Teiser, B., & Andersen, H. R. (2007). Irrigation of treated wastewater in Braunschweig, Germany: An option to remove pharmaceuticals and musk fragrances. *Chemosphere*, *66*(5), 894–904.

Tesi, J. N., Tesi, G. O., Ossai, J. C., & Agbozu, I. E. (2020). Organochlorine pesticides (OCPs) in agricultural soils of Southern Nigeria: Spatial distribution, source identification, ecotoxicological and human health risks assessment. *Environmental Forensics*, 1–13. https://doi.org/10.1080/15275922.2020.1850570

Thiombane, M., Petrik, A., Di Bonito, M., Albanese, S., Zuzolo, D., Cicchella, D., Lima, A., Qu, C., Qi, S., & De Vivo, B. (2018). Status, sources and contamination levels of organochlorine pesticide residues in urban and agricultural areas: A preliminary review in central–southern Italian soils. *Environmental Science and Pollution Research*, *25*(26), 26361–26382. https://doi.org/10.1007/s11356-018-2688-5

Tran, B. C., Teil, M. J., Blanchard, M., Alliot, F., & Chevreuil, M. (2015). Fate of phthalates and BPA in agricultural and non-agricultural soils of the Paris area (France). *Environmental Science and Pollution Research*, *22*(14), 11118–11126. https://doi.org/10.1007/s11356-015-4178-3

Tsunemi, K. (2009). Risk assessment of short-chain chlorinated paraffins in Japan. In J. de Boer, & T. El-Sayed Ali (eds.), *Chlorinated Paraffins. The Handbook of Environmental Chemistry* (pp. 56–100). Springer, Heidelberg.

Turgut, C., Atatanir, L., Mazmanci, B., Mazmanci, M. A., Henkelmann, B., & Schramm, K. W. (2012). The occurrence and environmental effect of persistent organic pollutants (POPs) in Taurus Mountains soils. *Environmental Science and Pollution Research*, *19*(2), 325–334. https://doi.org/10.1007/s11356-011-0561-x

Ukalska-Jaruga, A., Lewinska, K., Mammadov, E., Karczewska, A., Smreczak, B., & Medynska-Juraszek, A. (2020a). Residues of persistent organic pollutants (POPs) in agricultural soils adjacent to historical sources of their storage and distribution: The case study of Azerbaijan. *Molecules*, *25*(8). https://doi.org/10.3390/molecules25081815

Ukalska-Jaruga, A., Smreczak, B., & Siebielec, G. (2020b). Assessment of pesticide residue content in Polish agricultural soils. *Molecules*, *25*(3). https://doi.org/10.3390/molecules25030587

Ullah, R., Asghar, R., Baqar, M., Mahmood, A., Ali, S. N., Sohail, M., Schäfer, R. B., & Eqani, S. A. M. A. S. (2019). Assessment of organochlorine pesticides in the Himalayan riverine ecosystems from Pakistan using passive sampling techniques. *Environmental Science and Pollution Research International*, *26*(6), 6023–6037. https://doi.org/10.1007/s11356-018-3987-6

USEPA. (2024). *Regional Screening Level (RSL) – Summary Table (TR=1E-06, HQ=0.1)*. https://www.epa.gov/risk/regional-screening-levels-rsls-generic-tables

van Mourik, L. M., Gaus, C., Leonards, P. E., & de Boer, J. (2016). Chlorinated paraffins in the environment: A review on their production, fate, levels and trends between 2010 and 2015. *Chemosphere*, *155*, 415–428.

Vikelsøe, J., Thomsen, M., & Carlsen, L. (2002). Phthalates and nonylphenols in profiles of differently dressed soils. *Science of the Total Environment*, *296*(1–3), 105–116.

Wan, W., Zhang, S., Huang, H., & Wu, T. (2016). Occurrence and distribution of organophosphorus esters in soils and wheat plants in a plastic waste treatment area in China. *Environmental Pollution*, *214*, 349–353. https://doi.org/10.1016/j.envpol.2016.04.038

Wang, F., Shih, K., Ma, R., & Li, X. (2015a). Influence of cations on the partition behavior of perfluoroheptanoate (PFHpA) and perfluorohexanesulfonate (PFHxS) on wastewater sludge. *Chemosphere*, *131*, 178–183.

Wang, H., Liang, H., & Gao, D.-W. (2017). Occurrence and risk assessment of phthalate esters (PAEs) in agricultural soils of the Sanjiang Plain, northeast China. *Environmental Science and Pollution Research*, *24*(24), 19723–19732. https://doi.org/10.1007/s11356-017-9646-5

Wang, L., Xu, X., & Lu, X. (2015b). Phthalic acid esters (PAEs) in vegetable soil from the suburbs of Xianyang city, Northwest China. *Environmental Earth Sciences*, *74*(2), 1487–1496. https://doi.org/10.1007/s12665-015-4141-0

Wang, P., Zhang, Q., Wang, Y., Wang, T., Li, X., Li, Y., Ding, L., & Jiang, G. (2009). Altitude dependence of polychlorinated biphenyls (PCBs) and polybrominated diphenyl ethers (PBDEs) in surface soil from Tibetan Plateau, China. *Chemosphere*, *76*(11), 1498–1504. https://doi.org/10.1016/j.chemosphere.2009.06.045

Wang, X.-T., Jia, H.-H., Hu, B.-P., Cheng, H.-X., Zhou, Y., & Fu, R. (2019a). Occurrence, sources, partitioning and ecological risk of short-and medium-chain chlorinated paraffins in river water and sediments in Shanghai. *Science of the Total Environment*, *653*, 475–484.

Wang, X.-T., Zhang, Y., Miao, Y., Ma, L.-L., Li, Y.-C., Chang, Y.-Y., & Wu, M.-H. (2013a). Short-chain chlorinated paraffins (SCCPs) in surface soil from a background area in China: Occurrence, distribution, and congener profiles. *Environmental Science and Pollution Research International*, *20*(7), 4742–4749. https://doi.org/10.1007/s11356-012-1446-3

Wang, Y., Li, J., Cheng, Z., Li, Q., Pan, X., Zhang, R., Liu, D., Luo, C., Liu, X., Katsoyiannis, A., & Zhang, G. (2013b). Short- and medium-chain chlorinated paraffins in air and soil of subtropical terrestrial environment in the Pearl River Delta, South China: Distribution, composition, atmospheric deposition fluxes, and environmental fate. *Environmental Science & Technology*, *47*(6), 2679–2687. https://doi.org/10.1021/es304425r

Wang, X.-T., Xu, S.-Y., Wang, X.-K., Hu, B.-P., & Jia, H.-H. (2017). Occurrence, homologue patterns and source apportionment of short and mediumchain chlorinated paraffins in suburban soils of Shanghai, China. *Chemosphere, 180*, 302–311. https://doi.org/10.1016/j.chemosphere.2017.04.037

Wang, Y., Sun, H., Zhu, H., Yao, Y., Chen, H., Ren, C., Wu, F., & Kannan, K. (2018). Occurrence and distribution of organophosphate flame retardants (OPFRs) in soil and outdoor settled dust from a multi-waste recycling area in China. *Science of the Total Environment*, *625*, 1056–1064. https://doi.org/10.1016/j.scitotenv.2018.01.013

Wang, Y., Yao, Y., Li, W., Zhu, H., Wang, L., Sun, H., & Kannan, K. (2019b). A nationwide survey of 19 organophosphate esters in soils from China: Spatial distribution and hazard assessment. *Science of the Total Environment*, *671*, 528–535. https://doi.org/10.1016/j.scitotenv.2019.03.335

Washington, J. W., Rankin, K., Libelo, E. L., Lynch, D. G., & Cyterski, M. (2019). Determining global background soil PFAS loads and the fluorotelomer-based polymer degradation rates that can account for these loads. *Science of the Total Environment*, *651*, 2444–2449.

Wei, G.-L., Liang, X.-L., Li, D.-Q., Zhuo, M.-N., Zhang, S.-Y., Huang, Q.-X., Liao, Y.-S., Xie, Z.-Y., Guo, T.-L., & Yuan, Z.-J. (2016). Occurrence, fate and ecological risk of chlorinated paraffins in Asia: A review. *Environment International*, *92*, 373–387.

Wei, L., Li, Z., Sun, J., & Zhu, L. (2020). Pollution characteristics and health risk assessment of phthalate esters in agricultural soil and vegetables in the Yangtze River Delta of China. *Science of the Total Environment*, *726*, 137978. https://doi.org/10.1016/j.scitotenv.2020.137978

Wu, W., Hu, J., Wang, J., Chen, X., Yao, N., Tao, J., & Zhou, Y.-K. (2015). Analysis of phthalate esters in soils near an electronics manufacturing facility and from a non-industrialized area by gas purge microsyringe extraction and gas chromatography. *Science of the Total Environment*, *508*, 445–451.

Xia, Q. (2012). *The Research about SCCPs Level in the Zhejiang Province Soil and Sediments* (p. 71). Zhejiang University of Technology, Ningbo, Zhejiang, China.

Xia, X., Yang, L., Bu, Q., & Liu, R. (2011). Levels, distribution, and health risk of phthalate esters in urban soils of Beijing, China. *Journal of Environmental Quality*, *40*(5), 1643–1651.

Xu, C., Niu, L., Zou, D., Zhu, S., & Liu, W. (2019a). Congener-specific composition of polychlorinated biphenyls (PCBs) in soil-air partitioning and the associated health risks. *Science of the Total Environment*, *684*, 486–495. https://doi.org/10.1016/j.scitotenv.2019.05.334

Xu, C., Zhang, Q., Gao, L., Zheng, M., Qiao, L., Cui, L., Wang, R., & Cheng, J. (2019b). Spatial distributions and transport implications of short- and medium-chain chlorinated paraffins in soils and sediments from an e-waste dismantling area in China. *Science of the Total Environment*, *649*, 821–828. https://doi.org/10.1016/j.scitotenv.2018.08.355

Xu, J., Gao, Y., Zhang, H., Zhan, F., & Chen, J. (2016). Dispersion of short- and medium-chain chlorinated paraffins (CPs) from a CP production plant to the surrounding surface soils and coniferous leaves. *Environmental Science & Technology*, *50*(23), 12759–12766. https://doi.org/10.1021/acs.est.6b03595

Xu, J., Wu, L., Chen, W., Jiang, P., & Chang, A. C.-S. (2009). Pharmaceuticals and personal care products (PPCPs), and endocrine disrupting compounds (EDCs) in runoff from a potato field irrigated with treated wastewater in southern California. *Journal of Health Science*, *55*(2), 306–310.

Yadav, I. C., Devi, N. L., Li, J., & Zhang, G. (2018a). Organophosphate ester flame retardants in Nepalese soil: Spatial distribution, source apportionment and air-soil exchange assessment. *Chemosphere*, *190*, 114–123. https://doi.org/10.1016/j.chemosphere.2017.09.112

Yadav, I. C., Devi, N. L., Li, J., Zhang, G., & Covaci, A. (2018b). Concentration and spatial distribution of organophosphate esters in the soil-sediment profile of Kathmandu Valley, Nepal: Implication for risk assessment. *Science of the Total Environment*, *613–614*, 502–512. https://doi.org/10.1016/j.scitotenv.2017.09.039

You, G., Gan, S., Guo, H., & Dagestani, A. A. (2022). Public opinion spread and guidance strategy under COVID-19: A SIS model analysis. *Axioms*, *11*(6), 296.

Yuan, B., Fu, J., Wang, Y., & Jiang, G. (2017). Short-chain chlorinated paraffins in soil, paddy seeds (*Oryza sativa*) and snails (*Ampullariidae*) in an e-waste dismantling area in China: Homologue group pattern, spatial distribution and risk assessment. *Environmental Pollution*, *220*, 608–615.

Zeng, L., Wang, T., Han, W., Yuan, B., Liu, Q., Wang, Y., & Jiang, G. (2011). Spatial and vertical distribution of short chain chlorinated paraffins in soils from wastewater irrigated farmlands. *Environmental Science & Technology*, *45*(6), 2100–2106. https://doi.org/10.1021/es103740v

Zeng, Y.-H., Tang, B., Luo, X.-J., Zheng, X.-B., Peng, P.-A., & Mai, B.-X. (2016). Organohalogen pollutants in surface particulates from workshop floors of four major e-waste recycling sites in China and implications for emission lists. *Science of the Total Environment*, *569–570*, 982–989. https://doi.org/10.1016/j.scitotenv.2016.06.053

Zhang, Y., Liang, Q., Gao, R., Hou, H., Tan, W., He, X., Zhang, H., Yu, M., Ma, L., Xi, B., & Wang, X. (2015). Contamination of phthalate esters (PAEs) in typical wastewater-irrigated agricultural soils in Hebei, North China. *PLoS ONE*, *10*(9), 1–13. https://doi.org/10.1371/journal.pone.0137998

Zhang, Z., He, G., Peng, X., & Lu, L. (2014). Distribution and sources of phthalate esters in the topsoils of Beijing, China. *Environmental Geochemistry and Health*, *36*, 505–515.

Zhang, Z., Lin, G., Lin, T., Zhang, R., Jin, L., & Di, Y. (2020). Occurrence, behavior, and fate of organophosphate esters (OPEs) in subtropical paddy field environment: A case study in Nanning City of South China. *Environmental Pollution*, *267*, 115675.

Zhao, N., Cui, Y., Wang, P., Li, S., Jiang, W., Luo, N., Wang, Z., Chen, X., & Ding, L. (2019). Short-chain chlorinated paraffins in soil, sediment, and seawater in the intertidal zone of Shandong Peninsula, China: Distribution and composition. *Chemosphere*, *220*, 452–458. https://doi.org/10.1016/j.chemosphere.2018.12.063

Zhao, W., Lu, J., Lai, Y., Hou, Y., Zhao, X., Wei, Q., Zou, X., & Gou, Z. (2023). Occurrences, possible sources, and risk impacts of organochlorine pesticides in soil of Changchun Central Urban Area, Northeast China. *Sustainability*, *15*(24), 16801. https://doi.org/10.3390/su152416801

Zhou, B., Zhao, L., Wang, Y., Sun, Y., Li, X., Xu, H., Weng, L., Pan, Z., Yang, S., Chang, X., & Li, Y. (2020). Spatial distribution of phthalate esters and the associated response of enzyme activities and microbial community composition in typical plastic-shed vegetable soils in China. *Ecotoxicology and Environmental Safety*, *195*, 110495. https://doi.org/10.1016/j.ecoenv.2020.110495

Zhou, Q., Xu, C., Shen, C., Li, F., Liu, S., & Aamir, M. (2023). Congener profiles, air-soil exchange, and potential risks of short-and medium-chain chlorinated paraffins in demonstration zone of Yangtze River Delta. *Atmospheric Pollution Research*, *14*(1), 101639.

Zhu, Q., Qian, J., Huang, S., Li, Q., Guo, L., Zeng, J., Zhang, W., Cao, X., & Yang, J. (2022). Occurrence, distribution, and input pathways of per-and polyfluoroalkyl substances in soils near different sources in Shanghai. *Environmental Pollution*, *308*, 119620.

Zhu, Y. Y., Tian, J., Yang, H. B., Wei, E. Q., Yu, G., & Wei, F. S. (2010). Phthalate pollution in atmospheric particles and soils of Tianjin and their correlation. *Environmental Pollution Control*, *31*(10), 1535–1541.

4 Sorption of Emerging Organic Contaminants (EOCs) in Soil

Qandeel Zainab, Iqra Muhammad, Shahid Mahmood, Muzammil Anjum, Qaiser Hussain, and Azeem Khalid

4.1 INTRODUCTION

Emerging contaminants are substances that have been utilized in the past but have recently gained attention because of their potential negative effects or because they are newly produced pollutants. Synthetic polymers and plastics, medications, intimate hygiene items, plasticizers, and fire retardants are a few examples. For instance, ECs that are readily vaporized and released into the atmosphere have the potential to disperse widely and deposit (Liu et al. 2024). Emerging organic contaminants (EOCs) are any variety of substances as well as microorganisms that are not frequently observed in the environment, however, have the capacity to harm the environment and/or public health. Organic pollutants, or EOCs, have been detected or discovered in the environment as a result of recent analytical developments. They are composed of a diverse range of substances that enter, spread, and endure in the surroundings, together with their metabolites and transformation products (Mukhopadhyay et al. 2022). EOCs are gaining attention; however, the majority of research focuses on wastewater treatment, EOC incidence, and EOC fate in aquatic environments. It is unclear how active EOCs in agricultural soils are introduced by treated wastewater irrigation (Hurtado et al. 2017). Pesticides, industrial compounds, pharmaceuticals and personal care products (PPCPs) are often categorized as EOCs. One potential tactic to lessen the strain on water supplies is reusing wastewater to irrigate land. However, there is chance that EOCs might build up in the soil (Ferreira et al. 2020).

When EOCs are present in soil, composition of soil may change irreversibly, changing the soil's characteristics. According to Wan et al. (2019), microplastics significantly disturb soil structure and decrease water-holding capacity by affecting evaporation, which worsens soil moisture loss (Liu et al. 2024). Industrial and farming activities emit a variety of organic compounds, including pesticides, herbicides, public and livestock medications, biological and manufactured hormones, self-care items, surfactants as well as polymers, fire retardants, including other substances that

 DOI: 10.1201/9781003509523-4

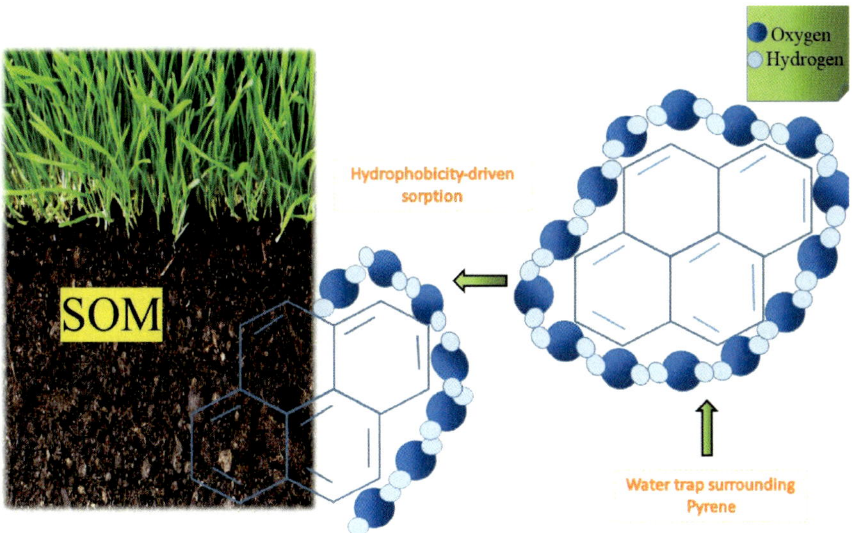

FIGURE 4.1 Formation of a water cage by a hydrophobic molecule disturbs the bulk liquid water structure. Pyrene absorption by hydrophobic components in soil organic matter lessens this disruption.

are believed to have an effect on ecosystem as well as on human health (Pignatello et al. 2010). One of the most significant chemical processes controlling the soil's ability for holding nutrients, contaminants, and other substances is probably sorption. The sorptive concentration may have the most important impact on the buildup of a sorbate on or inside a sorbent, even if the chemical makeup of sorbents (solids) varies. Le Chatelier's Principle of Chemical Equilibrium states that at low sorptive concentrations, rising or falling sorptive concentrations will cause sorbate concentrations to rise or fall. Fertilizers, for example, raise the concentration of potassium (K^+) in the solution, which in turn raises the quantity of K^+ that the solid phase absorbs. On the other hand, the sorbate K^+ will desorb from the soil as growing plants absorb K^+ from the soil solution. For plants and other soil creatures, the soil acts as a storehouse of nutrients. The bulk liquid water structure is altered when a hydrophobic molecule forms a water trap. This disturbance is reduced when pyrene is absorbed by hydrophobic components in soil organic matter (SOM). Hydrophobicity-driven sorption is illustrated in Figure 4.1.

Sorption affects organic molecules' mobility, chemistry, and biological activity. Sorption slows down the diffusion and advection of vapors or solutes. The sorbent may affect chemical reactivity by producing a reactive surface or altering the microscale chemical environment around reactive species. For example, when esters sorb to sediment particles, they decrease their reactivity to base hydrolysis. By preventing organic molecules from being directly absorbed by cells, sorption lowers bioavailability (Pignatello 1998).

4.2 ENVIRONMENTAL IMPLICATIONS OF EOC SORPTION

The distribution, persistence, and possible effects on ecological system and human health of EOCs are all significantly impacted by their sorption onto different environmental surfaces, including soils, sediments, and particulates (Hunt and Voulvoulis 2020).

4.2.1 Occurrence of EOCs in Soil

Pharmaceuticals are widely distributed and persistent biologically active compounds utilized in veterinary as well as human health care, plant development, and animal feeding. According to literature, veterinary medication is one of the main sources of frequently found pharmaceuticals in our surroundings. In general, the pharmaceutical industry is one of the major point sources of medicines and their metabolites, whereas sewage sludge and wastewater systems are examples of non-point sources. During Ghanaian examination, soils included following substances: diclofenac, chloramphenicol, doxycycline, and ciprofloxacin (0.90 mg/kg), caffeine (3.00 mg/kg), ibuprofen (1.50 mg/kg), amoxicillin (0.01 mg/kg), and paracetamol (2.10 mg/kg). Existence of drugs in soil, however, is understood little, especially in poor countries such as those in Africa (Masinga et al. 2024). Human and veterinary drugs and active ingredients in personal care items are a broad category of bioactive substances that are not given much consideration as possible environmental contaminants. Similarly, prescription drugs, biologics, diagnostic equipment, "nutritional supplements," fragrances, and sunscreens can also be EOCs in the environment (Daughton and Ternes 1999). Personal care items fall under the pseudo-persistent group since a significant percentage of them have characteristics with persistent organic pollutants (POPs), although not being officially categorized as POPs. Sewage, domestic waste, clinical effluents, and commercial and farming effluents are the main sources of PCPs (Masinga et al. 2024).

An important family of low molecular weight, nonvolatile substances that are frequently utilized as additives in polymer sector are plasticizing agents. Their primary purpose is to raise the elasticity and processing capability of polymers by lowering glass transition temperature, also known as second order transition temperature (Vieira et al. 2011). In terrestrial environments, soil is regarded as a significant sink for semi volatile organic compounds (SVOCs). Meanwhile, secondary contamination might result from SVOCs in soil volatilizing into the atmosphere as result of shifting sources or climate. Consequently, on a regional and global level, the exchange mechanism plays a crucial role in the transportation and environmental destiny of SVOCs (Wang et al. 2020b).

Groups that make up surfactants include fatty acid ethoxylates, fluorinated alkyl compounds, alcohol ethoxylates, alkylphenol ethoxylates, and alkylphenyl sulphonates such as linear alkylbenzene sulfonates (LAS) (Boontanon et al. 2013). Surfactants are a family of chemicals known for their ability to clean. Due to their excessive usage as ingredients, highly polluted wastewaters have been discharged into land and water from personal care products such as body washes and shampoos, as well as domestic cleaning goods including laundry, dishwashing, and hard-surface cleaners. Surfactants may degrade aerobically or anaerobically after entering various environmental compartments, such as lakes, rivers, soils, and sediments (Jardak et al. 2016). Crucially, by looking at distribution patterns of various surfactant homologs and their metabolites, it is feasible to distinguish between sources of sewage pollution that released treated and untreated wastewater.

Soil samples were found to contain gasoline additives at amounts ranging from 0.02 to 1.63 μg/kg, such as benzene, tert-butyl formate, methyl tert-butyl ether, diisopropyl ether, tert-butyl alcohol, ethyl tert-butyl ether, and xylenes (Rosell et al. 2006). Many polymeric products, such as plastics, foams, resins, and adhesives have used polybrominated diphenyl ethers, a class of brominated flame retardants, to prevent fires from spreading. Applying sewage sludge to soil as a nutrient amendment has been demonstrated to be a significant conduit of brominated flame retardants loading to agricultural land because of the high amounts of pollution found in sewage waste streams (McGrath et al. 2017). Septic tanks, hospital effluents, wastewater effluents from municipal treatment plants, livestock activities such as waste lagoons, manure application to soil, and subsurface storage of household and industrial waste are just a few of ways that these compounds are known to enter in environment. Furthermore, the groundwater–surface water exchange process is an indirect way for these substances to infiltrate the ecosystem (Lapworth et al. 2012).

4.2.2 Potential Toxic Effects and Bioavailability of EOCs in the Soil Ecosystem

The types of EOCs, exposure situations, biogeochemical endpoints, and experimental species all influence many ecological health impacts that EOCs cause (Gwenzi et al. 2022). Due to EOCs' mobility and ability to leach into groundwater as well as the possibility of accumulation within crop plants in agriculture, the reuse of treated wastewater for irrigation may pose health risks to humans (Ferreira et al. 2020). Different plant sections absorb and retain medications that are injected into the soil. Generally speaking, the generative components of plants, such as cereal grains, have the lowest accumulation coefficients, whereas the vegetative parts, in decreasing order, have the highest: roots, leaves, and stems (Gworek et al. 2021). Additionally, the trophic transfer of pharmaceuticals to and from food webs is frequently disregarded, which raises concerns about public health and highlights the urgent need to clarify the effects of today's diverse range of pharmaceuticals on the terrestrial system, their biophysical behaviors, and their interactions with soil metazoans (Zhang et al. 2024).

According to a few studies, drugs and PPCPs at trace levels negatively impact animals or soil microorganisms (destroying their variety and population structure). Crop development may be encouraged by the recovered water containing trace amounts of PPCPs. However, their buildup in crops may potentially have some morphological and physiological impacts. PPCPs have easier access to young plant seedlings than to mature ones. PPCPs can be absorbed by crops from the soil and transferred to humans through food. There may be adverse health impacts if edible parts of reclaimed watered crops are consumed. On average, 400 g of fresh weight crops per day may yield 0.4–20 μg of each target chemical. In humans, low PPCP levels can disrupt hormonal balance and metabolic functions (Qin et al. 2015).

Surfactants appear to profoundly change soil physics, and even at low concentrations, soil biology and chemistry are greatly influenced by sorption processes (Kuhnt 1993). Production of surfactants has a significantly negative impact on the environment, exacerbating issues with ozone layer depletion, climate change, global warming, and greenhouse gas emissions that are impossible to completely prevent. Both petrochemical and oleochemical-based surfactant manufactures produce solid waste, waterborne wastes, and atmospheric emissions (NO_x, CO_2, SO_2, hydrocarbons) that can lead to river and lake acidification and eutrophication. Surfactants mostly restrict or destroy plant roots, with relatively little inhibition in wheat seedling shoots. Bean plants' photosynthetic rate and chlorophyll concentration are decreased when detergent-contaminated water is used for cultivation (Rebello et al. 2014).

Once in the soil, pesticides undergo a variety of biodegradation mechanisms, such as biological breakdown (by the enzymes of microorganisms) and chemical degradation (by temperature, pH, and humidity). Pesticide residues and breakdown byproducts are absorbed by the roots through the xylem, which affects the soil and the plant in some negative ways. These consequences include excessive ROS generation, oxidative stress, DNA damage, photosynthetic obstruction, necrosis, chlorosis, twisted leaves, and plant death (Alengebawy et al. 2021).

Due to bioaccumulation, high plasticizer concentrations, particularly in urban and industrial areas, frequently surpass advised environmental limits, endangering both human health and ecological integrity. These substances' bioaccumulation in soil and water may have detrimental effects on microbial populations, nutrient cycling, and the stability of the ecosystem. The high concentrations above the threshold values raise concerns regarding their direct absorption by plants as well as possible hazards to human health and food safety (Manatunga et al. 2024).

4.2.3 The Role of Sorption in Mitigating EOC Bioaccumulation

Sorption has a major impact on the bioaccumulation of EOCs in environmental matrices such as soil, water, and sediment. Sorption can either lessen or increase the likelihood of bioaccumulation in organisms by changing mobility and bioavailability of these pollutants.

The rate at which organic pollutants move through porous media, including soils, sediments, and aquifers, is primarily determined by their solubility. Sorption can also dramatically change the pace of biotic and abiotic processes (Rao 1990). Plastics break down into tiny pieces called nanoplastics (NPs, smaller than 0.1 μm) when they are released into the environment and microplastics (MPs) ranging from 1 to 5000 μm. NPs and MPs have garnered a lot of attention in recent years. Because NPs and MPs have a large specific surface area and are hydrophobic, they may adsorb a variety of organic contaminants. Main mechanisms by which organic pollutants sorb to NPs/MPs include partitioning, surface sorption (hydrogen bonding, π–π interaction, electrostatic interaction, and van der Waals force), and pore filling. In addition to the properties of organic contaminants and NPs/MPs, chemistry of the soil solution, such as the pH, ionic strength, and dissolved organic matter, affects how organic pollutants sorb to NPs/MPs (Wang et al. 2020a). PPCPs, with half-lives of less than a month, can degrade rapidly in the environment, especially in aerobic settings, making them nonpersistent. However, new issues have emerged, such as mineralization not being observed in soils despite considerable degradation of endocrine disrupting chemicals (EDCs).

EDCs can change into other substances, such as estradiol to estrone, and metabolites may compete for sorption sites with parent substances (Pan et al. 2009). Furthermore, the detailed mechanism of sorption for various EOCs is discussed in next section.

4.3 PHYSICAL AND CHEMICAL PROPERTIES OF EOCs

Assessing the behavior of organic molecules in soil is challenging because it depends on the physical and chemical characteristics of the chemicals, the characteristics of the soil, and the surrounding environment (Wong and Wania 2011).

Physicochemical characteristics (pK_a, water solubility, partition coefficient, and vapor pressure) and degree of polarity of emerging organic pollutants are often hydrophobic by nature. Common EOCs (triclosan, carbamazepine, and diclofenac) have vapor pressure values of 6.9×10^{-5}, 1.17×10^{-5}, and 2.91×10^{-4} Pa, respectively. Depending on particular environmental conditions, these drugs can exist in a range of physical states. Chemical's flow between two or more environmental compartments can be described through equilibrium partitioning, which takes place between adjacent phases, as between solid and liquid (dissolution) or liquid and gas (volatilization). Partition coefficients may be used to quantify the net transfer of an organic chemical across phases, subjected to equilibrium restrictions (Figure 4.2).

Emerging contaminants are compounds, perhaps natural or manufactured, which are not often observed in the environment, yet have known or suspected negative impacts on ecosystems and people (Rodriguez-Narvaez et al. 2017). A broad class of compounds known as surfactants is intended to have solubilizing or cleaning qualities. They typically consist of a nonpolar hydrocarbon tail that is difficult to dissolve

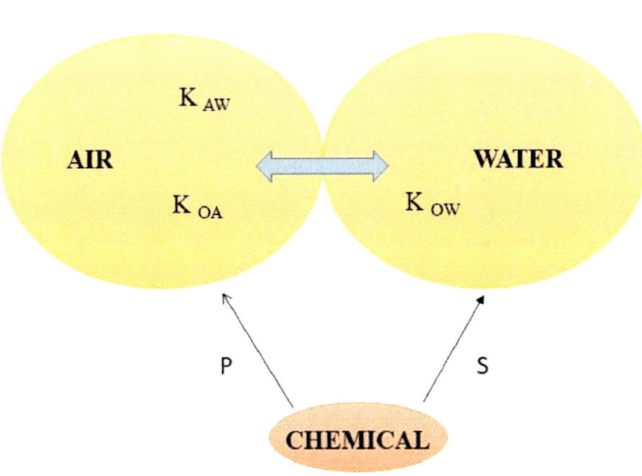

FIGURE 4.2 Physicochemical characteristics that regulate how EOCs are distributed throughout the environment. *S* stands for solubility, *K* for partition coefficient, and *P* for vapor pressure.

in water and a polar head group (either charged or uncharged) that is well solvated in water. As a result, surfactants are molecules that have both hydrophilic and hydrophobic characteristics (Ying 2006). According to reports, MNPs are vectors that transport hydrophobic organic microcontaminants in both terrestrial and aquatic environments. Polybrominated diethers, polyfluorinated alkyls, polychlorinated biphenyls, medication compounds (tetracycline, ciprofloxacin, sertraline, propranolol, and sulfamethoxazole), phthalates, surfactants, personal care items, polycyclic aromatic hydrocarbons, and pharmaceuticals contribute to their long-term presence in the environment, accumulation in living organisms, and causing possible harmful effects (Agboola and Nsikak 2021). Nevertheless, it has been discovered that a certain class of substances mimics or interferes with the normal operation of endocrine glands. In addition to being referred to as endocrine disruptive compounds, these substances are also known as endocrine disruptors, endocrine modulators, or endocrine disrupting chemicals. Endocrine disruptive substances can be either natural or synthetic substances that interact with estrogenic receptors to either boost or hinder the correct action of hormones. Endocrine disruptors are substances that prevent the body's natural hormones that are in charge of development, behavior, fertility, and the preservation of homeostasis (normal cell metabolism)—from being synthesized, secreted, transported, bound, acting, or being eliminated (Tijani et al. 2016).

Phenoxyalkanoic acid herbicides are the most researched chiral contaminants, used in forestry, industrial weed management, and agriculture. Due to their high water solubility, they are often mobile in the environment and leak into runoff and groundwater, contaminating water sources. Investigating their environmental occurrence and destiny is crucial to estimate their potential water resource danger. Mecoprop and dichlorprop are significant chiral members of the racemic medicine family, with only the R-enantiomers being herbicidal. Enantiopure products, containing the active enantiomer, have been created since the 1980s to replace or supplement original racemic medicines, reducing the total environmental load of xenobiotic chemicals (Wong 2006).

Acetamide pesticides are a significant family of agrochemicals that are frequently employed as fungicides and selective herbicides to control weeds and fungus in both agricultural and noncrop applications (such as lawn weed management). Like phenoxyalkanoic acids, acetamide insecticides frequently have a high water solubility and leach after being sprayed to surface and ground waters (Wong 2006).

Some of these herbicides have many stereogenic centers and many are chiral. As a result, diastereomers are created, which may be separated by achiral methods and have distinct physical and chemical characteristics. Two or more enantiomers make up each diastereomer. One such instance is metolachlor, which at normal temperature has an asymmetric carbon atom and axial chirality (atropisomerism) due to inhibited rotation of bonds (Müller et al. 2001). Studies show that polymeric properties, particle size, surface area, charge, crystallinity, compound-specific properties, and environmental elements including pH, salinity, and ionic strength all affect the adsorption of microplastics. Strong interactions between pollutants and microplastics are caused by major sorption mechanisms, which also control the fate, motion, or partitioning of organic contaminants between solid particles and water (Agboola and Nsikak 2021).

4.4 SOIL PROPERTIES AFFECTING EOC DISTRIBUTION

Numerous soil characteristics affect the allocation and destiny of EOCs in soils. These characteristics control how EOCs migrate through the soil, interact with soil particles, and either disintegrate or survive. Given that soil may serve as both a source and a sink for these pollutants, understanding these components is essential to evaluating environmental risks posed by EOCs.

4.4.1 Role of Organic Matter in EOC Retention and Sorption

Transport, reactivity, and bioavailability of organic contaminants in soil are all significantly influenced by the organic matter component of soils, which serves as the primary sorbent of organic compounds that are sparingly soluble (Pignatello 1998). In soil–water systems, SOM is frequently regarded as the primary sorptive phase for organic pollutants and pesticides.

The organic-matter-normalized sorption coefficients are often used to forecast pesticide soil-water dispersion, as they help estimate the extent to which a pesticide will adsorb to soil organic matter versus remaining in the aqueous phase (Sheng et al. 2001). Mineral and organic matter components of soil can either retain or sequester an organic contaminant, or it can build up within the soil biota or be removed by biodegradation, leaching, or volatilization after entering the soil (Figure 4.3).

Numerous factors, such as, type of soil (mineral and organic matter content) and the physicochemical characteristics of contaminant(s), such as water solubility, polarity, hydrophobicity, lipophilicity, and molecular structure, influence the destiny and behavior of EOCs in the soil (Semple et al. 2003).

SOM plays crucial role in controlling the movement and retention of organic pollutants in the subsurface environment. This is achieved through sorption onto SOM, which alters the chemistry of the soil surface and soil pore system to influence

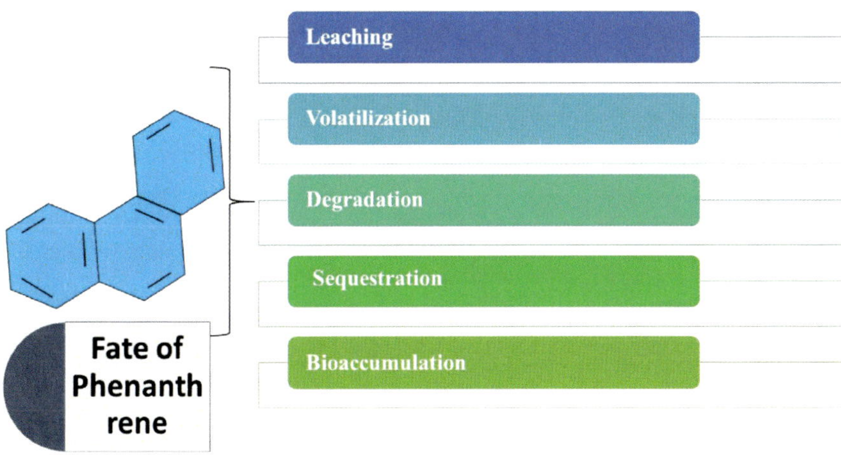

FIGURE 4.3 The behavior and possible fate of phenanthrene, a prototypical hydrophobic organic pollutant, in soil.

transport. Nonpolar SOM exhibits a greater sorption interaction with carbamazepine, a common PPCP found in soil, groundwater, and reclaimed water. These compounds, including bisphenol A, carbamazepine, and ibuprofen, are commonly used medications and potential endocrine disruptors. The presence of SOM in soils significantly influences the retention and sorption of substances such as carbamazepine, which has a greater affinity for sorption in soils with a depth of 0–5 cm (Qin et al. 2017).

4.4.2 Influence of pH, Ionic Strength, and Soil Buffering Capacity

Sorption kinetics, physicochemical characteristics (pH, cation exchange capacity, ionic strength, surface area, etc.), temperature, volatile compounds, and dissolved organic matter are some of variables that affect how organic pollutants are absorbed through natural sorbents such as soils, sediments, clays, and humic materials. Effects of these factors on the movement of contaminants through soil is crucial, as it influences the sorption of pollutants and their sorption coefficients for specific contaminants. Understanding these factors can help improve the management of organic pollutants in soil. There are variations in the influence of physicochemical properties of soil regarding different categories of EOCs.

A rise in pH (addition of NaOH) and an increase in the concentration of dissolved organic carbon are the opposing effects of different EOCs that raised Ca (addition of $Ca(NO_3)_2$). When a single dosage of NaOH was added, dissolved organic carbon increased more than when the same quantity of NaOH was supplied gradually (Oste et al. 2002).

The surface charge of soil minerals and organic matter, which frequently depends on the pH of the soil, causes them to retain pollutants, both biologically and inorganically, on their surfaces. Several processes include precipitation events, ligand exchange, ion exchange, complexation, and electrostatic attraction, which aid in retention of pollutants by soil particles. Soil minerals and organic matter can filter contaminants out of soil water by transferring them from the solution to solid phase. Therefore, soil with high clay concentration helps prevent diffuse contamination at nearby locations by acting as a physical barrier for the flow of contaminants. Furthermore, aerobic and anaerobic states of soil influence the movement, conversion, and bioavailability of contaminants as a function of the redox potential (Sarkar et al. 2021).

Leaching of EOCs may be impacted by the soil solution's ionic strength. While low ionic strength may increase the solubility and mobility of EOCs, increasing their likelihood of leaching into groundwater, high ionic strength may decrease solubility of some EOCs, leading them to precipitate or sorb more firmly to the soil. Phthalates, a group of plasticizers, are crucial for the elasticity and stability of polyvinyl chloride (PVC) processing. However, they can be harmful to aquatic organisms due to their endocrine-disrupting effects and toxicity. Understanding leaching kinetics of phthalates from PVC microplastics into aqueous environments is essential for assessing exposure concentrations. Environmental variables, such as salinity and flow conditions, can affect the process-specific parameters of leaching phthalates from PVC microplastics into aquatic systems (Henkel et al. 2023). Soils with high buffering capacity maintain conditions favorable for the sorption of contaminants to organic matter, reducing their mobility. Conversely, soils with low buffering capacity may experience rapid pH changes, enhancing leaching and causing EOCs to degrade more slowly. In contrast, soils with high buffering capacity allow microbial communities to thrive, resulting in longer persistence of contaminants due to stable conditions. Although concentrations in the soil–water continuum often decline with distance from the source of infiltration, the destiny of pharmaceuticals is heavily influenced by the soil's ability to attenuate pollutants as well as the chemical makeup of the drug (Barbagli et al. 2019).

4.4.3 Impact of Environmental Conditions like Moisture, Temperature, and Redox Potential

Environmental parameters such as temperature, moisture content, and redox potential have significant impacts on destiny, mobility, and decay of EOCs in soil and water systems.

The amount of moisture in soil has a direct impact on how pollutants leach. Because pollutants are more soluble in water, they are more likely to pass through the soil profile and into groundwater when the soil is wet or extremely damp. On the other hand, EOCs may become less mobile under dry circumstances as they become more firmly attached to soil particles and organic materials (Valentín et al. 2013). Contaminants may be dissolved in wetter circumstances, which increases their bioavailability to microbes, plants, and soil creatures. EOCs may bind to soil particles more firmly in drier environments, which would lower their bioavailability. Waterlogging, on the other hand, can result from excessive moisture, which may affect the rates at which some toxins degrade and increase their persistence.

The ability of soils to sorb organic pollutants can also be impacted by temperature. Higher temperatures tend to make water less viscous, which makes water-soluble EOCs more mobile. Additionally, warmer temperatures may make EOCs more mobile in the environment by decreasing their sorption affinity for minerals and SOM. As the temperature rose, most EOCs' detection frequency and concentration rose as well (Zhu et al. 2021).

Water affinity of soil particles is a significant factor in the case of organic contamination, since hydrophobic soils often retain more persistent organic contaminants than hydrophilic soils. Numerous human activities (such as the disposal of trash, the addition of fertilizers and organic matter, irrigation with tainted water, and mining operations) contribute to the soil's contamination in addition to natural causes. The speciation of pollutants, such as arsenic (As), chromium (Cr), selenium (Se), and vanadium (V), and organic contaminants are changed by the shift in soil redox potential, which also changes how hazardous they are to biological receptors. Organic molecules undergo change in a variety of redox regimes. When organic pollutants are present, high oxygen transfer rates can cause a redox interface to develop between the soil saturated zone and the subsurface vadose zone. Atrazine degraded more quickly when it was biodegraded under anaerobic settings with glucose and NO_3^- than when it was biodegraded under different redox conditions, with or without electron acceptors (O_2, NO_3^-), and glucose functioning as electron donors (Sarkar et al. 2021).

4.5 MECHANISM OF EOC SORPTION IN SOIL

Multiple sorption processes may be taking place at same time since soils include wide variety of particles and chemicals. However, since it is either the most prevalent solid phase in the soil or because compound in question has a strong attraction for sorption sites of minerals or organic matter, certain solid phase and sorption reaction mechanisms frequently dominate the sorption of a given chemical species. Two reactants are involved in a sorption reaction: a chemical compound in solution and solid (sorbent). A sorbed chemical's particular physicochemical interactions with the solid surface are determined by the reactant characteristics. The solid may have different specific surface areas, have a variety of reactive functional groups, and be permanently charged electrostatically. Hydrophobicity, polarizability, ionic and covalent bonding properties, hydration sphere, outer-shell electron configuration, ionic radius, steric effects, valence, molecular coordination, and redox properties are some of the unique properties of the chemicals in solution that will influence sorption mechanisms (Strawn 2021).

Many kinds of adsorption sites can exist even inside a single material. For instance, montmorillonite features unsatisfied bonds on its borders, which adsorb molecules by covalent attachment and electrostatic attraction, as well as persistent layer charge sites for adsorption on the inter-layers formed by isomorphic replacement. Each of the several distinct surface functional groups that make up montmorillonite's edges has a distinct adsorption affinity (Tournassat et al. 2016).

Another significant element that influences adsorption is the extent of the space that exists at the mineral surface. Because minerals are frequently porous solids, the physical limitations of their tiny holes exert forces on chemicals and water that alter the chemical reactions of the substances that they have absorbed (Zachara et al. 2016).

The organic matter in soils has comparatively large specific surface area and is rich in relatively weak acids and basic functional groups that are capable of adsorbing ions. Phenolic and carboxylic acids are weak acid functional groups that can adsorb cations through electrostatic interactions because they deprotonate and acquire a negative charge. As functional groups' outer-shell electrons are reactive and have the ability to exchange electrons with metal cations, inner-sphere complexation plays a significant role in SOM adsorption. Taking organic materials out of soil to separate fulvic and humic acids is a traditional technique for researching SOM adsorption capability. Cation exchange capacities of fulvic and humic acids range from 4850 to 14,000 mmol (+)/kg (Strawn 2021). Reactive sites for ion adsorption, including anions, are amine and thiol functional groups on SOM. For instance, X-ray absorption fine structure spectroscopy demonstrates that Hg^{2+} preferentially adsorbs to the thiol groups in humic acid. Other metals preferentially adsorb to amine and thiol functional groups (Xia et al. 1999).

Even though humic and fulvic acids are useful for separating SOM from soil minerals, physical and chemical characteristics of SOM after extraction differ (Kleber and Lehmann 2019). Therefore, it is best to study SOM's actual adsorption behavior in-situ (Strawn 2021). Helling et al. (1964) investigated the cation exchange capacity (CEC) of 60 soils in relation to pH and employed statistical analysis, and concluded that SOM possesses pH-dependent CEC, which is approximately 600 mmol (+)/kg at pH < 3 and approximately 3500 mmol (+)/kg at pH 8. Organic pollutants sorb to NPs/MPs primarily by partitioning, surface sorption (hydrogen bonding, π–π interaction, electrostatic interaction, and van der Waals force), and pore filling.

The specific surface area and number of sorption sites often increase as particle size decreases, improving sorption capacity. According to reports, NPs' sorption capability was larger than MPs', even going up to 1–2 orders of magnitude. It was discovered that as particle size decreased, PP MPs' sorption capacity for 1,3,6,8-tetrabromocarbazole and 3,6-dibromocarbazole increased. MPs' specific surface area rose as a result of the particle size decrease. BET specific surface area of PE, PS, and PVC were found to be significant in the sorption of pyrene (Pyr) by MPs. The three MPs' BET surface areas were in the following order: PE > PS > PVC. Their respective sorption capacities for Pyr were 164.5, 81.8, and 58.4 μg/g, respectively. (Wang et al. 2020a).

4.6 DISTRIBUTION CAPABILITIES AND CHARACTERISTIC ANALYSIS OF EOCs IN SOIL

Emerging organic pollutants have different polarity and physicochemical characteristics (pK_a, partition coefficient, water solubility, and vapor pressure), and they are inherently hydrophobic. Triclosan, carbamazepine, and diclofenac are typical EOCs that can exist in a range of physical states, depending on the surrounding environment. Their respective vapor pressure values are 6.9×10^{-5}, 1.17×10^{-5}, and 2.91×10^{-4} Pa (Peng et al. 2017).

The physicochemical characteristics of EOCs and surrounding environmental factors, such as pH of aqueous phase and makeup of soil, greatly influence their dispersal in environment. It is challenging to forecast the behavior of EOCs since they can also break down or change into metabolites, which may split and go to other stages in the environment. The transfer of chemicals between two or more environmental compartments, such as between a liquid and a gas (volatilization) or between a solid and a liquid (dissolution), can be referred to as equilibrium partitioning, which takes place between neighboring phases (Harrison 2007). Gas chromatography-mass spectrometry and liquid chromatography-mass spectrometry are two most used analytical methods for finding chemicals in environmental materials. The triple quadrupole mass detector (MS/MS), which enables more precise mass fragment characterization of micropollutants, has been employed in several investigations. Environmental samples must be pretreated in order to assess trace pollutants at mg/L or μg/L levels. This is especially important for getting rid of solid combinations. High-pressure solvent mixes are drawn through a metal sample cell using the selective and effective extraction method known as "accelerated solvent extraction" (Lapworth et al. 2012).

4.7 SORPTION OF EOCs FOR SOIL REMEDIATION USING BIOSORPTION

Many organic chemicals have a negative impact on natural ecosystems, while others are considered extremely harmful to humans. POPs include herbicides, insecticides, pesticides, polychlorinated biphenyls, and organochlorines. Some of these compounds are relatively new, while others have been discovered in the environment as analytical skills have improved, even though many of them have been known and utilized for a long time. These compounds, as explained above, are termed as EOCs (García et al. 2020; Lapworth et al. 2012). Because these chemicals pose major issues in ecological systems even at extremely low concentrations, it is critical to discover methods for their removal. Physicochemical procedures are not always effective, are more expensive, and may cause extra issues. In this regard, biosorption is an approach that avoids these disadvantages and is thus becoming more widely used for eliminating this type of chemical.

The process is comparable with metal biosorption, with similar parameters impacting efficiency. However, the response to these factors differs and must be investigated individually. The complexity of these compounds, unlike metals, allows for diverse functional groups with different levels of ionization and charge values according to the pH of the fluid. The outcomes for various compounds are more varied than those for metals, therefore it is important to optimize this parameter. Additionally, the process may be impacted by the varying levels of hydrophobicity and reactivity that these molecules possess. Despite not being easily soluble in water, hydrophobic chemicals can interact hydrophobically with biosorbent particles or even penetrate cell membranes when working with living biomass.

4.8 FUTURE DIRECTIONS

Research on the sorption of EOCs in soil has lot of promise to improve environmental science and develop sustainable treatment methods. The main goal should be to analyze sorption mechanisms at a molecular level. Understanding how EOCs interact with soil constituents, such as clay minerals, organic matter, and soil colloids, is crucial for predicting their mobility and environmental fate. Advanced methods including spectroscopy, computer modelling, and high-resolution chromatography can be employed to gain a deeper understanding of these interactions. Additional understanding of how enzymatic processes and soil microbial communities affect sorption and transformation dynamics of EOCs may also open up new avenues for combining physical and biological remediation techniques.

Another crucial direction is to assess how climate change affects the movement of EOCs in soils. By modifying soil structure, temperature, and precipitation, climate variability may alter the sorption, transport, and bioavailability of EOCs and increase the dangers they pose to ecosystems and human health. Future research should focus on the effects of climate-induced soil degradation or an increase in frequency of extreme weather events on the interactions between EOCs and soil. It is important to keep these changing factors in mind while creating adaptive management strategies for polluted environments.

Another intriguing area of research is boosting soil sorption capacity through creative additions. EOCs can be more effectively held and immobilized by using adsorptive materials such as charcoal, modified clays, and synthetic nanoparticles. Environmental suitability, possible adverse effects, and long-term stability of the changes need to be evaluated further. It is important to ensure their sustainability and broader applicability by conducting studies that consider their performance in variety of soil types and environmental conditions.

Combining sorption and bioremediation processes may provide a more thorough approach to addressing soil EOCs. When paired with microbial degradation mechanisms, sorption can entirely degrade EOCs into less dangerous metabolites, even if it is successful in immobilizing pollutants. Future research might identify and enhance microbial strains that are particularly adept in decomposing specific types of EOCs, as well as look at how soil amendments could enhance the conditions for microbial activity.

Finally, developing economically and environmentally sound soil remediation methods is essential. Predictive modelling of EOC behavior in various soil conditions may improve decision-making for remediation initiatives at specific sites. Future studies should focus on addressing knowledge gaps regarding the interactions between EOCs and environmental media—soil, water, and air—particularly investigating how EOCs migrate and transform across various soil types. To solve these problems and lessen the long-term risks associated with EOCs in the environment, a multidisciplinary approach integrating expertise in environmental engineering, microbiology, chemistry, and soil science is needed.

REFERENCES

Agboola, O.D., and U.B. Nsikak, 2021. Physisorption and chemisorption mechanisms influencing micro (nano) plastics-organic chemical contaminants interactions: A review. *Frontiers in Environmental Science*, 9, 678574.

Alengebawy, A., S.T. Abdelkhalek, S.R. Qureshi, and M.Q. Wang, 2021. Heavy metals and pesticides toxicity in agricultural soil and plants: Ecological risks and human health implications. *Toxics*, 9(3), 42.

Barbagli, A., B.N. Jensen, M. Raza, C. Schüth, and R. Rossetto, 2019. Assessment of soil buffer capacity on nutrients and pharmaceuticals in nature-based solution applications. *Environmental Science and Pollution Research*, 26(1), 759–74.

Boontanon, S.K., C. Kunacheva, N. Boontanon, N. Musirat, S. Fujii, and S. Tanaka, 2013. Occurrence of perfluorooctane sulfonate in the water environment of Bangkok, Thailand. *Journal of Environmental Engineering*, 139(4), 588–93.

Daughton, C.G., and T.A. Ternes, 1999. Pharmaceuticals and personal care products in the environment: Agents of subtle change. *Environmental Health Perspectives*, 107, 907–38.

Ferreira, A.R., P. Guedes, E.P. Mateus, A.B. Ribeiro, and N. Couto, 2020. Emerging organic contaminants in soil irrigated with effluent: Electrochemical technology as a remediation strategy. *Science of the Total Environment*, 743, 140544.

García, J., M.J. García-Galán, J.W. Day, R. Boopathy, J.R. White, S. Wallace, and R.G. Hunter, 2020. A review of emerging organic contaminants (EOCs), antibiotic resistant bacteria (ARB), and antibiotic resistance genes (ARGs) in the environment: Increasing removal with wetlands and reducing environmental impacts. *Bioresource Technology*, 307, 123228.

Gwenzi, W., T.T. Simbanegavi, J. Marumure, and Z. Makuvara, 2022. Ecological health risks of emerging organic contaminants. *Emerging Contaminants in the Terrestrial-Aquatic-Atmosphere Continuum*, 1, 215–242.

Gworek, B., M. Kijeńska, J. Wrzosek, and M. Graniewska, 2021. Pharmaceuticals in the soil and plant environment: A review. *Water, Air, & Soil Pollution*, 232, 145.

Harrison, R.M. 2007. *Principles of Environmental Chemistry*. Royal Society of Chemistry, Cambridge.

Helling, C.S., G. Chesters, and R.B. Corey, 1964. Contribution of organic matter and clay to soil cation-exchange capacity as affected by the pH of the saturating solution. *Soil Science Society of America Journal*, 28(4), 517–20.

Henkel, C., J. Lamprecht, T. Hüffer, and T. Hofmann, 2023. Environmental factors strongly influence the leaching of di (2-ethylhexyl) phthalate from polyvinyl chloride microplastics. *Water Research*, 242, 120235.

Hunt, C.F., and N. Voulvoulis, 2020. Chemical pollution of the aquatic environment and health. In *Environmental Pollutant Exposures and Public Health*, ed. R.M. Harrison, pp. 39–69. The Royal Society of Chemistry, London.

Hurtado, C., Y.N. Montano-Chávez, C. Domínguez, and J.M. Bayona, 2017. Degradation of emerging organic contaminants in an agricultural soil: Decoupling biotic and abiotic processes. *Water, Air, & Soil Pollution*, 228, 1–8.

Jardak, K., P. Drogui, and R. Daghrir, 2016. Surfactants in aquatic and terrestrial environment: Occurrence, behavior, and treatment processes. *Environmental Science and Pollution Research*, 23(4), 3195–216.

Kleber, M. and J. Lehmann, 2019. Humic substances extracted by alkali are invalid proxies for the dynamics and functions of organic matter in terrestrial and aquatic ecosystems. *Journal of Environmental Quality*, 48(2), 207–16.

Kuhnt, G. 1993. Behavior and fate of surfactants in soil. *Environmental Toxicology and Chemistry*, 12(10), 1813–20.

Lapworth, D.J., N. Baran, M.E. Stuart, and R.S. Ward, 2012. Emerging organic contaminants in groundwater: A review of sources, fate and occurrence. *Environmental Pollution*, 163, 287–303.

Liu, L., C. Liu, R.Z. Fu, W. Zuo, Y. Tian, and J. Zhang, 2024. Full-hain analysis on emerging contaminants in soil: Source, migration and remediation. *Chemosphere*, 363, 142854.

Manatunga, D.C., M. Sewwandi, K.I. Perera, M.D. Jayarathna, D.L. Peramune, R.S. Dassanayake, S. Ramanayaka, and M. Vithanage, 2024. Plasticizers: Distribution and impact in aquatic and terrestrial environments. *Environmental Science: Processes & Impacts*, 26(12), 2114–31.

Masinga, P., T.T. Simbanegavi, Z. Makuvara, J. Marumure, N. Chaukura, and W. Gwenzi, 2024. Emerging organic contaminants in the soil–plant-receptor continuum: Transport, fate, health risks, and removal mechanisms. *Environmental Monitoring and Assessment*, 196, 367.

McGrath, T.J., A.S. Ball, and B.O. Clarke, 2017. Critical review of soil contamination by polybrominated diphenyl ethers (PBDEs) and novel brominated flame retardants (NBFRs); concentrations, sources and congener profiles. *Environmental Pollution*, 230, 741–57.

Mukhopadhyay, A., S. Duttagupta, and A. Mukherjee, 2022. Emerging organic contaminants in global community drinking water sources and supply: A review of occurrence, processes and remediation. *Journal of Environmental Chemical Engineering*, 10(3), 107560.

Müller, R.H., S. Kleinsteuber, and W. Babel, 2001. Physiological and genetic characteristics of two bacterial strains utilizing phenoxypropionate and phenoxyacetate herbicides. *Microbiological Research*, 156(2), 121–31.

Oste, L.A., E.J.M. Temminghoff, and W.H. Van Riemsdijk, 2002. Solid-solution partitioning of organic matter in soils as influenced by an increase in pH or Ca concentration. *Environmental Science & Technology*, 36(2), 208–14.

Pan, B., P. Ning, and B. Xing, 2009. Part V: Sorption of pharmaceuticals and personal care products. *Environmental Science and Pollution Research*, 16, 106–16.

Peng, F.J., C.G. Pan, M. Zhang, N.-S. Zhang, R. Windfeld, D. Salvito, H. Selck, P.J. Van den Brink, and G.-G. Ying, 2017. Occurrence and ecological risk assessment of emerging organic chemicals in urban rivers: Guangzhou as a case study in China. *Science of the Total Environment*, 589, 46–55.

Pignatello, J.J., 1998. Soil organic matter as a nanoporous sorbent of organic pollutants. *Advances in Colloid and Interface Science*, 76, 445–67.

Qin, Q., X. Chen, and J. Zhuan, 2015. The fate and impact of pharmaceuticals and personal care products in agricultural soils irrigated with reclaimed water. *Critical Reviews in Environmental Science and Technology*, 45(13), 1379–408.

Qin, Q., X. Chen, and J. Zhuang, 2017. The surface-pore integrated effect of soil organic matter on retention and transport of pharmaceuticals and personal care products in soils. *Science of the Total Environment*, 599, 42–9.

Rao, P.S.C., 1990. Sorption of organic contaminants. *Water Science and Technology*, 22(6), 1–6.

Rebello, S., A.K. Asok, S. Mundayoor, and M.S. Jisha, 2014. Surfactants: Toxicity, remediation and green surfactants. *Environmental Chemistry Letters*, 12, 275–87.

Rodriguez-Narvaez, O.M., J.M. Peralta-Hernandez, A. Goonetilleke, and E.R. Bandala, 2017. Treatment technologies for emerging contaminants in water: A review. *Chemical Engineering Journal*, 323, 361–80.

Rosell, M., S. Lacorte, and D. Barceló, 2006. Simultaneous determination of methyl tert-butyl ether, its degradation products and other gasoline additives in soil samples by closed-system purge-and-trap gas chromatography–mass spectrometry. *Journal of Chromatography A*, 1132(1–2), 28–38.

Sarkar, B., R. Mukhopadhyay, S. Ramanayaka, N. Bolan, and Y.S. Ok, 2021. The role of soils in the disposition, sequestration and decontamination of environmental contaminants. *Philosophical Transactions of the Royal Society*, B, 376(1834), 20200177.

Semple, K.T., A.W.J. Morriss, and G.I. Paton, 2003. Bioavailability of hydrophobic organic contaminants in soils: Fundamental concepts and techniques for analysis. *European Journal of Soil Science*, 54(4), 809–18.

Sheng, G., C.T. Johnston, B.J. Teppen, and S.A. Boyd, 2001. Potential contributions of smectite clays and organic matter to pesticide retention in soils. *Journal of Agricultural and Food Chemistry*, 49(6), 2899–907.

Strawn, D.G., 2021. Sorption mechanisms of chemicals in soils. *Soil Systems*, 5(1), 13.

Tijani, J.O., O.O. Fatoba, O.O. Babajide, and L.F. Petrik, 2016. Pharmaceuticals, endocrine disruptors, personal care products, nanomaterials and perfluorinated pollutants: A review. *Environmental Chemistry Letters*, 14, 27–49.

Tournassat, C., J.A. Davis, C. Chiaberge, S. Grangeon, and I.C. Bourg, 2016. Modeling the acid–base properties of montmorillonite edge surfaces. *Environmental Science & Technology*, 50(24), 13436–45.

Valentin, L., A. Nousiainen, and A. Mikkonen, 2013. Introduction to organic contaminants in soil: Concepts and risks. In *Emerging Organic Contaminants in Sludges: Analysis, Fate and Biological Treatment. The Handbook of General Chemistry*, eds. T. Vicent, G. Caminal, E. Eljarrat, and D. Barceló, Vol. 24, pp. 1–29. Springer, Berlin, Heidelberg.

Vieira, M.G.A., M.A. da Silva, L.O. dos Santos, and M.M. Beppu, 2011. Natural-based plasticizers and biopolymer films: A review. *European Polymer Journal*, 47(3), 254–63.

Wang, F., M. Zhang, W. Sha, Y. Wang, H. Hao, Y. Dou, and Y. Li, 2020a. Sorption behavior and mechanisms of organic contaminants to nano and microplastics. *Molecules*, 25(8), 1827.

Wang, Y., Z. Li, F. Tan, Y. Xu, H. Zhao, and J. Chen, 2020b. Occurrence and air-soil exchange of organophosphate flame retardants in the air and soil of Dalian, China. *Environmental Pollution*, 265, 114850.

Wong, C.S., 2006. Environmental fate processes and biochemical transformations of chiral emerging organic pollutants. *Analytical and Bioanalytical Chemistry*, 386(3), 544–58.

Wong, F., and F. Wania, 2011. Visualising the equilibrium distribution and mobility of organic contaminants in soil using the chemical partitioning space. *Journal of Environmental Monitoring*, 13(6), 1569–78.

Wan, Z., C. Wang, J. Zhou, M. Shen, X. Wang, and Y. Jin, 2019. Effects of polystyrene microplastics on the composition of the microbiome and metabolism in larval zebrafish. *Chemosphere*, 217, 646–58.

Xia, K., U.L. Skyllberg, W.F. Bleam, P.R. Bloom, E.A. Nater, and P.A. Helmke, 1999. X-Ray absorption spectroscopic evidence for the complexation of Hg(II) by reduced sulfur in soil humic substances. *Environmental Science & Technology*, 33(2), 257–61.

Ying, G.G., 2006. Fate, behavior and effects of surfactants and their degradation products in the environment. *Environment International*, 32(3), 417–31.

Zachara, J., S. Brantley, J. Chorover, R. Ewing, S. Kerisit, C. Liu, E. Perfect, G. Rother, and A.G. Stack, 2016. Internal domains of natural porous media revealed: Critical locations for transport, storage, and chemical reaction. *Environmental Science & Technology*, 50(6), 2811–29.

Zhang, C., L.P. Barron, and S.R. Stürzenbaum, 2024. Pollution of soil by pharmaceuticals: Implications for metazoan and environmental health. *Annual Review of Pharmacology and Toxicology*, 65, 547–65.

Zhu, L., C. Jiang, S. Panthi, S.M. Allard, A.R. Sapkota, and A. Sapkota, 2021. Impact of high precipitation and temperature events on the distribution of emerging contaminants in surface water in the Mid-Atlantic, United States. *Science of the Total Environment*, 755, 142552.

5 Transport of Emerging Organic Contaminants (EOCs) in the Soil-Groundwater System

Noshin Masood, Nisbah Mushtaq, Mahtab Ahmad, and Abida Farooqi

5.1 INTRODUCTION

Emerging organic contaminants (EOCs) are a group of thousands of produced chemicals that primarily have unidentified toxicity hazards and are not subject to regulations (NORMAN 2019). These organic compounds are characterized as emergent since they were recently developed (Lapworth et al. 2012) or detected using advanced technical and monitoring methods (Moreau et al. 2019) as well as compounds that were only recently identified as contaminants. They may include pharmaceuticals and personal care products, veterinary products, pesticides, plasticizers, food additives, nanomaterials, metabolites, and their transformational products. These compounds are produced by different sources such as industries, agricultural, and livestock activities, e.g., waste lagoons and manure applications to the soil, septic tanks, hospital effluents, municipal treatment plants, or laboratory wastewater as well as through the groundwater and surface water exchange processes (Lapworth et al. 2012; Gogoi et al. 2018). Over the past few decades, EOCs have been identified in a variety of environmental compartments, such as soil, groundwater, surface water, atmosphere, and even glaciers (Falciglia et al. 2016; An et al. 2023). The groundwaters and soils of various countries such as Africa (Nasr et al. 2024), Austria (Brueller et al. 2018), China (Hao 2020; Lin et al. 2017; Zhong et al. 2022), Europe (Schulze et al. 2019), Germany (Gerhardt 2019), India (Brauns et al. 2024), Italy (Meffe and de Bustamante 2014), Mexico (Vázquez-Tapia et al. 2022), and the United States (Meena et al. 2024) are polluted with EOCs, where they are getting added every day altering the natural balance of the ecosystem. Being traced in almost every country's natural environment, they are considered as a global issue.

Once they enter the environment, EOCs may persist and accumulate within the water and soil compartment. EOC-contaminated groundwater may harm human and aquatic life that receives it (Moreau et al. 2019). In addition, the accumulation of EOCs in the soil system may affect its physicochemical properties and microbial population (Hou et al. 2024). Their persistent and nonbiodegradable nature causes

DOI: 10.1201/9781003509523-5

them to accumulate in living cells and transform via different trophic levels. On the other hand, sometimes the less persistent EOCs may degrade, and their daughter products are much more harmful as compared with the parent products (Michael et al. 2014), which can cause a significant threat to nontargeted organisms too (Gomes et al. 2017). However, the extent to which they harm human beings and the natural ecosystem is still unknown. This uncertainty is due to their vast range of classification and the different elements regulating their transport processes. These compounds, as the name implies, have not been thoroughly studied for their fate, toxicity, and transport pathways in soil-groundwater environments. As a result, the current chapter focuses on EOCs for their characteristics, routes, and transport mechanisms, as well as the variables influencing their fate and mobility in the soil-groundwater system. This chapter also addresses analytical approaches for investigating these EOCs, remedial methods as a solution as well as future dimensions to be considered for further investigations.

5.2 PROPERTIES AND CLASSIFICATION OF EOCs

EOCs encompass a wide array of substances with diverse chemical structures, uses, and environmental behaviors. Proper classification is critical for understanding their environmental fate, transport, and potential risks. The classification of emerging contaminants is still a part of an ongoing debate. In earlier studies on EOCs, classifications were often based on their concentrations and compositions in the environment (Bunting et al. 2021; Lapworth et al. 2012; Mahler and Musgrove 2019; Zhao et al. 2018). The designation of a compound as "emerging" varies by region and time of evolution. This is evident in cases where certain compounds were classified as 'emerging' in earlier studies but are no longer considered as emerging today — unlike pesticides, which continue to be recognized as persistent contaminants over the time. Thus, the term "emerging" can be taken as context-dependent and dynamic.

Another major challenge in listing emerging contaminants is the variation of legislation across different countries. For example, in Europe, microorganisms are regulated and governed at both national and regional levels instead of through global standards (Bunting et al. 2021). Likewise, in some European countries, pesticides are either regulated as whole compounds or sometimes as individual components, depending on the risks associated with their application (Reberski et al. 2022; Bunting et al. 2021).

Despite all these hurdles, EOCs can still be broadly classified and characterized into major groups depending on their source of origin, chemical characteristics, environmental persistence, application, usage, and transportation pathways. The following section elaborates on this concept, but the reader should note that often these groups would overlap with each other depending on their nature and application.

5.2.1 Based on Sources

Based on their source of origin, EOCs can be categorized into the following five classes:

a. **Industrial Chemicals:** This group includes industrial chemicals such as flame retardants, solvents, and plasticizers such as polybrominated diphenyl ether and bisphenol A (Lapworth et al. 2012).
b. **Pharmaceuticals and Personal Care Products:** This group includes drugs of human and veterinary origin, fragrances, and cosmetics (Silori et al. 2022; Dutta 2023).
c. **Agricultural Products:** This group includes all forms of agrochemicals including pesticides, herbicides, fungicides, algicides, and their metabolites (Shi et al. 2022).
d. **Nanomaterials:** This group includes all engineered nanoparticles utilized in industrial and medical applications (Ealias et al. 2024).
e. **Other Sources:** Apart from the abovementioned groups, chemicals from household cleaning products, dyes, and those used as food additives also make up a part of EOCs (Khan et al. 2022).

5.2.2 Based on Chemical Composition

Depending on their chemical nature and composition, EOCs are of three types:

a. **Polar Compounds:** These are the group of compounds that exhibit high solubility in water, including pharmaceuticals and some pesticides (Khan et al. 2023; Li et al. 2024).
b. **Nonpolar Compounds:** This group is characterized by their hydrophobic nature and their tendency to accumulate in sediments. Compounds such as hydrocarbons and persistent organic pollutants are some common examples of nonpolar EOCs (Meffe and de Bustamante 2014).
c. **Reactive Compounds:** This group of compounds is highly reactive and can easily be degraded in an environment producing toxic metabolites (Tong et al. 2022).

5.2.3 Based on Environmental Persistence

a. **Persistent EOCs:** As the name suggests, this group tends to persist in the environment for extended periods and are not easily biodegraded such as per- and polyfluoroalkyl substances (Johnson et al. 2022).
b. **Nonpersistent EOCs:** These are reactive groups of compounds that degrade relatively quickly but may yield toxic by-products as a result of degradation (Yu et al. 2024).

5.2.4 Based on Usage

EOCs based on their application and use can be classified as follows:

a. **Pharmaceuticals:** This group includes chemicals such as hormones, antibiotics, analgesics, antidepressants, and any human or veterinary drugs (Silori et al. 2022).

b. **Personal Care Products and Lifestyle Compounds:** This includes compounds that we use in daily life such as cosmetics, fragrances, stimulants, artificial sweeteners, and their metabolites (Silori et al. 2022; Dutta 2023).
c. **Industrial and Household Products:** As their name suggests, this group includes chemicals such as surfactants, plasticizers, flame retardants, and dyes used in daily life (Khan et al. 2022).
d. **Agricultural Products:** This group of EOCs can easily be transported through soil and groundwater matrix owing to their high water solubility (Zhao et al. 2018).

5.2.5 Based on Transport Mechanisms

The transportation mechanism divided them into two categories including

a. **Mobile EOCs:** Easily transported through soil and groundwater due to their high water solubility (Tong et al. 2022).
b. **Sorption-prone EOCs:** Due to their nature to bioaccumulate and high sorption capacity, this group of EOCs tends to bind to soil particles and organic matter, reducing their overall mobility (Yu et al. 2024).

5.3 EOC TRANSPORT IN SOIL AND GROUNDWATER

Several researchers have reported that EOCs have been detected in groundwater and soil systems in varying concentrations ranging between µg and ng/L (Lapworth et al. 2012; Sudarsan et al. 2024). Their transport within the soil and groundwater system can be determined by the source-pathway-receptor model (Jurado et al. 2012; Gomes et al. 2017). Thus, the pathway acts as a bridge between the source and receptor and is the key component in the model. Sources and pathways are associated with both natural and anthropogenic activities. The sources of EOCs are broadly classified into direct or point sources and nonpoint or diffuse source pathways (Chacón et al. 2022). Point source pathways, due to anthropogenic activities, may include landfills, mining, septic tanks, wastewater, and animal feeding facilities (James et al. 2016; Gomes et al. 2017). Diffused pollution pathways, on the other hand, may include land-based manure and biosolids applications, surface water, atmospheric deposition, and diffuse sewage leakage (Lapworth et al. 2012). Among the natural processes volcanic eruption, erosion, land sliding, and weathering are the major activities contributing towards emerging contaminant pollution. These pathways cause the emerging contaminants to source their way to receptors. For instance, these chemicals enter water systems via wastewater, rainfall, and runoff streams (Sun et al. 2013; Sudarsan et al. 2024). Conversely, terrestrial environments mainly receive these chemicals through agrochemical applications, manure applications, and untreated wastewater irrigation (Andreu and Picó 2019). The main processes that control EOC migration within the subsurface environment are sorption and desorption (to organic matter and clay minerals), advection and dispersion (through soil and air), infiltration (from surface runoff and precipitation), ion exchange (to soil and aquifers), and biodegradation/biotransformation (Lapworth et al. 2012; An et al. 2023).

5.4 FACTORS AFFECTING EOCs TRANSPORT

Once EOCs enter the terrestrial compartment, their fate mainly depends on various aspects. These aspects may include land use pattern, soil composition, soil organic matter, depth profile, molecular weight of the contaminant, soil leaching capacity, diffused aerial deposition, rainfall recharge, and surface water recharge (Yang et al. 2017).

Land use patterns, as a pathway, primarily control EOCs' contamination status in the soil and water systems. Urban land uses (industrial and residential areas) are among the hotspots and large-scale contamination sources (Chacón et al. 2022). For example, the landscape evolution of Tunisia in Africa, as a result of growing urbanization, provides a vivid image of how the extensive use of EOCs undermines groundwater quantitative and qualitative balance (Nasr et al. 2024).

Within the soil matrix, the contaminant follows two pathways, either horizontal movement or vertical movement, where the soil composition is associated with the contaminant's migration and remedial efficiency. For instance, the low clay composition of the soils facilitates EOC migration (Chacón et al. 2022). However, soils with high organic matter content can efficiently accumulate EOCs for years and transport them to water (Fabietti et al. 2010).

Once these contaminants have dispersed across the soil, they travel into the soil profiles and ultimately reach the aquatic environments through vertical and horizontal movement processes, such as atmospheric deposition, leaching, soil–water interactions, and surface runoff (Anawar and Strezov 2019). For example, they either volatilize/photodegrade and end up in the water system through erosion or soil runoff. These contaminants' fate is mostly determined by soil type, composition, and the pollutant itself. Some of them, when leached through adsorption/desorption, come into contact with the soil's organic and inorganic components and show vertical movement. Other substances may be biodegraded or chemically decomposed. Sometimes they may be absorbed by plants and accumulate within the food chain (Chacón et al. 2022). Vertical pollution context varies by pollutant characteristics as well as soil physicochemical features. Precipitation may also influence contaminant infiltration (Sverdrup et al. 2021; Chacón et al. 2022). Furthermore, EOCs are temperature-sensitive; their concentrations are higher in cooler months as compared with hotter ones (Biel-Maeso et al. 2018). On the other hand, the degradation rate and dilution patterns are mostly determined by the half-life of the pollutants.

The EOC's content within water environments is mainly influenced by the type of water catchment, its geology, water usage, water consumption patterns, and contaminant persistence (Meena et al. 2024; Kibuye et al. 2020). Porous aquifers facilitate pollutant transfer as compared to aquifers coated with clay and silt. Their usage patterns and geographical location also influence their content. For example, the use of pharmaceuticals is more prevalent in Africa, resulting in greater EOC contamination in African waters than in the Western world (Coronado-Apodaca et al. 2023). The same applies to personal care products, which are more commonly used in Europe than in third-world/developing countries, where water is less contaminated by these contaminants (Coronado-Apodaca et al. 2023). Similarly, the type and biodegradability of contaminants are known determinants of contamination. Due to their volatile nature, they are less concentrated in groundwater (Kibuye et al. 2020). Their fate and transport mechanism are complicated and highly dependent on the physicochemical and biological factors of groundwater. Physical factors may involve hydrodynamics conditions, temperature, rainfall, and solar radiation (Tong et al. 2022). Chemical characteristics may include pH, nutrients, and dissolved oxygen, whereas biological factors may include biological communities that trigger their biodegradation and transformation (Tong et al. 2022). For example, higher pH inhibited the sorption of acidic substances having hydroxyl and carboxyl groups (Wang et al. 2020).

Once they enter the soil and water ecosystems, these pollutants either agglomerate with soil and water chemicals, get diluted, or biodegrade into daughter products (Anawar and Strezov 2019). Such contaminants then enter the food chain, impacting humans, other species, and living things (receptors of the transport model). Therefore, predicting the EOC content in the receptor body requires a thorough understanding of emissions, residual time, dilution rate, and subsurface behavior.

5.5 ROLE OF BIOGEOCHEMICAL INTERACTIONS

The fate and transport of EOCs in soil and groundwater systems are largely dependent on their biogeochemical interactions within their existing environment. These interactions involving chemical, biological, and physical processes control their mobility, transport, biotransformation, and fate of EOCs in the environment. In groundwater, EOCs can easily be removed or eliminated due to microbial degradation, while sorption, in this case, is of less importance as it is a reversible process and only hinders the transport of contaminants instead of their elimination (Greskowiak et al. 2017).

The main processes involved in the biogeochemical transformations of EOCs include both microbial and abiotic degradation mechanisms and their interactions with natural organic matter (NOM).

5.5.1 Microbial Degradation and Transformation

Microbial degradation is the main pathway through which EOCs are eliminated or removed from soil and groundwater systems. In this process, indigenous microbial species and communities transform EOCs into less harmful by-products or metabolites through different metabolic and enzymatic processes (Joutey et al. 2013).

However, the effectiveness of microbial degradation depends on several different factors including the availability of nutrients, the presence of specific microbial strains necessary for metabolism, temperature, and the chemical nature and characteristics of the contaminant (Tong et al. 2022; Lapworth et al. 2012). For instance, compounds consisting of complex aromatic structures may resist microbial degradation, while compounds of simpler structures are more easily prone to undergo a series of biological transformations (Tang et al. 2019). Furthermore, the metabolites formed as a result of these biotransformations may exhibit either higher toxicity or environmental persistence as opposed to their parent compound (Michael et al. 2014).

5.5.2 Abiotic Degradation Mechanisms

Like microbial degradation, abiotic degradation of EOCs, either through hydrolysis or photolysis, is yet another mechanism that plays an important role in the removal of EOCs from soil-groundwater systems. In hydrolysis, EOCs are chemically broken down into their by-products in the presence of water. This process is highly variable as it depends on the pH and temperature of the existing water (Greskowiak et al. 2017). Similarly, photolysis uses solar radiation to degrade certain groups of EOCs present in surface water or shallow groundwater. However, in deeper aquifers where sunlight cannot penetrate, this process becomes ineffective for EOCs removal (Tong et al. 2022).

5.5.3 Interaction with Natural Organic Matter (NOM)

Another process that may influence or hinder the transport of EOCs in soil-groundwater systems is the presence of NOM. The NOM present in soil and aquifer systems can easily absorb contaminants onto its surface owing to its high sorption capacity. EOCs, either through hydrophobic or electrostatic interactions, often get adsorbed onto the surface of NOM, which hinders their mobility and transport (Wang et al. 2022). The EOCs that are adsorbed onto NOM can easily be released back into the environment once the conditions become favorable (change in pH, redox potential). This reversible sorption mechanism plays an important role in extending EOCs' persistence in the environment over time (Wilkinson et al. 2017).

5.6 MODELING EOC TRANSPORT IN SOIL-GROUNDWATER SYSTEMS

Understanding the transport pathways of EOCs in soil-groundwater systems can play an important role in developing effective remedial measures and management strategies associated with their risk of contamination. Numerous studies and research have employed a wide variety of modeling approaches ranging from empirical to mechanistic, for understanding and simulating the EOC behavior in environments.

5.6.1 Overview of Transport Modeling Approaches

The contaminant transport models are a set of tools employed by researchers to understand and describe the pathways taken and the biotransformation of EOCs in soil and groundwater systems. To predict the contaminant fate, transport modeling incorporates different processes such as advection, sorption, dispersion, and biodegradation into its application. Based on the review of studies, HYDRUS and MODFLOW are among the two most common modeling platforms used worldwide for the simulation of water flow and transport of solute in porous media (Locatelli et al. 2019; Tong et al. 2022). The following section briefly explains the different types of models used for predicting the fate and transport of EOCs in soil and groundwater systems.

a. **Analytical Models:** Analytical modeling is a much preferable approach for conducting risk evaluations and water management (Locatelli et al. 2019). This approach provides a simplified closed system solution for easy computation of preliminary assessments of contaminants present in the environment. However, this approach assumes a system with a completely homogeneous environment with steady-state conditions, which in turn, limits its application in cases with complex field scenarios (Pal et al. 2010; Lapworth et al. 2012). Some of the common examples of this modeling approach include RICS5 (for conventional backward and forward simulations), ROME, and CoronoScreen for studying horizontal and vertical transport mechanisms in 2D or 3D, and PLUME, which is based on the 3D analytical solution of the simple horizontal flow of ADE1 (Locatelli et al. 2019).
b. **Numerical Models:** For understanding complex hydrological processes, numerical models are much preferred as they consider the heterogeneous conditions of the environmental matrix, complex geometries and structures of contaminants, and time-variable inputs (Goyal et al. 2021). This data-driven model approach provides a detailed site-specific solution to a problem, but it requires comprehensive data inputs, computational resources, and understanding. MODLOW and MT3DMS are some of the most commonly used numerical tools for simulating contaminant transport in groundwater systems (Patil and Chore 2014). In addition, Goyal et al. (2021) suggested that the integration of geographic information system (GIS) applications with numerical modeling could offer a valuable advantage.
c. **Hybrid Models:** Hybrid models combine the strengths of both first principle-based (closed system) and data-driven models and are implemented in situations where neither approach can provide a solution to the problem. The main aim of the hybrid models is to balance computational efficiency for handling complex contaminant transport scenarios. They are primarily used in cases where the system requires a wide variety of details (Valhondo et al. 2018).

5.6.2 Key Parameters and Uncertainties in Models

The main parameters or factors that influence or contribute to model predictions include hydraulic conductivity, soil porosity, contaminant degradation rates, and partitioning coefficients. Nevertheless, uncertainties still exist in model predictions due to the variability of these parameters in their natural environments, along with limited data available for certain classes of EOCs.

a. **Sorption Coefficient (Kd):** Sorption coefficient (Kd) normally describes and explains the partitioning of contaminants between different media such as soil particles and pore water and is easily influenced by variability in mineral composition and soil organic matter content of the soil-groundwater system. For instance, the Kd values of personal care products and pharmaceuticals vary significantly with changing soil properties, affecting their mobility in aquifer systems (Roberts et al. 2014).

b. **Degradation Rates:** Both biotic and abiotic degradation processes reduce the availability and concentrations of contaminants over time in the environment. Microbial activity, pH, temperature, and redox conditions all influence the degradation rates of contaminants. Studies have shown that the transport behavior of EOCs in soil-aquifer treatment systems was largely impacted by their degradation rates, further influenced by site-specific existing conditions in treatment plants (Nham et al. 2015).
c. **Hydraulic Conductivity and Porosity:** Both hydraulic conductivity and soil porosity influence the transport of contaminants in water flow through soil-groundwater systems. Spatial heterogeneity of contaminants and their measurement challenges along with variability in overall soil structure affect the contaminant movement through the system (Alaoui et al. 2011)

Recent advancements in machine learning and sensitivity analysis, in conjunction with these parameters, are used to study uncertainties involved in such modeling approaches and ways to improve model reliabilities (Shu et al. 2023).

5.6.3 Application of Predictive Models in Assessing Contamination Risks

Predictive models are often used in studying the risk assessment of contaminants under varying conditions such as industrial discharges, wastewater (treatment plants, households), and agricultural and irrigational runoffs. Combining both site-specific information and variable inputs, these models provide a deeper understanding of the transport and fate of the contaminants and help in making informed decisions for ensuring sustainable water resource management (Ashraf et al. 2022; Chaudhry et al. 2021). Some of the main applications of these models include but are not limited to:

a. **Risk Assessment:** Predictive models can be used to estimate the concentration of contaminants at their point of origin (like drinking water wells) and can help in predicting potential health risks associated with it. Such an approach is used to assess and determine whether contaminant levels are within or above the regulatory threshold of a country (Ankley et al. 2021).
b. **Remediation Strategy Development:** Predictive models are yet another tool that assists researchers in designing and developing effective remedial measures for a certain group of contaminants. These measures include pump and treat systems, natural attenuation or in-situ removal and treatment of contaminants, constructed wetlands, or biochar amendments to soil. The effectiveness of these measures in reducing EOC concentrations can be assessed through the application of such predictive models (Chaves et al. 2022).
c. **Regulatory Compliance:** These models can be used to ensure compliance with the national regulatory framework of contaminants by predicting contaminant behavior over time and any monitoring programs in place for their regulation (Bunting et al. 2021; Reberski et al. 2022).

5.7 CASE STUDIES OF EOC TRANSPORT IN SOIL-GROUNDWATER SYSTEMS

Real-world case studies provide critical insights into the intricate processes that govern the transport of EOCs as indicated in Table 5.1.

The studies highlight the use of analytical techniques, field investigations, and modeling approaches to better understand and mitigate contamination risks. Globally, most research has focused on pharmaceuticals, personal care products, and agrochemicals within soil and groundwater systems. The table summarizes region-specific studies, detailing EOC transport pathways and their environmental impacts. These case studies emphasize the necessity of a multidisciplinary approach to comprehensively understanding EOC transport and devising effective remediation strategies.

5.8 ANALYTICAL TECHNIQUES FOR STUDYING EOC TRANSPORT

The precise and accurate determination of EOCs in soil and groundwater necessitates robust analytical methods capable of measuring even small concentrations in the environment.

These methods are a group of techniques ranging from sampling, laboratory-based experiments, monitoring, and advanced analytical tools employed for contaminant detection, characterization, and quantification in soil-groundwater systems. The current set of techniques that govern or impact the transport and fate of contaminants include: (i) in-situ measurement and monitoring, (ii) lab-based experiments for determining their kinetic coefficients, and (iii) numerical modeling (Tong et al. 2022).

5.8.1 In-Situ Measurement and Monitoring in Soil and Groundwater

a. **Soil Sampling:** The most basic and common methods for determining contaminant concentrations in soil are core sampling, pore sampling, and lysimeters. With recent advancements in analytical techniques, advanced sampling designs can ensure spatial and temporal coverage to account for contaminant heterogeneity in the soil matrix (Khan et al. 2023).
b. **Groundwater Monitoring:** For monitoring groundwater, piezometers and multilevel sampling wells are often employed to check EOC concentrations in aquifers over time. These wells are often equipped with sensor-based devices for the continuous monitoring of EOC concentrations that provide real-time information and data on contaminant dynamics in groundwater (Sintim et al. 2021).

5.8.2 Laboratory-Based Methods

In laboratory-based methods, the commonly employed approach is to use column experiments as a primary setup for studying and simulating contaminant transport through soil-water media under controlled settings. This allows for the effective study of underlying processes such as adsorption, degradation, and distribution or

TABLE 5.1
The Relevant Complexities and Environmental Impacts of EOCs in Various Regions

Case Studies	EOCs Type	Environmental Consequences	Source and Pathways	Transporting Agents	References
Africa (sub-Saharan cities)	Pharmaceuticals and personal care products, lifestyle compounds, and industrial compounds	Surface water and groundwater EOC concentration reaches 61,273 ng/L, versus 16,677 ng/L in groundwater.	Land use patterns domestic, hospital, and brewery industry wastewaters, landfills, and pit latrines.	Sandy nature of the aquifer, and the high rainfall	Djieugoue et al. (2024)
China (Xiong'an New Area)	Organochlorine pesticides (OCPs), polychlorinated biphenyls (PCBs), and volatile organic compounds (VOCs)	Higher groundwater EOC levels were detected along historical sewage reservoirs.	Historically input from multiple industries, including paper, chemical, textiles, and printing and dyeing	Rapid urbanization and population growth in the Tanghe River basin	An et al. (2023)
China (Northeastern China)	16 different EOCs	EOC concentrations ranging from 1.25 to 1471 ng/L were detected in groundwater monitoring wells around the landfills.	Landfills	The low permeability of the aquifer	Zhang et al. (2025)
Europe	Pharmaceuticals such as carbamazepine and diclofenac were frequently detected in groundwater samples.	Carbamazepine and caffeine had groundwater concentrations up to 2.3 and 14.8 μg/L, respectively.	Agricultural practices and urban wastewater discharges	Low biodegradability and strong sorption to soil organic matter.	Bunting et al. (2021)
Mexico	Pharmaceuticals, hormones, plasticizers, personal care products, sweeteners, drugs, and pesticides	The levels of EOCs and EDs varied from ng/L to 140 mg/L in surface and groundwater.	Cropland irrigation with raw wastewater was the principal activity.	Biodegradation and soil-water exchange	Vázquez-Tapia et al. (2022)

(Continued)

TABLE 5.1 (*Continued*)
The Relevant Complexities and Environmental Impacts of EOCs in Various Regions

Case Studies	EOCs Type	Environmental Consequences	Source and Pathways	Transporting Agents	References
New Zealand (Waikato region)	Pesticides, pharmaceuticals, industrial, preservatives/food additives, and personal care products	Groundwater EOC concentrations ranging from 0.1 to 11,000 ng/L	Geographical location and land use	Organic matter and clay minerals, ion exchange (soil and aquifer), and microbial degradation define EOC content in the groundwater	Moreau et al. (2019)
North Africa (Tunisia)	Pharmaceuticals and personal care products, industrial compounds, and lifestyle compounds	EOC concentrations reached extreme levels (up to 24,000 ng/L)	Domestic sewage from on-site sanitation systems, industrial and hospital waste	Higher concentrations of EOCs are mainly connected to the use patterns of particular classes of pharmaceuticals and lifestyle products in Tunisia as well as the hydrogeology of the area.	Nasr et al. (2024)
USA (Susquehanna River Basin (SRB))	Pharmaceuticals and agrochemicals	Higher EOC concentrations were measured in surface water indicating medium-to-high risk to aquatic organisms.	Surface runoff and wastewater effluent	EOC occurrences are generally lower in the reservoir than in riverine sources likely due to higher forested land use and increased residence time	Kibuye et al. (2020)
USA (Midwestern United States)	Pesticides and veterinary pharmaceuticals	Significant contamination of shallow aquifers	Heavy rainfall	Tile drainage systems are facilitating contaminant transport to groundwater.	Delgado et al. (2023)

disruption with time. These are the best methods to estimate and calculate retardation factors or degradation rates, used in transport modeling. Likewise, batch column experiments can be used to determine equilibrium coefficients, including sorption and kinetic factors, under varying conditions (Saeidi et al. 2023).

5.8.3 Analytical Tools

The analytical methods mentioned in Table 5.2 play a significant role in the detection, determination, identification, and quantification of EOCs in soil-groundwater systems. Spectroscopic techniques such as high-resolution mass spectrometry, Fourier transform infrared spectroscopy, ultra-performance liquid chromatography coupled with tandem mass spectrometry, and ultraviolet–visible spectroscopy are best suitable for the analysis of organic pollutants, while atomic absorption and inductively-coupled plasma spectroscopy provide efficient and reliable results for detecting metals and metalloids. Likewise, liquid chromatography offers robust separation and identification capabilities, while molecular techniques such as next-generation sequencing and polymerase chain reaction offer microbial and genetic marker analysis required for biodegradation studies.

5.9 MITIGATION STRATEGIES FOR EOC CONTAMINATION

Reducing the risks posed by EOCs in the environment requires a multi-faceted approach that incorporates prevention, natural remedial processes, and advanced treatment technologies. Effective approaches can lessen and reduce concentrations of EOC released into the environment, enhance and improve natural attenuation of contaminants, and improve the effectiveness of engineered remediation systems. The following section describes briefly the strategies that can be used to reduce and prevent the introduction and release of EOCs into the environment by proactively managing the contamination risk at its source.

5.9.1 Source Reduction

As a first step in preventing the release of contaminants into the environment, it is important to upgrade existing wastewater treatment plants (WWTPs) either with membrane bioreactors or advanced oxidation processes (AOPs) to improve the efficiency of WWTPs in the removal of EOCs from discharge (Roberts et al. 2014). Similarly, promoting and incentivizing industries to adopt green and sustainable environment principles, and replacing harmful EOCs with environmentally benign alternatives (Ealias et al. 2024). Also, encouraging and implementing take-back programs in hospitals and pharmacies for unused pharmaceuticals is a way to prevent and reduce improper disposal and release of pharmaceuticals into the environment (Chaves et al. 2022).

5.9.2 Agricultural Interventions

The release of EOCs can be reduced in agricultural activities by encouraging the use of organic fertilizers or biochar amendments that can immobilize and reduce

TABLE 5.2
Analytical Tools and Their Significance in Analyzing EOCs

Analytical Technique	Specification	Analytical Principle	Contaminant Analyzed	Significance	References
Spectroscopy	Modified QuEChERS method coupled to UPLC-MS/MS and high-resolution mass spectrometry (HRMS)	To screen for a wider range of organic contaminants at trace levels using wide-scope target, suspect, and nontarget screening approaches	Fully validated for 90 EOCs for both soil and water matrices	High sensitivity, specificity, and accuracy	Hollender et al. (2017)
	FTIR and UV–Vis	UV–Vis' spectroscopy analyzes chromophores and conjugated systems in aqueous solutions. FTIR detects functional groups and chemical bonds in solid and liquid samples	Contaminants' chemical structure and functional groups	Used for qualitative study and identification with excellent resolution and sensitivity	Enyoh et al. (2020)
	AAS and ICP-OES	NA	Environmental metals and metalloids	High throughput and multi-element detection	King et al. (2022)
Chromatography	*Liquid chromatography* coupled with *electrospray ionization tandem mass spectrometry*	Detect new contaminants by separating complicated mixtures of chemicals by chemical characteristics	Organic pollutants, including pesticides, medicines, and industrial chemicals	This technique requires a great investment, and high operational costs associated with skilled maintenance and highly specialized operating workers	Vázquez-Tapia et al. (2022) and Coronado-Apodaca et al. (2023)
Molecular methods	PCR and NGS	NA	Identify and quantify microbial communities, antibiotic-resistance genes, and genetic markers.	Useful for microbial ecology and biodegradation potential in natural settings	Solangi et al. (2023)

contaminant transport in groundwater systems (Shi et al. 2022). In the same way, adopting precision farming, controlled application and usage of pesticides, and limiting the use of veterinary antibiotics as part of integrated pest management measures (Delgado et al. 2023).

5.9.3 Policy and Public Awareness

Strengthening regulatory framework, such as setting maximum allowable concentrations for EOCs in effluents and natural water bodies, ensures better environmental protection and mitigates the potential human and ecological risks. Likewise, raising public awareness about EOC pollution and encouraging environmentally responsible behaviors, including proper disposal of household chemicals and pharmaceuticals, promotes sustainable wastewater resource management (Khan et al. 2022).

5.9.4 Chemical Treatment Methods

Chemical methods such as coagulation-flocculation, adsorption, and aqueous phase oxidation also play an important role in the removal of EOCs from groundwater systems. Coagulation-flocculation/lime softening simplifies the removal of colloidal particles by producing aggregates that settle down at the bottom of treatment plants. Owing to this property, this method is preferred in treating hospital wastewater containing personal care products (Kumar et al. 2022). Likewise, aqueous oxidation utilizing chlorine or ozone has proven 90%–100% efficiency in the removal of EOCs under optimal conditions (Noutsopoulos et al. 2015). Similarly, materials such as activated carbon, zeolites, or biochar have been proven effective in the removal of EOCs from wastewater through adsorption. This is because recent advancements in nanomaterials have improved their adsorption capacities and selectivity for specific contaminants (Valhondo et al. 2018).

Studies on biochar-amended soils highlight their potential for immobilizing EOCs in agricultural runoff, reducing leaching into groundwater (Pal et al. 2010). However, harnessing natural processes and enhancing their effectiveness through intervention can significantly mitigate EOC contamination in soil and groundwater systems.

5.10 NATURAL ATTENUATION AND BIOREMEDIATION APPROACHES

Harnessing natural processes and enhancing their effectiveness through intervention can significantly mitigate EOC contamination in soil and groundwater systems. In natural attenuation, for reducing EOC concentrations over time, a total reliance on naturally occurring processes such as microbial degradation, dilution, and sorption is being made. To ensure its effectiveness, a proper monitoring system has been designed to ensure that these processes meet the necessary remediation goals under controlled site-specific conditions (Close et al. 2021). This is yet another way, where the influence and impact of different environmental factors such as redox conditions, temperature, pH, nutrient availability, and soil textures can be evaluated (Tong et al. 2022). Different natural attenuation practices may involve:

Biological Methods: Biological remediation may include phytoremediation and bioremediation, offering environmentally friendly solutions. Phytoremediation uses plants to absorb and break down pollutants, proving effective in soil remediation. Bioremediation leverages microbial processes for organic contaminant degradation, ensuring sustainability and cost-efficiency (Sudarsan et al. 2024; Liang et al. 2022).

Biological Treatments: Among the biological treatment technologies available, both aerobic (such as membrane bioreactors, active sludge, and a sequencing batch reactor) and anaerobic (sludge reactors and anaerobic film) treatment methods are found to be environmentally sustainable (Ahmed et al. 2021).

Bioaugmentation: It is another bioremedial measure in which microbial strains with high degradation potential are introduced to contaminant environments. For instance, studies have shown that species of *Pseudomonas* are quite effective in degrading pharmaceuticals such as carbamazepine (Hu et al. 2023).

Biostimulation: In this process, microbial activity is enhanced through the addition of electron donors, or nutrients. As is shown in studies, nitrate and sulfate amendments to soils have been proven effective in promoting the biodegradation of certain classes of pesticides (Chen et al. 2023).

5.11 ROLE OF ADVANCED WATER TREATMENT METHODS IN MINIMIZING GROUNDWATER CONTAMINATION

Advanced water treatment technologies provide a critical line of defense against EOC contamination by removing them from wastewater before they infiltrate the environment. These methods, ranging from physical to chemical and biological treatments, offer diverse mechanisms for effective contaminant removal from the soil-groundwater systems.

5.11.1 Advanced Oxidation Processes (AOPs)

AOPs, including techniques such as ozonation, UV/H_2O_2, and photocatalysis are highly effective in degrading a wide range of EOCs by generating highly reactive hydroxyl radicals that break down contaminants into nontoxic byproducts (Kumar et al. 2022; Pandis et al. 2022). For instance, ozonation has been successfully applied in municipal WWTPs for the removal of pharmaceuticals, including diclofenac and sulfamethoxazole (Titchou et al. 2021).

Similarly, photocatalysis utilizing light energy and catalysts such as hydrogen peroxide can achieve up to 99%–100% removal efficiency in the case of pesticides (Prieto-Rodríguez et al. 2013).

5.11.2 Membrane Filtration

Membrane technologies including reverse osmosis, nanofiltration, ultrafiltration, and forward osmosis are effective in physically removing EOCs, especially those resistant to degradation (Shahid et al. 2021).

Studies have shown that reverse osmosis can effectively remove particles as fine as 10 Å with nearly 99% effectiveness, while ultrafiltration, with pore sizes ranging between 0.001 and 0.1 μm, effectively targets biological and inorganic contaminants (Kumar et al. 2022; Predieri et al. 2022; Sudarsan et al. 2024). Integrating membrane systems with biological treatment has been shown to further enhance contaminant removal while reducing energy consumption (Tang et al. 2019).

5.11.3 Constructed Wetlands

Engineered wetlands provide a sustainable and cost-effective option for EOC mitigation. Processes such as sorption, microbial degradation, and photodegradation occur naturally in these systems, making them particularly suitable for decentralized wastewater treatment (Shi et al. 2022). Research indicates that hybrid wetlands combining horizontal and vertical flow designs achieve higher removal efficiencies for pharmaceuticals such as ibuprofen and paracetamol (Reberski et al. 2022).

5.11.4 Integrated Approach to Mitigation

Despite the availability of the abovementioned remedial techniques, emerging contaminants within soil and water systems are still difficult to remediate. Their recalcitrant, persistent, and complex nature requires scalable and cost-effective solutions. Addressing EOC contamination requires combining multiple strategies to achieve optimal outcomes. For example, integrating advanced treatment technologies in WWTPs with natural attenuation processes in downstream ecosystems can provide a holistic solution. Similarly, regulatory measures and public participation are essential for the sustained success of technical interventions. By adopting such integrated approaches, stakeholders can mitigate the environmental and health risks posed by EOCs effectively and sustainably.

5.12 CONCLUSIONS, CHALLENGES, AND FUTURE PERSPECTIVE

For many years, academia, health professionals, and policymakers have been concerned about EOCs because of their specific characteristics (i.e., persistent and nonbiodegradable) and direct impact on all aspects of the environment. However, their ever-emerging lists and advances in technologies have increased their profound impacts. These pollutants and contaminants released through various natural and anthropogenic activities exhibit multifaceted transport and fate pathways influenced directly by different physiochemical, climatic, and biological factors. The persistent nature of EOCs and their presence in soil and groundwater systems emphasize the urgent need to address this widespread global problem. Their effective and efficient management requires both advancements in analytical methods for their determination and also rapid methods for their removal from the soil–water ecosystem. This collaborative approach needs to be adopted addressing regulatory, public awareness, and interdisciplinary research frameworks in place to reduce and mitigate the risks associated with EOCs and ensure the long-term sustainability of the soil-groundwater system.

Keeping this in mind, all future work should be prioritized:

- Large-scale risk assessments
- The development of eco-friendly mitigatory interventions along with solutions for combatting EOC pollution comprehensively.
- Regular monitoring for EOCs above acceptable limits (70 ng/L) in wastewater in developing and developed countries.
- Regularly update a thorough inventory of various EOC classes.
- Devise sustainable and robust analytical methods to monitor both qualitative and quantitative measurements of EOCs and their precursors.
- Undertake toxicological and epidemiological studies on EOCs in underdeveloped nations.
- Address the EOC issues using upstream solutions by combining the technologies and policy interventions.
- Risk assessment studies to determine ecosystem and human health impacts.

REFERENCES

Ahmed, S.F, M. Mofijur, S. Nuzhat, et al. 2021. Recent developments in physical, biological, chemical, and hybrid treatment techniques for removing emerging contaminants from wastewater. *Journal of Hazardous Materials*, 416:125912.

Alaoui, A., J. Lipiec, and H.H Gerke, 2011. A review of the changes in the soil pore system due to soil deformation: A hydrodynamic perspective. *Soil and Tillage Research*, 115:1–15.

An, R., B. Li, S. Zhong, G. Peng, J. Li, R. Ma, Q. Chen, and J. Ni, 2023. Distribution, source identification, and health risk of emerging organic contaminants in groundwater of Xiong'an New Area, Northern China. *Science of the Total Environment*, 893:164786.

Anawar, H.M., and V. Strezov, 2019. Transport, fate, and toxicity of the emerging and nanomaterial contaminants in aquatic ecosystems: Removal by natural processes. In *Emerging and Nanomaterial Contaminants in Wastewater*, eds. A.K. Mishra, H.M.D. Anawar, and N. Drouiche, pp. 43–62. Elsevier, Amsterdam.

Andreu, V., and Y. Picó, 2019. Pressurized liquid extraction of organic contaminants in environmental and food samples. *Trends in Analytical Chemistry*, 118:709–721.

Ankley, G.T., P. Cureton, R.A. Hoke, et al. 2021. Assessing the ecological risks of per-and polyfluoroalkyl substances: Current state-of-the science and a proposed path forward. *Environmental Toxicology and Chemistry*, 40(3):564–605.

Ashraf, M., S.Z. Ahammad, and S. Chakma, 2022. Recent advances in the occurrence, transport, fate, and distribution modeling of emerging contaminants: A review. In *Soil-Water, Agriculture, and Climate Change*, eds. S.K. Dubey, P.K. Jha, P.K. Gupta, A. Nanda, and V. Gupta, pp. 185–203. Water Science and Technology Library, vol. 113. Springer, Cham.

Biel-Maeso, M., C. Corada-Fernández, and P.A. Lara-Martín, 2018. Monitoring the occurrence of pharmaceuticals in soils irrigated with reclaimed wastewater. *Environmental Pollution*, 235:312–321.

Brauns, B., S. Chandra, W. Civil, et al. 2024. Presence of emerging organic contaminants and microbial indicators in surface water and groundwater in urban India. *Environmental Pollution*, 362:124983.

Brueller, W., N. Inreiter, T. Boegl, et al. 2018. Occurrence of chemicals with known or suspected endocrine disrupting activity in drinking water, groundwater and surface water, Austria 2017/2018. *Die Bodenkultur: Journal of Land Management, Food and Environment*, 69(3):155–173.

Bunting, S.Y, D.J Lapworth, E.J Crane, et al. 2021. Emerging organic compounds in European groundwater. *Environmental Pollution*, 269:115945.

Chacón, L., L. Reyes, L. Rivera-Montero, and K. Barrantes, 2022. Transport, fate, and bioavailability of emerging pollutants in soil, sediment, and wastewater treatment plants: Potential environmental impacts. In *Emerging Contaminants in the Environment*, eds. H. Sarma, D.C. Dominguez, and W.Y. Lee, pp. 111–136. Elsevier, Amsterdam.

Chaudhry, A.K., K. Kumar, and M.A. Alam, 2021. Groundwater flow and contaminant transport models: A short review. *Desalination and Water Treatment*, 211:80–91.

Chaves, M.J.S., J. Kulzer, P.R.P. Lima, S.C. Barbosa, and E.G. Primel, 2022. Updated knowledge, partitioning and ecological risk of pharmaceuticals and personal care products in global aquatic environments. *Environmental Science: Processes and Impacts*, 24(11):1982–2008.

Chen, L., X. Zheng, K. Zhang, et al. 2023. Sustained-release nitrate combined with microbial fuel cell: A novel strategy for PAHs and odor removal from sediment. *Journal of Hazardous Materials*, 455:131610.

Close, M.E, B. Humphries, and G. Northcott, 2021. Outcomes of the first combined national survey of pesticides and emerging organic contaminants (EOCs) in groundwater in New Zealand 2018. *Science of the Total Environment*, 754:142005.

Coronado-Apodaca, K.G., S.E.R.D. Luna, R.G. Araújo, et al. 2023. Occurrence, transport, and detection techniques of emerging pollutants in groundwater. *MethodsX*, 10:102160.

Delgado, N., J. Orozco, S. Zambrano, J.C. Casas-Zapata, and D. Marino, 2023. Veterinary pharmaceutical as emerging contaminants in wastewater and surface water: An overview. *Journal of Hazardous Materials*, 460:132431.

Djieugoue, B., B. Nlend, S.N. Boum-Nkot, et al. 2024. Emerging organic compounds in surface and groundwater reflect the urban dynamics in sub-Saharan cities. *Science of the Total Environment*, 956:177217.

Dutta, S. 2023. Pharmaceuticals and Personal Care Products in Water: Occurrence, Removal, and Impacts on Microbiomes and Invertebrates. Doctoral Thesis, Kent State University.

Ealias, M.A., G. Meda, and K. Tanzil, 2024. Recent progress in sustainable treatment technologies for the removal of emerging contaminants from wastewater: A review on occurrence, global status and impact on biota. *Reviews of Environmental Contamination and Toxicology*, 262(1):16.

Enyoh, C.E., A.W. Verla, W. Qingyue, et al. 2020. An overview of emerging pollutants in air: Method of analysis and potential public health concern from human environmental exposure. *Trends in Environmental Analytical Chemistry*, 28:e00107.

Fabietti, G., M. Biasioli, R. Barberis, and F. Ajmone-Marsan, 2010. Soil contamination by organic and inorganic pollutants at the regional scale: The case of Piedmont, Italy. *Journal of Soils and Sediments*, 10:290–300.

Falciglia, P.P, G.D. Guidi, A. Catalfo, and F.G.A Vagliasindi, 2016. Remediation of soils contaminated with PAHs and nitro-PAHs using microwave irradiation. *Chemical Engineering Journal*, 296:162–172.

Gerhardt, A. 2019. Plastic additive bisphenol A: Toxicity in surface-and groundwater crustaceans. *Journal of Toxicology and Risk Assessment*, 5(1):017.

Gogoi, A., P. Mazumder, V.K. Tyagi, G.G.T. Chaminda, A.K. An, and M. Kumar, 2018. Occurrence and fate of emerging contaminants in water environment: A review. *Groundwater for Sustainable Development*, 6:169–180.

Gomes, A.R., C. Justino, T. Rocha-Santos, A.C. Freitas, A.C. Duarte, and R. Pereira, 2017. Review of the ecotoxicological effects of emerging contaminants to soil biota. *Journal of Environmental Science and Health, Part A*, 52(10):992–1007.

Goyal, D., A.K. Haritash, and S.K. Singh, 2021. A comprehensive review of groundwater vulnerability assessment using index-based, modelling, and coupling methods. *Journal of Environmental Management*, 296:113161.

Greskowiak, J., E. Hamann, V. Burke, and G. Massmann, 2017. The uncertainty of biodegradation rate constants of emerging organic compounds in soil and groundwater: A compilation of literature values for 82 substances. *Water Research*, 126:122–133.

Hao, P.P. 2020. Determination of bisphenol A in barreled drinking water by a SPE–LC–MS method. *Journal of Environmental Science and Health, Part A*, 55(6):697–703.

Hollender, J., E.L. Schymanski, H.P. Singer, and P.L. Ferguson, 2017. Nontarget screening with high resolution mass spectrometry in the environment: Ready to go? *Environmental Science & Technology*, 51:11505–11512.

Hou, R., J. Zhang, Q. Fu, T. Li, S. Gao, R. Wang, S. Zhao, and B. Zhu, 2024. The boom era of emerging contaminants: A review of remediating agricultural soils by biochar. *Science of the Total Environment*, 931:172899.

Hu, F., P. Wang, Y. Li, et al. 2023. Bioremediation of environmental organic pollutants by *Pseudomonas aeruginosa*: Mechanisms, methods and challenges. *Environmental Research*, 239:117211.

James, C.A., J.P. Miller-Schulze, S. Ultican, A.D. Gipe, and J.E. Baker, 2016. Evaluating contaminants of emerging concern as tracers of wastewater from septic systems. *Water Research*, 101:241–251.

Johnson, G.R., M.L. Brusseau, K.C. Carroll, G.R. Tick, and C.M. Duncan, 2022. Global distributions, source-type dependencies, and concentration ranges of per-and polyfluoroalkyl substances in groundwater. *Science of the Total Environment*, 841:156602.

Joutey, N.T., W. Bahafid, H. Sayel, and N. El Ghachtouli, 2013. Biodegradation: Involved microorganisms and genetically engineered microorganisms. *Biodegradation-Life of Science*, 1:289–320.

Jurado, A., E. Vàzquez-Suñé, J. Carrera, M. López de Alda, E. Pujades, and D. Barceló, 2012. Emerging organic contaminants in groundwater in Spain: A review of sources, recent occurrence and fate in a European context. *Science of the Total Environment*, 440:82–94.

Khan, N.A., E.A. López-Maldonado, A. Majumder, et al. 2023. A state-of-art-review on emerging contaminants: Environmental chemistry, health effect, and modern treatment methods. *Chemosphere*, 344:140264.

Khan, S., M. Naushad, M. Govarthanan, J. Iqbal, and S.M. Alfadul, 2022. Emerging contaminants of high concern for the environment: Current trends and future research. *Environmental Research*, 207:112609.

Kibuye, F.A., H.E. Gall, T.L. Veith, K.R. Elkin, H.A. Elliott, J.P. Harper, and J.E. Watson, 2020. Influence of hydrologic and anthropogenic drivers on emerging organic contaminants in drinking water sources in the Susquehanna River Basin. *Chemosphere*, 245:125583.

King, M.E., C. Wang, M.V.F. Guzman, and M.B. Ross, 2022. Plasmonics for environmental remediation and pollutant degradation. *Chemical Catalysis*, 2(8):1880–1892.

Kumar, R., M. Qureshi, D.K. Vishwakarma, N. Al-Ansari, A. Kuriqi, A. Elbeltagi, and A. Saraswat. 2022. A review on emerging water contaminants and the application of sustainable removal technologies. *Case Studies in Chemical and Environmental Engineering*, 6:100219.

Lapworth, D.J., N. Baran, M.E. Stuart, and R.S. Ward, 2012. Emerging organic contaminants in groundwater: A review of sources, fate and occurrence. *Environmental Pollution*, 163:287–303.

Li, X., X. Shen, W. Jiang, Y. Xi, and S. Li, 2024. Comprehensive review of emerging contaminants: Detection technologies, environmental impact, and management strategies. *Ecotoxicology and Environmental Safety*, 278:116420.

Liang, W., G. Wang, C. Peng, et al. 2022. Recent advances of carbon-based nano zero valent iron for heavy metals remediation in soil and water: A critical review. *Journal of Hazardous Materials*, 426:127993.

Lin, Z., L. Wang, Y. Jia, Y. Zhang, Q. Dong, and C. Huang, 2017. A study on environmental bisphenol A pollution in plastics industry areas. *Water, Air, and Soil Pollution*, 228:1–9.

Locatelli, L., P.J. Binning, X. Sanchez-Vila, G.L. Søndergaard, L. Rosenberg, and P.L. Bjerg, 2019. A simple contaminant fate and transport modelling tool for management and risk assessment of groundwater pollution from contaminated sites. *Journal of Contaminant Hydrology*, 221:35–49.

Mahler, B., and M.L. Musgrove, 2019. Emerging contaminants in groundwater, karst, and the Edwards (Balcones Fault Zone) Aquifer. In *The Edwards Aquifer: The Past, Present, and Future of a Vital Water Resource*, eds. J.M. Sharp, R.T. Green, and G.M. Schindel, Geological Society of America, Boulder, CO.

Meena, V., D. Swami, A. Chandel, N. Joshi, and S.O. Prasher, 2024. Selected emerging contaminants in water: Global occurrence, existing treatment technologies, regulations and associated risk. *Journal of Hazardous Materials*, 483:136541.

Meffe, R., and I. de Bustamante, 2014. Emerging organic contaminants in surface water and groundwater: A first overview of the situation in Italy. *Science of the Total Environment* 481:280–295.

Michael, I., M..I. Vasquez, E. Hapeshi, T. Haddad, E. Baginska, K. Kümmerer, and D. Fatta-Kassinos, 2014. Metabolites and transformation products of pharmaceuticals in the aquatic environment as contaminants of emerging concern. In *Transformation Products of Emerging Contaminants in the Environment*. eds. D.A. Lambropoulou, and L.M.L. Nollet, pp. 413–458, John Wiley and Sons Ltd, Hoboken, NJ.

Moreau, M., J. Hadfield, J. Hughey, F. Sanders, D.J. Lapworth, D. White, and W. Civil, 2019. A baseline assessment of emerging organic contaminants in New Zealand groundwater. *Science of the Total Environment*, 686:425–439.

Nasr, W.B., F. Huneau, R. Trabelsi, K. Zouari, E. Garel, and T. Leydier, 2024. Emerging organic compounds as markers of the degradation of groundwater qualitative and quantitative equilibrium in a context of rapid urban expansion. *Science of the Total Environment*, 915:170068.

Nham, H.T.T., J. Greskowiak, K. Nödler, et al. 2015. Modeling the transport behavior of 16 emerging organic contaminants during soil aquifer treatment. *Science of the Total Environment*, 514:450–458.

NORMAN, Network. 2019. *List of Emerging Substances, Network of Reference Laboratories, Research Centres and Related Organisations for Monitoring of Emerging Environmental Substances*. Available online at www.norman-network.com. Accessed June.

Noutsopoulos, C., E. Koumaki, D. Mamais, M.C. Nika, A.A. Bletsou, and N.S. Thomaidis, 2015. Removal of endocrine disruptors and non-steroidal anti-inflammatory drugs through wastewater chlorination: The effect of pH, total suspended solids and humic acids and identification of degradation by-products. *Chemosphere*, 119:S109–S114.

Pal, A., K.Y.H. Gin, A.Y.C. Lin, and M. Reinhard, 2010. Impacts of emerging organic contaminants on freshwater resources: Review of recent occurrences, sources, fate and effects. *Science of the Total Environment*, 408(24):6062–6069.

Pandis, P.K, C. Kalogirou, E. Kanellou, et al. 2022. Key points of advanced oxidation processes (AOPs) for wastewater, organic pollutants and pharmaceutical waste treatment: A mini review. *Chemical Engineering Journal*, 6(1):8.

Patil, S.B., and H.S. Chore, 2014. Contaminant transport through porous media: An overview of experimental and numerical studies. *Advances in Environmental Research*, 3(1):45–69.

Predieri, B., C.A.D. Alves, and L. Iughetti, 2022. New insights on the effects of endocrine-disrupting chemicals on children. *Jornal de Pediatria*, 98(suppl 1):73–85.

Prieto-Rodríguez, L., I. Oller, N. Klamerth, A. Agüera, E.M. Rodríguez, and S. Malato, 2013. Application of solar AOPs and ozonation for elimination of micropollutants in municipal wastewater treatment plant effluents. *Water Research*, 47(4):1521–1528.

Reberski, J.L., J. Terzić, L.D. Maurice, and D.J. Lapworth, 2022. Emerging organic contaminants in karst groundwater: A global level assessment. *Journal of Hydrology*, 604:127242.

Roberts, S., C. Higgins, and J. McCray, 2014. Sorption of emerging organic wastewater contaminants to four soils. *Water*, 6(4):1028–1042.

Saeidi, N., F. Harnisch, V. Presser, F.D. Kopinke, and A. Georgi, 2023. Electrosorption of organic compounds: State of the art, challenges, performance, and perspectives. *Chemical Engineering Journal*, 471:144354.

Schulze, S., D. Zahn, R. Montes, R. Rodil, J.B. Quintana, T.P. Knepper, T. Reemtsma, and U. Berger, 2019. Occurrence of emerging persistent and mobile organic contaminants in European water samples. *Water Research*, 153:80–90.

Shahid, M.K., A. Kashif, A. Fuwad, and Y. Choi, 2021. Current advances in treatment technologies for removal of emerging contaminants from water: A critical review. *Coordination Chemistry Reviews*, 442:213993.

Shi, Q., Y. Xiong, P. Kaur, N.D. Sy, and J. Gan, 2022. Contaminants of emerging concerns in recycled water: Fate and risks in agroecosystems. *Science of the Total Environment*, 814:152527.

Shu, X, Y. Wu, X. Zhang, and F. Yu, 2023. Experiments and models for contaminant transport in unsaturated and saturated porous media: A review. *Chemical Engineering Research and Design*, 192:606–621.

Silori, R., V. Shrivastava, A. Singh, P. Sharma, M. Aouad, J. Mahlknecht, and M. Kumar, 2022. Global groundwater vulnerability for pharmaceutical and personal care products (PPCPs): The scenario of second decade of 21st century. *Journal of Environmental Management*, 320:115703.

Sintim, H.Y., S. Bandopadhyay, M.E. English, et al. 2021. Four years of continuous use of soil-biodegradable plastic mulch: Impact on soil and groundwater quality. *Geoderma*, 381:114665.

Solangi, N.H., R.R. Karri, S.A. Mazari, et al. 2023. MXene as emerging material for photocatalytic degradation of environmental pollutants. *Coordination Chemistry Reviews*, 477:214965.

Sudarsan, J.S., K. Dogra, R. Kumar, et al. 2024. Tricks and tracks of prevalence, occurrences, treatment technologies, and challenges of mixtures of emerging contaminants in the environment: With special emphasis on microplastic. *Journal of Contaminant Hydrology*, 265:104389.

Sun, P., D. Barmaz, M.L. Cabrera, S.G. Pavlostathis, and C.H. Huang, 2013. Detection and quantification of ionophore antibiotics in runoff, soil and poultry litter. *Journal of Chromatography A*, 1312:10–17.

Sverdrup, L.E., T.A. Sogn, T. Källqvist, et al. 2021. Risk assessment of contaminants in sewage sludge applied on Norwegian soils. *European Journal of Nutrition and Food Safety*, 13(11):1–7.

Tang, Y., M. Yin, W. Yang, et al. 2019. Emerging pollutants in water environment: Occurrence, monitoring, fate, and risk assessment. *Water Environment Research*, 91(10):984–991.

Titchou, F.E., H. Zazou, H. Afanga, et al. 2021. Removal of organic pollutants from wastewater by advanced oxidation processes and its combination with membrane processes. *Chemical Engineering and Processing-Process Intensification*, 169:108631.

Tong, X., S. Mohapatra, J. Zhang, N.H. Tran, L. You, Y. He, and K.Y.H. Gin, 2022. Source, fate, transport and modelling of selected emerging contaminants in the aquatic environment: Current status and future perspectives. *Water Research*, 217:118418.

Valhondo, C., L. Martinez-Landa, J. Carrera, C. Ayora, K. Nödler, and T. Licha, 2018. Evaluation of EOC removal processes during artificial recharge through a reactive barrier. *Science of the Total Environment*, 612:985–994.

Vázquez-Tapia, I., T. Salazar-Martínez, M. Acosta-Castro, et al. 2022. Occurrence of emerging organic contaminants and endocrine disruptors in different water compartments in Mexico: A review. *Chemosphere*, 308:136285.

Wang, T., L. Wang, Q. Chen, N. Kalogerakis, R. Ji, and Y. Ma, 2020. Interactions between microplastics and organic pollutants: Effects on toxicity, bioaccumulation, degradation, and transport. *Science of the Total Environment*, 748:142427.

Wang, X., D. Liang, Y. Wang, et al. 2022. A critical review on the biological impact of natural organic matter on nanomaterials in the aquatic environment. *Carbon Research*, 1(1):13.

Wilkinson, J., P.S. Hooda, J. Barker, S. Barton, and J. Swinden, 2017. Occurrence, fate and transformation of emerging contaminants in water: An overarching review of the field. *Environmental Pollution*, 231:954–970.

Yang, Y., Y.S. Ok, K.H. Kim, E.E. Kwon, and Y.F. Tsang, 2017. Occurrences and removal of pharmaceuticals and personal care products (PPCPs) in drinking water and water/sewage treatment plants: A review. *Science of the Total Environment*, 596:303–320.

Yu, Y., Z. Wang, B. Yao, and Y. Zhou, 2024. Occurrence, bioaccumulation, fate, and risk assessment of emerging pollutants in aquatic environments: A review. *Science of the Total Environment*, 923:171388.

Zhang, Z., N. Zhang, M. Zhao, Y. Zhang, W. Yang, and B. Liu, 2025. Occurance and pollution risk assessment of emerging contaminants in groundwater in the vicinity of a typical municipal landfill in northeastern China. *Journal of Hydrology*, 648:132408.

Zhao, Y., L. Ye, and X.X. Zhang, 2018. Emerging pollutants: Part I: Occurrence, fate and transport. *Water Environment Research*, 90(10):1301–1322.

Zhong, M., T. Wang, W. Zhao, et al. 2022. Emerging organic contaminants in Chinese surface water: Identification of priority pollutants. *Engineering*, 11:111–125.

6 The Uptake of EOCs from Soil by Plants is Essential for Agricultural Food Safety and Phytoremediation

Saher Shahid and Mahtab Ahmad

6.1 INTRODUCTION

An enormous range of anthropogenic chemical compounds have been synthesized for many applications since the start of industrialization. Chemical products are frequently released into the environment after they have fulfilled their intended purpose (Drakvik et al. 2020). Emerging organic contaminants (EOCs) consist of a large group of compounds, including pharmaceuticals and personal care products (PPCPs), antibiotics, per-fluoro-alkyl sulfonate (PFAS), flame retardants, endocrine disruptors, pesticides, and surfactants (Priya et al. 2022; Rogowska et al. 2020). These EOCs are observed in soil, air, and water bodies (Masinga et al. 2024).

One of the main repositories of waste compounds, which can be harmless, toxic, or nontoxic to living things, is soil (Wołejko et al. 2021). Industrial products and chemicals, after use, come into contact with soil and serve as a major source of contamination in soil (Figure 6.1). Personal care products and industrial items are the primary sources of soil waste chemicals (Bahuguna et al. 2022). The soil biota, including nematodes, microorganisms, and plants, may be in danger from the careless or uncontrolled disposal of the EOCs released into the environment (Bender and van der Heijden 2015). As plants play a special function among living things (Masinga et al. 2024), any disruption to the environment in which plants grow affects living organisms, such as consumers (humans and animals) (Hashimi et al. 2020). Plants have the ability to uptake the chemicals existing in the soil (Pullagurala et al. 2018).

As EOCs are susceptible to volatilization, microbial degradation, and photodegradation, their destiny in the soil medium is uncertain (Reddy and Kim 2015). Apart from these procedures, external elements, including air currents, soil erosion, and surface runoff, also have significant effects. In addition to their negative influence on the surrounding environment, there are still unanswered questions on how these

 DOI: 10.1201/9781003509523-6

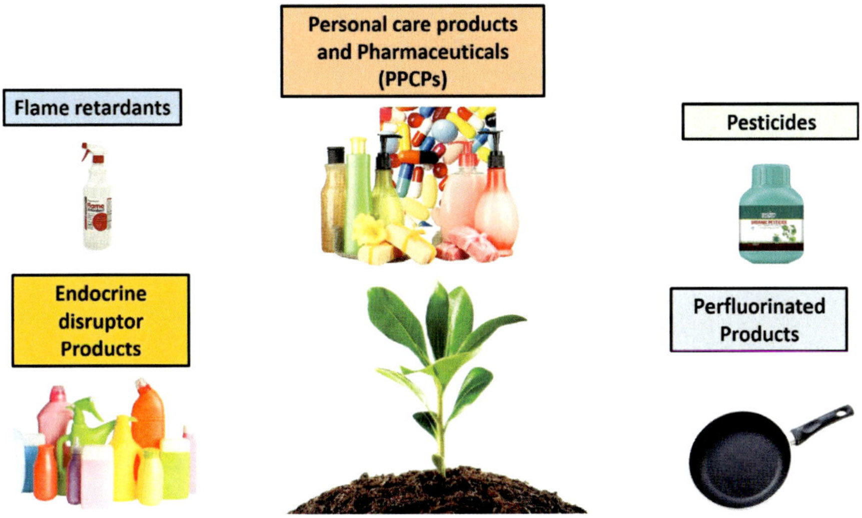

FIGURE 6.1 Potential emerging organic contaminant (EOC) products.

EOCs are exposed and affect the terrestrial food chain. Determining their absorption and possible impact on plants, which are the main contributors to a large portion of terrestrial food chains, is therefore crucial.

6.2 ROLE OF PLANTS AS GREEN LIVER

The translocation and transfer of EOCs by the plants from the soil is largely dependent on their physicochemical characteristics. Some EOCs are very stable and remain in their pure form regardless of their interaction that builds up during the general translocation or uptake and transfer processes. An essential tool for determining the total amount of EOCs absorbed by plants from the soil into their xylem is the transpiration stream concentration factor (TSCF), which is defined as "the ratio of chemical concentration in the transpiration stream to the concentration found in an external solution" (Miller et al. 2016). The TSCF might change depending on the variable properties of the chemicals.

Plants have certain enzymes essential for digesting contaminant compounds like the detoxification processes in livers of animals (Pullagurala et al. 2018). These enzymes include *N*-malonyltransferases, carboxylesterases, glutathione transferases, and cytochrome P450. The enzyme detoxification processes in plants allow them to detoxify pollutants, including EOCs. The term "green livers" refers to this plant detoxification process (Pullagurala et al. 2018). Also, research has revealed that some of the pollutants are lipophilic, meaning they possess the ability to enter and move across different plant organs and tissues. Phases I, II, and III of the pollutants' chemical changes are aided by enzymes once absorbed (Kurade et al. 2021). In phase 1, reactive groups are involved, which leads to the hydrolysis and redox-based conversion of pollutants into metabolites. During phase II, the conjugation of metabolites with the assistance of glutathione, sugars, or amino acids takes place. Lastly, during phase III, metabolites that are persistent and incapable of undergoing additional transformation are either integrated into cell walls or retained in vacuoles (Kurade et al. 2021). The three metabolic phases are crucial in plants lacking an excretory system (Pullagurala et al. 2018).

6.3 THE INTERACTIONS OF EOCs IN PLANTS

In the past 10 years, there has been a significant rise in the scientific interest in processes and mechanisms by which plants detoxify and get rid of organic pollutants (Bartrons and Peñuelas 2017). In this section, we will discuss the in-depth methods of absorption and transfer of PPCPs in the plant structure. Therefore, the mechanism for uptake is also applicable to other emerging contaminants. The absorption of PPCPs by plants depends on the multiple properties (physical and chemical) of plants. As a result, these critical properties influencing absorption and translocation must be considered for each emerging contaminant and the type of plant with which it is interacting. The physicochemical properties of the organic contaminant also influence the transportation mechanism. For instance, the transport of organic contaminants to the plant system through the root is affected by the size of the molecule. Molecules with minute sizes can pass through membranes and enter and exit the phloem and xylem with ease. Whereas, substances having greater molecular sizes and limited membrane permeability have a difficult time moving through the phloem (Shahid et al. 2024; Kvesitadze et al. 2015). The PPCPs frequently affect the plant growth, development, and germination processes, but they can also have toxic effects on plant composition and functioning (Bártíková et al. 2016). The three major biotic processes—bioadsorption, bioaccumulation, and biodegradation—are responsible for removing the organic chemicals from the soil (Kurade et al. 2016; Xiong et al. 2016).

6.3.1 Route of EOCs for Plant Uptake

There are multiple routes by which the PPCPs can enter the plant: movement via root channels, diffusion, ionic fraction uptake, and intake through aerial parts by the deposition of aerosols and volatilized chemicals (Trapp and Legind 2011). Diffusion of dissolved organics or mass stream entrance are the two ways for PPCPs to enter roots and aerial tissues (Miller et al. 2016). While organic contaminants can be absorbed by plants, the majority of PPCPs are mainly exposed to plants either through passive or active mode in contact with water or soil (Zhang et al. 2017). As a result, the roots are the main entry point for PPCPs into plants. Plants exhibit a mechanism of active and passive transport that allows the uptake of foreign particles and nutrients and their storage in the vacuoles. There are only few chemicals (such as phenoxy acid herbicides) that follow active uptake for adsorption; the majority of organic pollutants are absorbed through a general process of passive uptake, which is controlled by transpiration current passing through the tracheids and xylem (Madikizela et al. 2018). The porous cell wall of the root cortex allows the contaminants to enter readily and pass through before reaching the endodermis. Therefore, two potential routes for the movement of pollutants in plants are identified: (i) conducting tissue transportation, also referred to as short-distance transport; and (ii) intercellular and intracellular transport, also known as long-distance transport (Zhang et al. 2017).

6.3.2 Mechanism(s) Involved in the Uptake of EOCs by Plants

As transpiration creates a pressure gradient, the transferred pollutants from the external aqueous phase move upward into the aerial parts through mass flow through the xylem. However, the contaminants need to go through the epidermis, cortex, endodermis, and pericycle to get to the xylem from the roots. The translocation due to evaporation from the leaf surface causes a continuous flow of transpiration. Therefore, water evaporating off leaves produces a suction force and capillary action in the xylem vessels. Also, the osmotic pressure process allows the movement of water and PPCPs from the roots to the leaves, which is then followed by a "transpiration-cohesion-adhesion" mechanism. Compared with the xylem sap, the phloem has a comparatively higher concentration of dissolved organics, but the phloem's vessels are mainly focused on movement of the photosynthesis products from leaves to the roots (Madikizela et al. 2018). The pollutants' dissolution in the aqueous phase and lipid-rich cell membrane of the plant determine whether or not the pollutants can pass through it. For the pollutants to move through the plant roots, they must get to the vascular tissue (Miller et al. 2016). The cell walls of the newborn root hairs, just behind the root tips, allow the organic contaminants to enter the root along with water and other nutrients. Subsequently, compounds move through three main channels in the root into the xylem: symplastic (via plasmodesmata across cells), apoplastic (free intercellular space), and transmembrane (via cell membranes) (Zhang et al. 2017; Kvesitadze et al. 2015; Shahid et al. 2024).

6.4 FACTORS AFFECTING THE UPTAKE OF EOCs IN PLANTS

6.4.1 Environmental Factors

Numerous physicochemical traits of the contaminants, biotic traits of the plant, and the environmental conditions all impact plants' absorption and transport of EOCs (Madikizela et al. 2018; Figure 6.2). Plant cell membranes lack a dedicated transport mechanism to uptake artificial chemicals. Consequently, it has been postulated that a direct diffusion process determined by the chemical characteristics of the compounds drives plants' uptake of EOCs (Zhang et al. 2014). A contaminant's ionic strength, molecular mass, hydrophobicity, solubility, pK_a, K_{OW}, solution pH, biodegradation, and sorption capacity are the crucial factors for their transport from the external environment to the plant tissues (Malchi et al. 2014; Adeel et al. 2017). Several studies reveal that the uptake of EOCs by crops is influenced by their bioavailability and bioaccessibility in the pore water of the soil (Christou et al. 2019). The movement or translocation of EOC inside the plant, i.e., through the xylem and phloem, highly depends on the physicochemical characteristics of the contaminants, including lipophilicity and electrical charge, along with the physiological structure of the plant and transpiration rate (Dodgen et al. 2015). The variable environmental

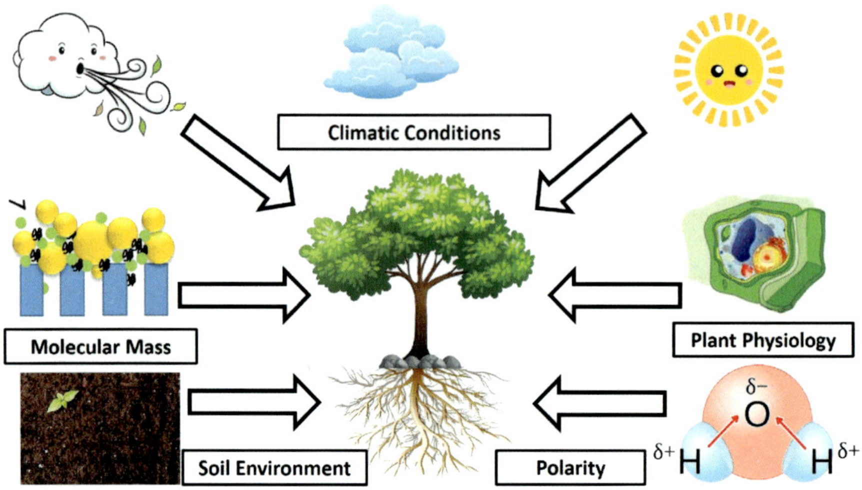

FIGURE 6.2 Factors affecting the uptake of EOCs by plants.

conditions, such as drought, also impact the uptake of contaminants in plants (Zhang et al. 2016). However, rather than affecting PPCPs' translocation in plants, these factors mostly affect adsorption or chelate formation. While the concentration of contaminants determines the passive uptake, the plant's lipid content and metabolism affect the active uptake (Christou et al. 2019). Absorption is negatively impacted by acidity and organic matter (Goldstein et al. 2014).

6.4.2 Physical and Chemical Factors

6.4.2.1 Hydrophobicity and Molecular Mass

The hydrophobicity and molecular mass of organic compounds are the main factors that affect plant absorption. The molecular mass and hydrophobicity are determined by K_{OW} and K_{OA}, i.e., the partition coefficients between octanol and water and octanol and air, respectively. Plants efficiently absorb lipophilic particles with log K_{OW} values between 0.5 and 3.5 because of their capacity to move across membranes and translocate into cell fluids (Weyens et al. 2009). Roots absorb organic compounds having molecular masses less than 1000 and high K_{OW} values. On the other hand, substances with low K_{OW} are easily absorbed by the leaves by selective absorption from the air. The PPCPs are generally divided into membrane lipids and detained in the roots due to anionic and hydrophobic nature, indicating that the interaction of PPCPs in plants depend upon their pK_a values (Madikizela et al. 2018). Herklotz et al. (2010) presented that PPCPs with low pK_a values show more accumulation than other PPCPs. According to Miller et al. (2016), the EOCs with p$K_a < 7$ and log $K_{OW} < 3$ can move to fruits due to ion-trapping mechanisms in the phloem. Whereas, the EOCs with p$K_a > 7$ and log $K_{OW} < 0$ translocate in both xylem and phloem; and the EOCs with log $K_{OW} > 0$ and log $K_{OW} < 4$ translocate only in the xylem tissues of the plant. Contaminants carried through the xylem diffuse into the next layers during the translocation, along with nutrients and water.

6.4.2.2 Polarity

According to a predicted model developed by Dettenmaier et al. (2009), extremely hydrophilic substances show high affinity for the uptake and translocation of EOCs by roots. For example, caffeine was greatly captivated by *Scirpus validus* and in artificial wetlands by *Salvinia molesta* and *Elodea canadensis* (Zhang et al. 2014). Nonionic PPCPs travel across cell membranes, moving mostly in the route of the transpiration current to finally gather in plant leaves. Hydrophilic compounds translocate into the xylem with the water (Adeel et al. 2017). The nonionic pollutants are absorbed in two stages: first, the amount of contaminant in the root water phase and the external soil solution reach equilibrium. Later, the pollutant is chemically sorbed onto the lipophilic entities. Lipids make up entities, which adhere to cell walls and membranes (Collins et al. 2011).

Also, due to the repulsion between the cytoplasm and the cell walls, ionic PPCPs are restricted in the phloem and accumulate in numerous plant sections, including fruit (Goldstein et al. 2014). Certain extremely hydrophilic compounds, however, accrue specifically in the roots (Tanoue et al. 2012). This may be due to the absence of particular protein transporters that could help the compounds move into other organs. The PPCP in the roots were greater than those in the leaves and stem for sulfamethazine (Kurwadkar et al. 2017), salbutamol, trimethoprim, sulfamethoxazole, carbamazepine (Herklotz et al. 2010), ciprofloxacin, metformin, and narasin (Eggen et al. 2011). The amount of lipids extracted from roots also affects how well organic contaminants are absorbed (Dodgen et al. 2015).

6.4.2.3 Soil Environment

Crops that grow in agricultural soils deficient in organic matter, such as sandy soil, generally show higher potential for EOC uptake compared with clay soils or soils rich in organic matter (Malchi et al. 2014). Malchi et al. (2014) studied the pharmaceutical contaminants in carrots and sweet potatoes in three soils, including sandy clay, loamy sand, and sandy loam. They observed that the contaminant concentrations were higher in the plants grown in sandy soil compared with those grown in loamy soil. Also, nonionic pharmaceutical uptake and accumulation were higher in the cucumber and tomato plants grown in soil with low organic matter and clay content (Goldstein et al. 2014). The antibiotics sulfamethoxazole, lincomycin, and oxytetracycline were exposed to the lettuce plant. The plant was grown in three soils of different textures, i.e., loam, sandy loam, and sand. It was observed that only sulfamethoxazole was taken up in the lettuce, indicating a positive association between the contaminant uptake percentages and the sand content in the soil (Zhang et al. 2016). The EOCs in the soil undergo sorption/desorption and biotic and/or abiotic transformation processes. These mechanisms influence their concentrations for biodegradation, runoff and leaching into soil, and uptake by plants (Grossberger et al. 2014).

The ionized and polarized EOCs undergo several contacts with soil organic matter to dissolve organic matter on the mineral surface. These interactions include hydrogen bonding, cation–anion exchange, protonation, hydrophobic partitioning, water binding, surface complexion, and cation binding (Christou et al. 2019). The good aerated soils support the uptake of EOCs by plants as compared with water-logged or compacted soil (nonaerated) as aeration facilitates the activity of roots, respiration, water and nutrients uptake, and plant development (Christou et al. 2019).

6.4.2.4 Climatic Conditions

The evapotranspiration rate of plants, and their uptake of EOCs, water, and nutrients are affected by climatic conditions such as wind speed, humidity in the air, and ambient temperature. The uptake of EOCs, nutrients, and water is enhanced in the plant under high wind speeds and low humidity conditions. Crops grown in the Mediterranean region (hot and dry climate) show a high uptake of EOCs. Plants in cold and humid agricultural sites exhibit low uptake of EOCs. Environmental conditions, such as drought, also influence the uptake of EOCs in plants. The drought conditions caused higher translocation of lincomycin and decreased movement of oxytetracycline and sulfamethoxazole in the lettuce (Zhang et al. 2016). Also, the transpiration rate is one of the factors in the uptake and translocation of EOCs, especially in arid and hot climates (Dodgen et al. 2015). The evapotranspiration in crops depends on climate and crop coefficient K_C that serve as virtuous indicators for crop potential for uptake and accumulation of EOCs in the plant tissues. Plants with a high evapotranspiration and net irrigation rate show high potential for EOC in hot and dry climates. These plants include citrus fruits, bananas, walnuts, cucumbers, peppers, eggplant, peanuts, and tomatoes. The seasonal plants during the summer season, greenhouse plants, and perennial crops grown in treated wastewater show a high probability for EOC uptake. Leafy vegetables (cabbage, spinach, broccoli, and celery), often extensively grown all year round, hoard more amounts of EOCs in the plant tissues, whereas the plants of autumn and winter seasons, including succulent plants, are all categorized as plants with decreased EOC uptake (agave and aloe vera) (Christou et al. 2019).

6.4.2.5 Plant Physiology

Plant physiology greatly affects uptake, translocation, and accumulation of EOCs in plants. Plants, when interact with xenobiotic substances, induce different coping and adaptive mechanisms including antioxidant defense mechanism, stomatal closing, osmoregulation, ion sequestration and exclusion, hormone and metabolic regulation, and transcription and post-translation alterations (Christou et al. 2019). It has been studied that plants grown under stress conditions show a hindrance to the uptake of EOCs in comparison to plants under nonstressed conditions. Therefore, the nonstressed plants have a greater potential for EOC uptake as plants under stress conditions show lower photosynthetic efficiency and growth rate (Hepworth et al. 2015).

The type and physiology of the plant is also responsible for the translocation of EOCs. For instance, Pi et al. (2017) showed that *E. horemanii*, a submerged species exposed to aqueous particulate matter pollutants through direct contact with the leaves, predominately accumulated atenolol, carbamazepine, ibuprofen, and triclosan in its leaf as opposed to the roots. Conversely, it was found that the majority of the examined substances tended to concentrate more on the roots like *E. crassipes*. The concentrations of dissolved organic carbon, plant physiology, organic matter content, electrolyte type, and pH are among the most important environmental parameters; temperature, on the other hand, has a very small impact on the pollutants' uptake in plants (Goldstein et al. 2014).

6.5 FOOD SECURITY AND POTENTIAL TOXICITIES OF EOCs

Worldwide, starvation and hunger have affected more than 690 million people on Earth. Any component of the diet, when lacking in the diet may pose a detrimental outcome on the human health (Naresh et al. 2021). The presence of nutrients in the soil estimates the mineral content of plant-based food. Soil contamination due to EOCs, metals, and organic chemicals directly damage human health by consuming the contaminated crops (Masinga et al. 2024; Zhao et al. 2022). The EOCs, due to their toxic effects on agriculture, may pose high food security risks. It has been studied that EOCs such as fluoroquinolones significantly reduce reproduction rates and root and shoot length, causing chlorosis in plants. Pharmaceuticals including antibiotic ciprofloxacin present in macrophyte *Eichhornnia crassipes* are associated with a decrease in chlorophyll content showing faded leaf color and an effect on enzyme activity (Yan et al. 2019). Humans are affected directly or indirectly when exposed to EOCs. Therefore, considering the level of exposure, EOCs can cause toxic effects, microbial infection, endocrine disruption, and skin illnesses (Wang et al. 2022).

6.5.1 Effect of EOCs on Soil Organisms

The EOCs, present in soil, causes environmental health concern to soil microbiota and crops grown therein. The pharmaceuticals bioaccumulate in the bodies of soil microorganisms and show toxic effects (Zhao et al. 2022). For example, pharmaceuticals ivermectin, fuazuoron, fipronil, and closantel decrease the propagative capacity of *Folsomia candida* and soil-dwelling arthropods (Zortéa et al. 2017). The oxidative stress and disrupting neurotransmissions were observed in *F. candida* when exposed to carbamazepine and fluoxetine in the concentrations of 0.04, 0.4, 4, and 40 mg/kg (Oliveira et al. 2018). The reproduction capacity of earthworms was decreased due to bisphenol-S at concentration levels of 10, 100, and 1000 mg/kg (Marcos et al. 2021). Similarly, when exposed to *Dendrobaena veneta* and *Eisenia fetida*, another endocrine disruptor, bisphenol-A, disrupted its endocrine pathways and gene expression of the male organ, causing reduction in the growth and reproduction level (Novo et al. 2018). In earthworms, halogenated flame retardants caused negative impacts on enzymatic activities and DNA (Navarro et al. 2017). Some studies also revealed the toxic impact of PFAS on soil and soil organisms. The polycyclic musks caused cellular oxidative injury, increased lipid peroxidation levels, and gene expression of superoxide dismutase enzymes in earthworms. In another study, the cocoon production in *Eisenia fetida* was decreased when exposed to tonalide and galaxolide at a concentration of 100 μg/g (Masinga et al. 2024).

6.5.2 Effect of EOCs on Plants

Soil contaminated with EOCs also poses a negative impact on crops grown in these soils. The presence of pharmaceuticals, such as sulfamethoxazole and trimethoprim, in tomatoes decreased the carbohydrate content (Christou et al. 2019). The tissue biomass and yield were decreased in peanut plants due to veterinary antibiotics and tetracyclines (Zhao et al. 2022). Various edible crops were damaged due to the presence of analgesics such as paracetamol and ibuprofen. For example, exposure to analgesics affected the development of *Lactuca sativa* as the plant indicated toxicity (Rede et al. 2019). The endocrine-disrupting pesticides were detected in the lettuce, onion, radish, spinach, garlic, and turnip in the concentration range of 14.7–28.9 μg/kg (Ali et al. 2019). Likewise, the plant growth of the *Vigna radiata* was affected when exposed to bisphenol-A contamination (Kim et al. 2018). Also, doses of 250 and 500 mg/kg caused reduction in the chlorophyll content and stomata size (Zhang et al. 2016). The exposure of plants to PPCPs, including sulfonamide, negatively affected the root length, growth rate, fluorescence of chloroplast, chlorophyll content, and leaf length in the *Vallisneria natans* plant (Zhu et al. 2020). The PPCPs, tetracyclines, lincosamides, β-lactams, and macrolides showed toxic effects in plants of diverse categories (Bártíková et al. 2016). Plants show resistance to toxic stress by initiating detoxification processes such as H_2O_2 and O_2 (reactive oxygen species) and antioxidant enzymes (Sun et al. 2018).

6.5.3 Effect of EOCs on Humans

Humans' health is at serious risk when they consume edible plants or vegetables contaminated with EOCs. The exposure usually occurs through inhalation, ingestion, and dermal exposure (Pawar et al. 2017). Different EOCs pose a multiple health threats. For example, a flame-retardant tris(2-butoxyethyl) was observed to cause hepatotoxicity and carcinogenicity in human liver cells in the concentration range of 100–400 μM (Saquib et al. 2022). In another study, tris(2-butoxyethyl) upregulated 12% of cancer genes and 0.3% indicating carcinogenicity (Liu et al. 2022). Endocrine disruptors and flame retardants are associated with endocrine damage (Rosenmai et al. 2021). Flame retardants also cause ovarian dysfunction (Wang et al. 2021). Long-term exposure to phthalates causes abnormalities (Qadeer et al. 2022). Also, phthalates causes asthma in childhood, decreased semen quality, impaired neurodevelopment, autism, breast or uterine cancer, low birth weight, obesity, low testosterone level, endometriosis, reduced estradiol, thyroid, and hearing function (Eales et al. 2022). Few EOCs are genotoxic, risking life (Billing et al. 2021).

6.6 PHYTOREMEDIATION OF EOCs

A low-risk, environmentally friendly, plant-based method powered by solar energy, *phytoremediation*, gained attention for removing toxins through biotic processes that pose no danger (Rane et al. 2016; Chandanshive et al. 2017; Gikas et al. 2018). Due to their widespread distribution across the globe and ability to absorb both organic and inorganic substances from soil and water, plants serve as excellent pollution tracers. With the advantages of habitat restoration and biomass production, which causes feedstock for producing renewable bioenergy, phytoremediation might thus be developed as an in-situ treatment for removing EOCs with ease (Khandare and Govindwar 2015).

Many contaminants have been removed from soil by using plants. For instance, different plant species have been used to remove a variety of heavy metals

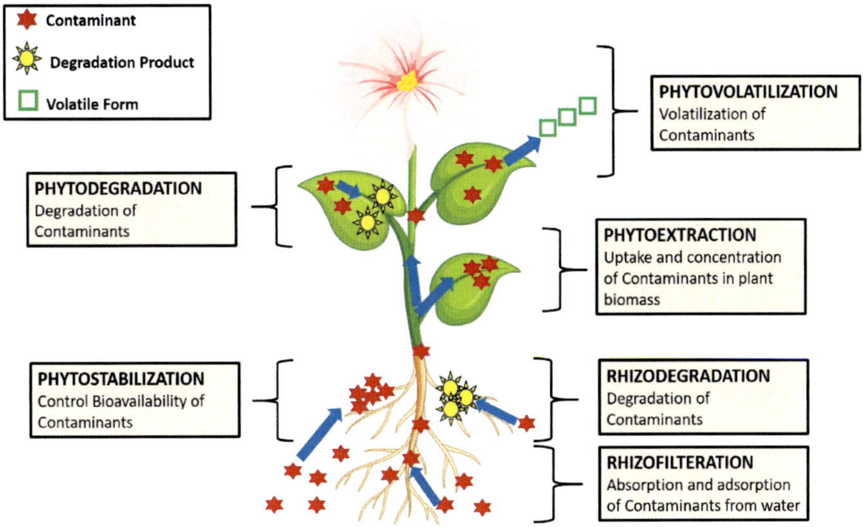

FIGURE 6.3 Phytoremediation of EOCs.

successfully (Wang et al. 2017). Plants such as *Aster amellus, Glandularia pulchella, Portulaca grandiflora, Petunia grandiflora*, and *Zinnia angustifolia*, are used due to their ability to degrade dyes (Khandare and Govindwar 2015). Also, rhizodegradation and uptake in plants are effective in removing polychlorinated biphenyls (Passatore et al. 2014).

The mechanisms including phytostabilization, rhizoid grading, rhizofiltration, phytodegradation, phytoextraction, phytoaccumulation, and phytovolatilization have all been used to remediate different pollutants in plants (Wang et al. 2017; Figure 6.3). Additionally, plant–microbe pairings such as those between endophytic or rhizospheric plants might promote phytoremediation. Endophytes emit phytohormones and biosurfactants, inhibiting phytopathogens plants, enhancing plant uptake of nutrients and water. They can also improve plant growth and the activity of antioxidant enzymes in plants while repairing atmospheric damage due to organic pollutants (Feng et al. 2017).

Rhizoremediation also improves phytoremediation by encouraging the growth of microbial communities. This is because root exudation can induce the production of enzymes that cause the breakdown of contaminants. Therefore, plants secrete secondary metabolites (Didier et al. 2012).

6.7 MECHANISM(S) OF PHYTOREMEDIATION OF EOCs

Phytoremediation is a process in which plants and microorganisms interact for the exclusion, stabilization, transmission, and annihilation of toxic contaminants from the environment (Kristanti et al. 2023). In the process of phytoremediation, plants are grown at a site contaminated with pollutants, and the plants uptake these contaminants through their root systems. The roots are garnered and become drenched with contaminants (Abdel-Shafy and Mansour 2018). The fast-growing plants with broad roots and high transpiration rates are recommended for use (Srivastav et al. 2018). Plant growth-promoting bacteria can help make phytoremediation more effective, i.e., with a synergistic effect, in which they can release phytohormones and biosurfactants, resist abiotic stress, and lessen plant toxicity through biosorption or bioaccumulation. This is explained by the large surface area to volume ratio of the bacteria's cells, which can release a variety of hormones, organic acids, and antibiotics to promote plant growth (Saxena et al. 2020). Numerous mechanisms are engaged in phytoremediation, based on the pollutant type and the polluted substrate.

6.7.1 Phytostabilization

Phytostabilization works by immobilizing pollutants in soil or groundwater by precipitation in the rhizosphere, adsorption onto roots, or accumulation. It seeks to stop the pollutants' migration and re-entry into the environment by lowering their mobility (Jeevanantham et al. 2019). Due to wind and rain, the covered plants in the contaminated region help prevent soil erosion and surface runoff (Shmaefsky 2020). Plants secrete root exudates, which can draw a wide variety of microorganisms from the rhizosphere and serve as sources of nutrients. Additionally, root exudates can neutralize pH and cleanse the soil of impurities (Jeevanantham et al. 2019).

6.7.2 Phytodegradation

Phytodegradation or phytotransformation is another plant-based remediation technology in which metabolic processes occurring in plants' body break the contaminants. The metabolic activities, such as photosynthesis, oxidation, and plant enzymatic activity, are responsible for phytodegradation of the contaminants (Jeevanantham et al. 2019). In this context, the role of enzymes is crucial. Both oxidases and hydrolases are involved in the hydrolysis of contaminants from large and complex forms to small and simple forms (Shmaefsky 2020). Enzymes that are effective in the degradation of contaminants are: nitrilase, peroxidase, nitroreductase, laccase, and dehalogenase (Srivastav et al. 2018). This technique has been successfully employed for the treatment of insecticides, herbicides, and personal care chemicals in the soil (Shmaefsky 2020). However, the technique is restricted to specific and limited contaminants depending on their hydrophobic nature.

6.7.3 Phytoextraction

The process of using hyper-accumulating plants to translocate pollutants from the environment by roots and transfer them to above-ground biomass is known as phytoextraction, sometimes known as phytoaccumulation, phytosequestration, phytoadsorption, or phytomining (Abdel-Shafy and Mansour 2018). Certain carrier proteins, transporters, and enzymes may also be able to metabolically break down the pollutants (Shmaefsky 2020). Plants that may absorb metals or pollutants at a pace 100 times faster than non-hyper-accumulating plants are referred to as hyper-accumulating plants (Abdel-Shafy and Mansour 2018). The chosen plants can support huge amounts of biomass, absorb a high concentration of contaminants, have broad root systems, have high tolerance for the toxin's concentration and toxicity, and show rapid growth (Srivastav et al. 2018).

6.7.4 Phytovolatilization

In phytovolatilization, plants take up the contaminants from the soil, convert them into volatile forms, and release them into the atmosphere through stomata during transpiration process. During this process, the contaminants are detoxified and their water solubility is increased (Jeevanantham et al. 2019; Shmaefsky 2020). In addition to stomatal release, the contaminants can also diffuse through the plant stem. However, the volatility of the compounds is a key factor regulating the efficiency of this technique. Greater is the volatility of the contaminants, greater will be their release through the hyper-accumulating plants without adversely affecting the plant itself during accumulation (Jeevanantham et al. 2019). Phytovolatilization has been utilized effectively for the removal of organic pollutants from the soil (Shmaefsky 2020).

6.7.5 Phytostimulation

Rhizodegradation, another name for phytostimulation, uses the mutualism between microorganisms and plants to break down pollutants in the rhizosphere through metabolic processes (Mishra et al. 2020). Plant roots increase microbial activity by promoting aerobic transformation, removing soil to expose more surface area, moving water, raising the bioavailability of organic carbon, and providing nutrients and energy to microorganisms in the form of exudates. Microbes also encourage the biological activity of plant roots in the interim. Treatment of organic pollutants such as pesticides, aromatic compounds, and poly-aromatic hydrocarbons can be achieved with this technique (Abdel-Shafy and Mansour 2018).

6.7.6 Rhizofiltration

The process that draws impurities from contaminated water and absorbs, concentrates, and precipitates them through the root system is called rhizofiltration. Rhizofiltration is also known as phytofiltration. Although it works in water instead of soil, it is comparable with phytoextraction. Filtration by the root system results in the elimination of hazardous contaminants from the water. Similarly, the selected plants should possess high levels of root biomass, accumulation capacity, adsorption surface, and pollution resistance (Kristani et al. 2023; Abdel-Shafy and Mansour 2018).

6.8 POTENTIAL OF PHYTOREMEDIATION OF EOCs FOR FOOD SAFETY

Phytoremediation is an effective technology for soil purification and groundwater due to heavy metal and organic contamination (Shi et al. 2023). There are several advantages of phytoremediation including: (i) there are several organic and inorganic compounds that are amenable to phytoremediation process (Moosavi and Seghatoleslami 2013); (ii) phytoremediation is applicable process in both in situ and ex situ conditions; (iii) in situ application is more effective as it reduces the chances of soil disturbance and controls the spread of contamination through air and waterborne waste; (iv) a green technology, if implemented effectively, is both environmental friendly and aesthetically pleasing (Abdel-Shafy and Mansour 2018); (v) this technology is suitable for large areas where other conventional technologies become ineffective or are very expensive; (vi) it does not require any expensive equipment and is easily applicable; (vii) the biggest advantage of phytoremediation is the cost-effectiveness as compared with other conventional clean up technologies (Moosavi and Seghatoleslami 2013); (viii) the phytoremediation process does not require disposal sites and also does not require excavation and transport of the polluted soil or wastewater, therefore reducing the risk of spread of contamination; (ix) it has the potential to treat the affected site contaminated with more than one pollutant; (x) lastly, the extension of contamination is controlled by preventing leaching and soil erosion caused by wind and water activity (Abdel-Shafy and Mansour 2018).

According to Pan et al. (2014), several antibiotics such as tetracycline, sulfamethazine, norfloxacin, erythromycin, and chloramphenicol were found in the edible parts of different crops. The leafy vegetables and root vegetables, such as carrots, show higher concentrations of EOCs in their edible parts compared with fruits and vegetables including cucumber, bell pepper, and tomato (Christou et al. 2017). The definition of acceptable daily intake is: "an estimate of the amount of a residue, expressed on a bodyweight basis that can be ingested daily over a lifetime without appreciable health risks" (Bueno et al. 2022). In all the above cases, the level of EOCs was 100–1000 times lower than the levels established by the European Commission. Also, the pesticides detected in the fruit reflected no risks to human life.

6.9 CONCLUSION

The EOCs have become an integral part of our daily lives due to modernization. The contamination of soil with EOCs has raised concerns due to the substantial ecological adverse impacts on the environment. The EOCs are taken up by the plants from soil and transformed to become parts of the plant. The contaminants can be taken up by plants either through the root or foliar systems. The common pathways that contaminants follow to reach the xylem of the plant are the active transport or passive transport mechanisms. The enzymes present in the plants transform the contamination during different phases. The uptake of contaminants in plants depends on numerous factors, including molecular weight, hydrophobicity, polarity, climatic conditions, soil environment, and plant physiology. The EOCs, once taken up by the plant, have the potential to become part of the edible part of the plant and enter the food chain. These toxic contaminants not only become toxic for plant growth but also for soil organisms. Phytoremediation is an environmentally friendly, plant-based method that can remove EOCs through biotic processes that pose no danger. Due to their widespread distribution across the globe and ability to absorb contaminants from environment, plants serve as excellent pollution tracers. Phytoremediation might thus be established for removing EOCs from the soil.

REFERENCES

Abdel-Shafy, H.I., and M.S.M. Mansour, 2018. Phytoremediation for the elimination of metals, pesticides, PAHs, and other pollutants from wastewater and soil. In *Phytobiont and Ecosystem Restitution*, eds. V. Kumar, M. Kumar, and R. Prasad, pp. 101–136. Springer Nature, Singapore.

Adeel, M., X. Song, Y. Wang, D. Francis, and Y. Yang, 2017. Environmental impact of estrogens on human, animal and plant life: a critical review. *Environment International*, 99:107–119.

Ali, N., S. Khan, M.A. Khan, M. Waqas, and H. Yao, 2019. Endocrine disrupting pesticides in soil and their health risk through ingestion of vegetables grown in Pakistan. *Environmental Science and Pollution Research*, 26:8808–8820.

Bahuguna, A., S.K. Singh, S. Sharma, et al. 2022. Soil pollution due to sewage sludge and industrial effluents. In *Soil Health and Environmental Sustainability: Application of Geospatial Technology*, eds. P.K. Shit, P.P. Adhikari, G.S. Bhunia, and D. Sengupta, pp. 345–367. Springer International Publishing, Cham.

Bártíková, H., R. Podlipná, and L. Skálová, 2016. Veterinary drugs in the environment and their toxicity to plants. *Chemosphere*, 144:2290–2301.

Bartrons, M., and J. Peñuelas, 2017. Pharmaceuticals and personal-care products in plants. *Trends in Plant Science*, 22:194–203.

Bender, S.F., and M.G.A. van der Heijden, 2015. Soil biota enhance agricultural sustainability by improving crop yield, nutrient uptake and reducing nitrogen leaching losses. *Journal of Applied Ecology*, 52:228–239.

Billings, A., K.C. Jones, M.G. Pereira, and D.J. Spurgeon, 2021. Plasticisers in the terrestrial environment: sources, occurrence and fate. *Environmental Chemistry*, 18:111–130.

Bueno, M.J.M., M.G. Valverde, M.M. Gómez-Ramos, et al. 2022. Fate, modeling, and human health risk of organic contaminants present in tomato plants irrigated with reclaimed water under real-world field conditions. *Science of the Total Environment*, 806:150909.

Chandanshive, V.V., N.R. Rane, A.S. Tamboli, et al. 2017. Co-plantation of aquatic macrophytes *Typha angustifolia* and *Paspalum scrobiculatum* for effective treatment of textile industry effluent. *Journal of Hazardous Materials*, 338:47–56.

Christou, A., A. Agüera, J.M. Bayona, et al. 2017. The potential implications of reclaimed wastewater reuse for irrigation on the agricultural environment: the knowns and unknowns of the fate of antibiotics and antibiotic resistant bacteria and resistance genes: a review. *Water Research*, 123:448–467.

Christou, A., G. Papadavid, P. Dalias, et al. 2019. Ranking of crop plants according to their potential to uptake and accumulate contaminants of emerging concern. *Environmental Research*, 170:422–432.

Collins, C., I. Martin, and W.J. Doucette, 2011. Organic xenobiotics and plants. In *Plant Uptake of Xenobiotics*, eds. P.C. Schröder, and D. Christopher, p. 311. Springer Publishing Company: New York.

Dettenmaier, E.M., W.J. Doucette, and B. Bugbee, 2009. Chemical hydrophobicity and uptake by plant roots. *Environmental Science & Technology*, 43:324–329.

Didier, T., L.G. Philippe, H. Sonia, B. Amar, M.C. Claudia, and F. Jairo, 2012. Prospects of *Miscanthus x giganteus* for PAH phytoremediation: a microcosm study. *Industrial Crops and Products*, 36:276–281.

Dodgen, L.K., A. Ueda, X. Wu, D.R. Parker, and J. Gan, 2015. Effect of transpiration on plant accumulation and translocation of PPCP/EDCs. *Environmental Pollution*, 198:144–153.

Drakvik, E., R. Altenburger, Y. Aoki, et al. 2020. Statement on advancing the assessment of chemical mixtures and their risks for human health and the environment. *Environment International*, 134:105267.

Eales, J., A. Bethel, T. Galloway, P. Hopkinson, K. Morrissey, R.E. Short, and R. Garside, 2022. Human health impacts of exposure to phthalate plasticizers: an overview of reviews. *Environment International*, 158:106903.

Eggen, T., T.N. Asp, K. Grave, and V. Hormazabal, 2011. Uptake and translocation of metformin, ciprofloxacin and narasin in forage-and crop plants. *Chemosphere*, 85:26–33.

Feng, N.X., J. Yu, H.M. Zhao, et al. 2017. Efficient phytoremediation of organic contaminants in soils using plant–endophyte partnerships. *Science of the Total Environment*, 583:352–368.

Gikas, G.D., M. Pérez-Villanueva, M. Tsioras, et al. 2018. Low-cost approaches for the removal of terbuthylazine from agricultural wastewater: constructed wetlands and biopurification system. *Chemical Engineering Journal*, 335:647–656.

Goldstein, M., M. Shenker, and B. Chefetz, 2014. Insights into the uptake processes of wastewater-borne pharmaceuticals by vegetables. *Environmental Science & Technology*, 48:5593–5600.

Grossberger, A., Y. Hadar, T. Borch, and B. Chefetz, 2014. Biodegradability of pharmaceutical compounds in agricultural soils irrigated with treated wastewater. *Environmental Pollution*, 185:168–177.

Hashimi, M.H., R. Hashimi, and Q. Ryan, 2020. Toxic effects of pesticides on humans, plants, animals, pollinators and beneficial organisms. *Asian Plant Research Journal*, 5:37–47.

Hepworth, C., T. Doheny-Adams, L. Hunt, D.D. Cameron, and J.E. Gray, 2015. Manipulating stomatal density enhances drought tolerance without deleterious effect on nutrient uptake. *The New Phytologist*, 208:336.

Herklotz, P.A., P. Gurung, B.V. Heuvel, and C.A. Kinney, 2010. Uptake of human pharmaceuticals by plants grown under hydroponic conditions. *Chemosphere*, 78:1416–1421.

Jeevanantham, S., A. Saravanan, R.V. Hemavathy, P.S. Kumar, P.R. Yaashikaa, and D. Yuvaraj, 2019. Removal of toxic pollutants from water environment by phytoremediation: a survey on application and future prospects. *Environmental Technology & Innovation*, 13:264–276.

Khandare, R.V., and S.P. Govindwar, 2015. Phytoremediation of textile dyes and effluents: current scenario and future prospects. *Biotechnology Advances*, 33:1697–1714.

Kim, S., C.M. Park, M. Jang, et al. 2018. Aqueous removal of inorganic and organic contaminants by graphene-based nanoadsorbents: a review. *Chemosphere*, 212:1104–1124.

Kristanti, R.A., R. Tirtalistyani, Y.Y. Tang, et al. 2023. Phytoremediation mechanism for emerging pollutants: a review. *Tropical Aquatic and Soil Pollution*, 3:88–108.

Kurade, M.B., J.R. Kim, S.P. Govindwar, and B.H. Jeon, 2016. Insights into microalgae mediated biodegradation of diazinon by Chlorella vulgaris: microalgal tolerance to xenobiotic pollutants and metabolism. *Algal Research*, 20:126–134.

Kurade, M.B., Y.H. Ha, J.Q. Xiong, S.P. Govindwar, M. Jang, and B.H. Jeon, 2021. Phytoremediation as a green biotechnology tool for emerging environmental pollution: a step forward towards sustainable rehabilitation of the environment. *Chemical Engineering Journal*, 415:129040.

Kurwadkar, S., G. Struckhoff, K. Pugh, and O. Singh, 2017. Uptake and translocation of sulfamethazine by alfalfa grown under hydroponic conditions. *Journal of Environmental Sciences*, 53:217–223.

Kvesitadze, G., G. Khatisashvili, T. Sadunishvili, and E. Kvesitadze, 2015. Plants for remediation: uptake, translocation and transformation of organic pollutants. In *Plants, Pollutants and Remediation*, eds. M. Öztürk, M. Ashraf, A. Aksoy, M. Ahmad, and K. Hakeem, pp. 241–308. Springer, Dordrecht.

Liu, Y., Y. Le, M. Xu, W. Wang, H. Chen, Q. Zhang, and C. Wang, 2022. Remodeling on adipocytic physiology of organophosphorus esters in mature adipocytes. *Environmental Pollution*, 305:119287.

Madikizela, L.M., S. Ncube, and L. Chimuka, 2018. Uptake of pharmaceuticals by plants grown under hydroponic conditions and natural occurring plant species: a review. *Science of the Total Environment*, 636:477–486.

Malchi, T., Y. Maor, G. Tadmor, M. Shenker, and B. Chefetz, 2014. Irrigation of root vegetables with treated wastewater: evaluating uptake of pharmaceuticals and the associated human health risks. *Environmental Science & Technology*, 48:9325–9333.

Marcos, A., D. Trigo, A.B. Muñiz-González, N. Tilikj, J.L. Martínez-Guitarte, and M. Novo, 2021. Effects of bisphenol S on the life cycle of earthworms and its assessment in the context of climate change. *Science of the Total Environment*, 781:146689.

Masinga, P., T.T. Simbanegavi, Z. Makuvara, J. Marumure, N. Chaukura, and W. Gwenzi, 2024. Emerging organic contaminants in the soil–plant-receptor continuum: transport, fate, health risks, and removal mechanisms. *Environmental Monitoring and Assessment*, 196:367.

Miller, E.L., S.L. Nason, K.G. Karthikeyan, and J.A. Pedersen, 2016. Root uptake of pharmaceuticals and personal care product ingredients. *Environmental Science & Technology*, 50:525–541.

Mishra, A., S.P. Mishra, A. Arshi, A. Agarwal, and S.K. Dwivedi, 2020. Plant-microbe interactions for bioremediation and phytoremediation of environmental pollutants and agro-ecosystem development. In *Bioremediation of Industrial Waste for Environmental Safety*, eds. R. Bharagava, and G. Saxena, pp. 415–436. Springer, Singapore.

Moosavi, S.G., and M.J. Seghatoleslami, 2013. Phytoremediation: a review. *Advance in Agriculture and Biology*, 1:5–11.

Naresh, R.K., M.S. Chandra, N.C. Mahajan, P.K. Singh, A. Baliyan, P. Ahlawat, and P. Shalini, 2021. Neglected and underutilized crop species the key to improving soil nutritional security for fighting poverty, hunger and malnutrition in north-western IGP: a review. *The Pharma Innovation Journal*, 10:1833–1839.

Navarro, I., A. De la Torre, P. Sanz, et al. 2017. Uptake of perfluoroalkyl substances and halogenated flame retardants by crop plants grown in biosolids-amended soils. *Environmental Research*, 152:199–206.

Novo, M., I. Verdú, D. Trigo, and J.L. Martínez-Guitarte, 2018. Endocrine disruptors in soil: effects of bisphenol A on gene expression of the earthworm *Eisenia fetida*. *Ecotoxicology and Environmental Safety*, 150:159–167.

Oliveira, M., D.N. Cardoso, A.M.V.M. Soares, and S. Loureiro, 2018. Toxic effects of human pharmaceuticals to *Folsomia candida*: a multigeneration approach. *Science of the Total Environment*, 625:1225–1233.

Pan, M., C.K.C. Wong, and L.M. Chu, 2014. Distribution of antibiotics in wastewater-irrigated soils and their accumulation in vegetable crops in the Pearl River Delta, Southern China. *Journal of Agricultural and Food Chemistry*, 62:11062–11069.

Passatore, L., S. Rossetti, A.A. Juwarkar, and A. Massacci, 2014. Phytoremediation and bioremediation of polychlorinated biphenyls (PCBs): state of knowledge and research perspectives. *Journal of Hazardous Materials*, 278:189–202.

Pawar, G., M.A.E. Abdallah, E.V. de Sáa, and S. Harrad, 2017. Dermal bioaccessibility of flame retardants from indoor dust and the influence of topically applied cosmetics. *Journal of Exposure Science & Environmental Epidemiology*, 27:100–105.

Pi, N., J.Z. Ng, and B.C. Kelly, 2017. Bioaccumulation of pharmaceutically active compounds and endocrine disrupting chemicals in aquatic macrophytes: results of hydroponic experiments with *Echinodorus horemanii* and *Eichhornia crassipes*. *Science of the Total Environment*, 601:812–820.

Priya, A.K., L. Gnanasekaran, S. Rajendran, J. Qin, and Y. Vasseghian, 2022. Occurrences and removal of pharmaceutical and personal care products from aquatic systems using advanced treatment: a review. *Environmental Research*, 204:112298.

Pullagurala, V.L.R., S. Rawat, I.O. Adisa, J.A. Hernandez-Viezcas, J.R. Peralta-Videa, and J.L. Gardea-Torresdey, 2018. Plant uptake and translocation of contaminants of emerging concern in soil. *Science of the Total Environment*, 636:1585–1596.

Qadeer, A., K.L. Kirsten, Z. Ajmal, X. Jiang, and X. Zhao, 2022. Alternative plasticizers as emerging global environmental and health threat: another regrettable substitution? *Environmental Science & Technology*, 56:1482–1488.

Rane, N.R., S.M. Patil, V.V. Chandanshive, et al. 2016. *Ipomoea hederifolia* rooted soil bed and *Ipomoea aquatica* rhizofiltration coupled phytoreactors for efficient treatment of textile wastewater. *Water Research*, 96:1–11.

Reddy, P.V.L., and K.H. Kim, 2015. A review of photochemical approaches for the treatment of a wide range of pesticides. *Journal of Hazardous Materials*, 285:325–335.

Rede, D., L.H.M.L.M. Santos, S. Ramos, F. Oliva-Teles, C. Antão, S.R. Sousa, and C. Delerue-Matos, 2019. Individual and mixture toxicity evaluation of three pharmaceuticals to the germination and growth of *Lactuca sativa* seeds. *Science of the Total Environment*, 673:102–109.

Rogowska, J., M. Cieszynska-Semenowicz, W. Ratajczyk, and L. Wolska, 2020. Micropollutants in treated wastewater. *Ambio*, 49:487–503.

Rosenmai, A.K., S.B. Winge, M. Möller, et al. 2021. Organophosphate ester flame retardants have antiandrogenic potential and affect other endocrine related endpoints in vitro and in silico. *Chemosphere*, 263:127703.

Saquib, Q., A.M. Al-Salem, M.A. Siddiqui, S.M. Ansari, X. Zhang, and A.A. Al-Khedhairy, 2022. Organophosphorus flame retardant TDCPP displays genotoxic and carcinogenic risks in human liver cells. *Cells*, 11:195.

Saxena, G., S.P. Goutam, A. Mishra, S.I. Mulla, and R.N. Bharagava, 2020. Emerging and eco-friendly technologies for the removal of organic and inorganic pollutants from industrial wastewaters. In *Bioremediation of Industrial Waste for Environmental Safety*, ed. B.R. Saxena, pp. 113–126. Springer, Singapore.

Shahid, S., H. Imtiaz, J. Rashid, M. Xu, M. Vithanage, and M. Ahmad, 2024. Uptake, translocation, and nutrient efficiency of nano-bonechar as a plant growth regulator in hydroponics and soil systems. *Environmental Research*, 251:118695.

Shi, Q., P. Kaur, and J. Gan, 2023. Harnessing the potential of phytoremediation for mitigating the risk of emerging contaminants. *Current Opinion in Environmental Science & Health*, 32:100448.

Shmaefsky, B.R., 2020. Principles of phytoremediation. In *Phytoremediation. Concepts and Strategies in Plant Sciences*, ed. B.R. Shmaefsky, pp. 1–26. Springer, Cham.

Srivastav, A., K.K. Yadav, S. Yadav, N. Gupta, J.K. Singh, R. Katiyar, and V. Kumar, 2018. Nano-phytoremediation of pollutants from contaminated soil environment: current scenario and future prospects. In *Phytoremediation*, eds. A. Ansari, S. Gill, R. Gill, G. Lanza, and L. Newman, pp. 383–401. Springer, Cham.

Sun, C., S. Dudley, J. Trumble, and J. Gan, 2018. Pharmaceutical and personal care products-induced stress symptoms and detoxification mechanisms in cucumber plants. *Environmental Pollution*, 234:39–47.

Tanoue, R., Y. Sato, M. Motoyama, S. Nakagawa, R. Shinohara, and K. Nomiyama, 2012. Plant uptake of pharmaceutical chemicals detected in recycled organic manure and reclaimed wastewater. *Journal of Agricultural and Food Chemistry*, 60:10203–10211.

Trapp, S., and C.N. Legind, 2011. Uptake of organic contaminants from soil into vegetables and fruits. In *Dealing with Contaminated Sites*, ed. F. Swartjes, pp. 369–408. Springer, Dordrecht.

Wang, F.X., C.C. Wang, X. Du, Y. Li, F. Wang, and P. Wang, 2022. Efficient removal of emerging organic contaminants via photo-Fenton process over micron-sized Fe-MOF sheet. *Chemical Engineering Journal*, 429:132495.

Wang, L., B. Ji, Y. Hu, R. Liu, and W. Sun, 2017. A review on in situ phytoremediation of mine tailings. *Chemosphere*, 184:594–600.

Wang, M., D.Q. Zhang, J.W. Dong, and S.K. Tan, 2017. Constructed wetlands for wastewater treatment in cold climate: a review. *Journal of Environmental Sciences*, 57:293–311.

Wang, X., B.F. Hales, and B. Robaire, 2021. Effects of flame retardants on ovarian function. *Reproductive Toxicology*, 102:10–23.

Weyens, N., D. van der Lelie, S. Taghavi, and J. Vangronsveld, 2009. Phytoremediation: plant–endophyte partnerships take the challenge. *Current Opinion in Biotechnology*, 20:248–254.

Wołejko, E., A. Jabłońska-Trypuć, A. Butarewicz, and U. Wydro, 2021. Heavy metals, hydrocarbons, radioactive materials, xenobiotics, pesticides, hazardous chemicals, explosives, pharmaceutical waste and dyes bioremediation. In *Rhizomicrobiome Dynamics in Bioremediation*, ed. V. Kumar, pp. 269–298. CRC Press, Boca Raton, FL.

Xiong, J.Q., M.B. Kurade, R. AI Abou-Shanab, et al. 2016. Biodegradation of carbamazepine using freshwater microalgae *Chlamydomonas mexicana* and *Scenedesmus obliquus* and the determination of its metabolic fate. *Bioresource Technology*, 205:183–190.

Yan, Y., X. Xu, C. Shi, W. Yan, L. Zhang, and G. Wang, 2019. Ecotoxicological effects and accumulation of ciprofloxacin in *Eichhornia crassipes* under hydroponic conditions. *Environmental Science and Pollution Research*, 26:30348–30355.

Zhang, C., F.E.N.G. Yao, Y.W. Liu, H.Q. Chang, Z.J. Li, and J.M. Xue, 2017. Uptake and translocation of organic pollutants in plants: a review. *Journal of Integrative Agriculture*, 16:1659–1668.

Zhang, D., R.M. Gersberg, W.J. Ng, and S.K. Tan, 2014. Removal of pharmaceuticals and personal care products in aquatic plant-based systems: a review. *Environmental Pollution*, 184:620–639.

Zhang, Y., J.B. Sallach, L. Hodges, et al. 2016. Effects of soil texture and drought stress on the uptake of antibiotics and the internalization of Salmonella in lettuce following wastewater irrigation. *Environmental Pollution*, 208:523–531.

Zhao, F., L. Yang, G. Li, L. Fang, X. Yu, Y.T. Tang, M. Li, and L. Chen, 2022. Veterinary antibiotics can reduce crop yields by modifying soil bacterial community and earthworm population in agro-ecosystems. *Science of the Total Environment*, 808:152056.

Zhu, L., H. Xu, W. Xiao, et al. 2020. Ecotoxicological effects of sulfonamide on and its removal by the submerged plant *Vallisneria natans* (Lour.) Hara. *Water Research*, 170:115354.

Zortéa, T., J.C. Segat, A.P. Maccari, J.P. Sousa, A.S. Da Silva, and D. Baretta, 2017. Toxicity of four veterinary pharmaceuticals on the survival and reproduction of *Folsomia candida* in tropical soils. *Chemosphere*, 173:460–465.

7 Emerging Organic Contaminants in Soil Biota

Ruwandi Nilakshi, Nalin Madusanka Gunathilaka, Sahan Nishada Perera, Meththika Vithanage, and Anushka Upamali Rajapaksha

7.1 EMERGING ORGANIC CONTAMINANTS

In recent years, emerging organic contaminants (EOCs) such as pesticides (Mali et al. 2023), pharmaceuticals, personal care products (Gomes et al. 2017; Osuoha et al. 2023), antibiotics (García et al. 2020), perfluoroalkyl sulfonate (Arrieta-Cortes et al. 2017), flame retardants (Navarro et al. 2016), and endocrine disruptors (Chaturvedi et al. 2023) have posed a high risk to the natural ecosystem. The cumulative increase occurs due to the use and production of these substances in farming activities, industrial manufacturing, domestic activities, and healthcare services. This surge has led to contaminating soil and other ecosystems with alarming levels of pollutants, posing threats to living organisms (Gomes et al. 2017; Houida et al. 2024). Pesticides are crucial agrochemicals that are used in agricultural systems to protect crops against pests, which are frequently sprayed multiple times, and a portion of their droplets contaminates the soil (Arora and Sahni 2016). Perfluoroalkane sulfonates are utilized in a range of sectors, including firefighting substances, inks, lubricants, and water proofing for leather, paper, and fabric production (Fabregat-Palau et al. 2021). They are primarily stored in soil and accessed through a variety of means. These include irrigation with water contaminated with perfluoroalkane sulfonates, applying treated sewage sludge on land, leachates from landfills, increased usage of aqueous firefighting foam to quench fire from petroleum oils, and industrial atmospheric emissions (Li and Koosaletse-Mswela 2023). Flame retardants get into the soil through atmospheric dispersion, accumulation of wet and dry methods, runoffs, irrigation of contaminated water, and municipal sludge dumping. The highest amounts of flame retardants are found in waste treatment regions and present variable toxicity against animals, plants, and microbes (Tian et al. 2023). Endocrine disrupters including di (2-ethylhexyl) phthalate, methyl paraben triclocarban, 4-nonylphenol, 4-tert octylphenol, butyl paraben, bisphenol-A, propyl paraben, triclosan, benzyl butyl phthalate, and di-*n*-butyl phthalate are extraneous pollutants that disrupt the activities of endocrine system (Saha et al. 2022). Among them, bisphenol A (BPA) recognized as a leading endocrine disruptor and it reveals high estrogenic activity and acute

 DOI: 10.1201/9781003509523-7

toxicity (Eltoukhy et al. 2020). Abundance of BPA has grown in recent years, since mass manufacturing and utilization of BPA based plastic, epoxy resin food containers, and polyvinylchloride films is increased (Novo et al. 2018).

The soil environment is exposed to pharmaceuticals through wastewater discharge by homes, hospitals, and other healthcare facilities. Natural fertilizers are also a significant source of pharmacological release. Physical, chemical, and biological features of the soil and properties of pharmaceuticals control these contaminants in the soil (Gworek et al. 2021). Numerous target and nontarget living organisms are at risk due to the persistence of these emerging pollutants in the environment. Figure 7.1 depicts the origin sources of several emerging organic pollutants and their route into the soil ecosystem.

There are few data on the toxicity and the ability of bioaccumulation in biota (Gomes et al. 2017). However, the investigation of the consequences of the surge of organic pollutants in soil biota is crucial. Small invertebrates, including earthworms, are at the first place of food chains; therefore, when the pollutant chemicals are consumed by earthworms, they transferred those chemicals to the upper level and into the food webs through bioaccumulation and biomagnification (Carter et al. 2016). Hence, this chapter specifically focuses on: (i) the absorption of EOCs by small animals, such as earthworms and soil insects, (ii) the influences of EOCs on the soil microbiological diversity, and (iii) the eco-risks of soil emerging organic pollutants.

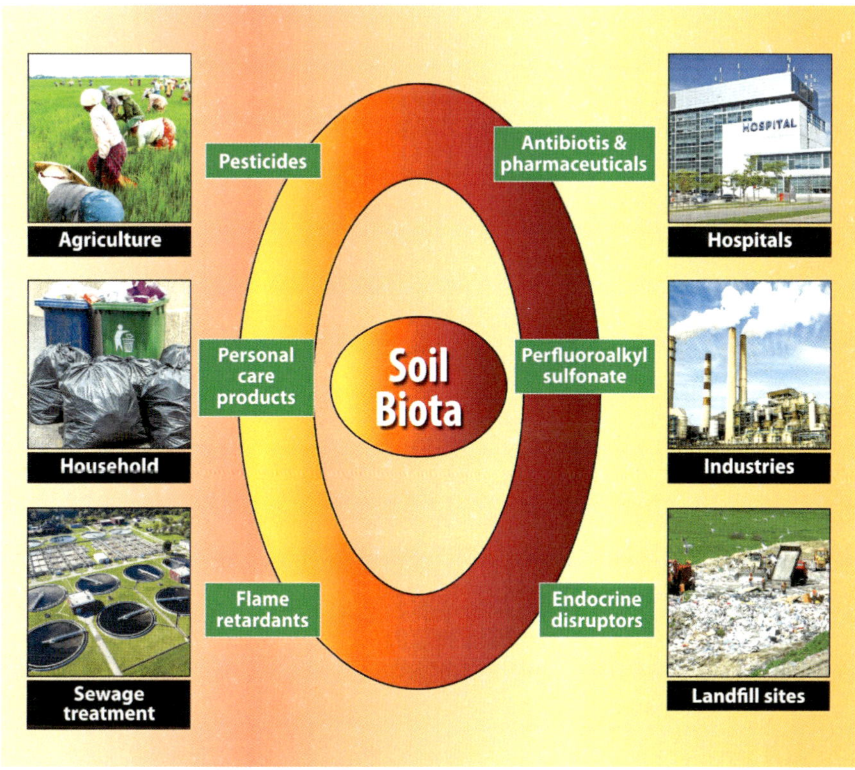

FIGURE 7.1 Sources of different emerging organic pollutants in soil (Gomes et al. 2017).

7.1.1 Soil Biota Function in a Sustainable Soil Ecosystem

Soil organisms can be classified as decomposing organisms and nutrients transforming organisms (e.g., soil microorganisms), biocontrollers (e.g., nematodes, mites), and ecosystem engineers (e.g., earthworms and termites) based on their functions, and they are interconnected to form and regulate ecosystem functions. Nematodes perform a variety of ecological roles, including the breakdown of organic materials, predation and pest management, soil bioturbation, biogeochemical cycles such as the nitrogen and carbon cycles, and interactions with plants and other animals such as bacteria, fungi, and earthworms (van den Hoogen et al. 2019; Wang et al. 2024). Further, microarthropods contribute to soil structure and humus production, and through their interactions with microbial and fungal populations, they regulate the rate of decomposition and nutrient cycling (Marchi 2015). Earthworms can change the properties of the soil to affect its habitat. They might greatly enhance the soil physical attributes, such as bulk density, infiltration, hydraulic conductivity, porousness of soil structure, and aggregate stability (Kim 2016; Ojha and Devkota 2014). Microbiological populations strengthen the earthworm cast. Nutrient mineralization and release are facilitated by increased microbial activity in the soil. Phosphorus and other immobile macronutrients that plants absorb through root interception become more easily accessible when earthworm activity increases root dispersion. Microbial activity and other biological components are encouraged in the soil by earthworm activity (Ojha and Devkota 2014). Soil biodiversity influences biological soil water consumption, but numerous life forms also aid in chemical and physical interactions between water and the soil structure. Burrowing animals, such as earthworms and ground beetles, alter the water infiltration rate in the soil by forming mega and macropores. Boosted termite feeding in regenerated forests boosted water penetration by several folds compared with agricultural areas and bare land. This can help prevent floods after heavy rains, minimize soil erosion, and enhance hydration in subsurface soil. To achieve a sustainable soil environment, consideration of features such as ecosystem processes and functions of soil biota is essential.

As compared to the invertebrates, the soil microorganisms are responsible for main soil ecosystem activities and functions including biogeochemical cycling, eliminating toxic pollutants, maintenance of soil texture, repairs soil structure, and breakdown of organic matter. When considering plant detritus and organic matter decomposition, earthworms, ants, fungi, and bacteria are particularly involved. Physical fragmentation, chemical degradation, and organic substrate leaching are all critical stages of the decomposition process (Bach et al. 2020).

7.2 ABSORPTION OF EMERGING ORGANIC CONTAMINANTS TO SOIL BIOTA

EOCs represent a newly discovered source comprising novel exposure pathways inside soil (Ailijiang et al. 2022). The specific proportion and the exact form of the emerging pollutants in the soil are not yet completely discovered (Gomes et al. 2017). Also, much less work has been done regarding the absorption of EOCs into soil organisms (Carter et al. 2014). Earthworms have usually been used as a testing organism for soil toxicity studies. *Eisenia fetida*, a soil organism, was tested for the prevalence of toxicity of 18 pharmaceuticals and the results showed acute toxicity toward three blood lipid-reducing drugs and four anti-inflammatory nonsteroidal drugs (Pinto et al. 2024). Concentrations of the tested drugs are within the permissible limits for fatality indicating hindered toxic effects (Pino et al. 2015). In earthworms, the pharmaceutical compound trimethoprim was found at a higher level (127 μg/kg dry weight) in biosolids-amended soils, compared to a lower level (61 μg/kg dry weight) when manure-amended soils were utilized (Carter et al. 2014; Kinney et al. 2008). The absorption and fate of diclofenac, ^{14}C-labeled carbamazepine, orlistat, and fluoxetine in soil earthworms have been investigated (Carter et al. 2014). It is shown that the sorption coefficients were raised according to carbamazepine < diclofenac < fluoxetine < orlistat, respectively. Pesticides influence both macro and microorganisms in the soil. Earthworms are also highly vulnerable to pesticides, which may lead to their immobility, biomass decrease, and disturbance of growth and reproduction by disrupting numerous physiological activities (Sandeep et al. 2024). The pathway and uptake portion of one ionizable substance in personal care product, such as triclosan, and two ionizable pharmaceuticals including fluoxetine and diclofenac, for *Lumbriculus variegatus* (sediment dwelling worm) was investigated (Karlsson et al. 2015). Studies were conducted using ^{14}C-isotope ingredients while using worms without the head ("nonfeeding") and complete worms ("feeding"). However, based on uptake of ^{14}C, biota sediment accumulation factors (BSAF) for feeding worms increased according to the concentrations fluoxetine (0.3) < diclofenac (0.5) < triclosan (9), respectively. Several investigations have determined the BSAF values of several substances to organisms (Table 7.1). BSAF stands for the ratio of lipid-normalized concentration of a hydrophobic organic chemical (HOC) in an organism and organic carbon-normalized concentration of the HOC in the sediment to which the organism has been exposed (Burkhard and Votava 2023). BSAF can be calculated according to the following equation:

$$\mathrm{BSAF} = \frac{C_O / f_l}{C_s / f_{soc}}$$

C_O The HOC concentration in the organism (wet or dry weight)
f_l The lipid concentration (fraction) in the organism (wet or dry weight)
C_s The HOC concentration in the sediment (generally dry weight)
f_{soc} The total organic carbon content (fraction) of the sediment (generally dry weight)

A study shows that the significant differences in the uptake of diclofenac and fluoxetine in feeding and nonfeeding worms were not observed, however the uptake of the more hydrophobic antimicrobial triclosan was significantly greater in feeding worms than in nonfeeding worms. The 48-h BSAF for feeding worms tested as 36% higher than that of nonfeeder. The above findings suggest that the uptake of triclosan is facilitated by food uptake, which could be due to the compound's high hydrophobicity

TABLE 7.1
Biota Sediment Accumulation Factors of Different Contaminants in Soil

Type of Emerging Organic Contaminant	Chemical Name	Organism	Biota Sediment Accumulation Factor (BSAF)	References
Polyfluoroalkyl substances (PFAS)	Perfluorooctanoic acid (PFOA)	Earthworm sp.	9.54×10^{-3}	(Burkhard and Votava 2023, Zhao et al. 2017)
	Perfluoro tridecanoic acid (PFTrDA)	Earthworm sp.	3.04	
	Perfluoro butane sulfonic acid (PFBS)	Earthworm sp.	5.00×10^{-2}	
	Perfluorohexacene sulfonic acid (PFDS)	Earthworm sp.	3.39×10^{-1}	
Flame retardants	Polybrominated diphenyl ethers (PBDEs)	Earthworm sp.	3.33 (freshly spiked soil)	(Liang et al. 2010)
	Polychlorinated biphenyls (PCBs)	Earthworm sp.	4.80×10^{-1}	(Blankenship et al. 2005)

(Karlsson et al. 2015). Another study revealed that fungal mycelium can efficiently transport chemicals in soil due to their long-distance transport and ability to cross air-filled pores. Cytoplasmic streaming enables the active transport of vesicle-associated polyaromatic hydrocarbons, resulting in more efficient distribution than diffusion mechanisms. Tri-*n*-butyl phosphate (TnBP) is a type of endocrine disruptor (a typical alkyl organophosphate ester), which has been heavily used as a plasticizer for decades (Furuno et al. 2012). One of the studies investigated TnBP, induced intestinal toxicity activities on earthworm *E. fetida* (Yang et al. 2020). High TnBP bioconcentrations were found in earthworm intestinal paths. Degradation of the digestive tract, disruption of the digestive enzyme activity, and disturbance in nutrient absotpion were observed in earthworms due to destruction of osmotic equilibrium between intestinal epithelial cells. The levels of endocrine-disrupting chemicals such as dibutylphthalate, dioctylphthalate, BPA, and 17b-estradiol were determined using tissue samples taken from earthworm, *E. fetida*, mostly found in farming soil and sewage filtering beds. The most available substances in the earthworms from both the sites were the phthalates (especially the dioctylphthalate). Phthalates are vastly used as plasticizers (Markman et al. 2007). The insecticide chlorpyrifos and the fungicide tebuconazole on the *Eisenia andrei* were measured in two soil samples for bioaccumulation ability and results showed different organic carbon contents of 1.02% and 1.93%, respectively (Svobodová et al. 2018).

7.2.1 Routes of Access for Soil Creatures

Soil-burrowing organisms are contacted by chemicals through various pathways. In most organisms, the contribution of different pathways varies. The water repulsion of the chemical and environment, such as soil type and climate, all contribute to its ecological impact. Terrestrial species can be exposed to potentially toxic contaminants through uptake pathways. These pathways include: (i) pore water, (ii) ingesting food and soil, and (iii) inhaling air from soil pores. The importance of each uptake route depends on morphology (e.g., epidermis structure), physiology (e.g., water uptake method [drinking vs. uptake through the skin], oxygen uptake method, feeding habits), and behavioral characteristics. Hard-bodied animals include arthropods such as spiders, mites, insects, collembolans, millipedes, centipedes, harvestman, isopods, and some crab species, which have specialized organs for absorbing oxygen and water, while soft-bodied animals, including nematodes, earthworms, enchytraeids, and insect larvae, primarily absorb water and oxygen through their skin. When assessing the uptake pathway of soil biota, their ecology and physiology must be known. The skin structure, water uptake mode, oxygen uptake method, and feeding habits are all crucial factors in this regard. This highlights the significance of organism behavior (Peijnenburg et al. 2012).

7.2.2 Bioaccumulation and Biomagnification of EOCs in Soil Biota

Bioaccumulation refers to "a process by which a chemical is taken up by an organism either directly from contaminated medium or by consumption of food containing the chemical". Biomagnification refers to increasing concentration within an organism in comparison with its food (Janssen and Hogervorst, 1993). Biomagnification relies on the consumer's ability to absorb the substance from the food source and store or remove it (Mortensen et al. 2018). To estimate the risk of toxicants in the environment, bioaccumulation is a commonly employed criterion. Pollutants absorbed by plants and other organisms can be transported to higher trophic levels through ingestion (Chormare and Kumar 2022). Because of biomass loss during respiration and excretion, there is a risk of bioaccumulation at higher trophic levels (Mortensen et al. 2018). EOCs can accumulate, alter, and rise in concentrations inside biological systems. Small sizes promote toxicity. Bioaccumulation involves the accumulation of EOC substances in an organism, followed by biomagnification in the predatory creature. Biomagnification causes the highest risk to the soil ecosystem and human well-being. A growing number of stakeholders are curious in the possible toxicity and other hazards linked to nanoparticles production, including development, production, usage, and disposal. Nanomaterials can have cumulative harmful effects on living organisms, which can be mitigated through storage or excretion (Uddin et al. 2021). Persistent organisms may be more vulnerable to pollutants due to increased exposure duration. Surface uptake occurs when organisms with a high surface-to-volume ratio come into direct contact with contaminated environments, increasing exposure.

Microorganisms are buried in the substrate and have a high surface-to-volume ratio. In nematodes, ectodermal tissue thickness and collagen density can affect permeability and drug intake (Mortensen et al. 2018). Several environmental protection authorities have established standards to measure the bioaccumulative potential of man-made substances, aimed at managing commercial chemical production, usage, and release into the environment. The bioconcentration factor is the calculation of ratio of chemical concentration (mg/kg) in the ecosystem to the surrounding water (mg/L). It can be estimated by exposing the organism to the contaminated water. If no data are available, the octanol–water partition coefficient (K_{OW}) (1–4) is used instead (Armitage and Gobas 2007; Kelly et al. 2007). Biomagnification or trophic magnification factors are important metrics for quantifying bioaccumulation in terrestrial food webs since they consider food exposure and may be applied to air and water breathing animals. Trophic magnification factors can be challenging to identify due to short trophic ranges or a lack of measurable concentrations in the environment, particularly at lower levels of the food chain (Fremlin et al. 2020). Bioaccumulation rates vary depending on the soil matrix. Scientists created different models to identify the behaviors of accumulation rates (Gobas et al. 2016). A study (Carter et al. 2016) investigated the uptake of pharmaceuticals such as carbamazepine, diclofenac, fluoxetine, and orlistat into the earthworm *E. fetida*. The result of the study shows that earthworm interaction with many soil ecosystems may lead to the variability of chemical metabolism in the soil or *E. fetida*. Fluoxetine treatment had the greatest uptake rate in all soils (0.96–2.35 mL/g/day), and carbamazepine treatment had the fastest depuration rate in all five soils (0.16–0.24 per day). The bioaccumulation factors of each compound were found to deviate across soil types, with the highest variability observed for diclofenac (7.02–69.57) and orlistat (30.50–115.88), and lower variability of the bioaccumulation factor was noted for fluoxetine (14.09–20.42) and carbamazepine (1.05–1.61). The overall experiment mortality was less than 4% across all exposure situations (Carter et al. 2016).

7.3 CONSEQUENCES OF EOCs ON SOIL MICROBES AND THEIR FUNCTION

Soil microorganisms are affected differently by various contaminants. Currently, the consequences of polyfluoroalkyl substance (PFAS) on soil microbial communities have gained vast attention. The organisms that live in contaminated soil can take up organic pollutants and can store large amounts of PFAS into their bodies. Perfluorooctane sulfonate (10 mg/kg), which belongs to the polyfluoroalkyls, can interfere with soil enzyme function and cause damage to DNA, RNA, cellular structure, and the immune system of soil bacteria under greenhouse conditions (Ehsan et al. 2024). Also, perfluorohexanesulfonic acid (PFHxS) caused a considerable conveyance in a group of bacteria than in fungi. The phylum Ascomycota can be identified as a pollution indicator because of its greater endurance to the pollution caused by humans and its ability to succeed in polluted environments. Therefore, it shows a huge relative abundance usually in the soil. The relative abundance of Ascomycota increased by more than 20% at a the highest PFHxS level (250 mg/kg) in soil (Samarasinghe et al. 2023). Pharmaceuticals give rise to the stimulation and inhibition of several steps in nitrogen cycle such as ammonification, nitrification, and denitrification in the soil. For instance, sulfamethazine, clarithromycin, erythromycin, amoxicillin, chlortetracycline, ciprofloxacin, enrofloxacin, and norfloxacin had substantial impacts. Hence, even at low concentrations, pharmaceuticals exhibit much greater responsiveness during the nitrogen cycle than the other nitrogen transformations. In contrast, the existing information about the effect of pharmaceuticals on nitrogen fixation and ammonification is insufficient (Pashaei et al. 2022). The usage of pesticides has a deleterious impact on microbial functions. When exposed to pesticides, the generation of CO_2 was drastically reduced, indicating that the insecticides are poisonous to soil bacteria (Pashaei et al. 2022). An endocrine disruptor in soil is nonylphenol, which adversely affects microbial nitrogen and carbon mineralization processes; it also reduces breakdown rates and bioavailability while causing detrimental effects on soil microbes. However, it increases respiration activity (Stenrød et al. 2013). Picoxystrobin is a fungicide that impedes cellular respiration inside mitochondria by suppressing electron transport. This fungicide has high acute toxicity on earthworms, with field experiments exhibiting immediate toxic effects even at permitted concentrations (Stenrød et al. 2013). However, contamination can only be recovered if the toxicity of the molecules does not impede microbial activity (Barra Caracciolo et al. 2015). The microbes in agricultural soils can respond to pharmaceuticals by stress, dormancy, stimulation, and inhibition. Sulfamethoxazole can cause stimulatory effects on fungi and bacteria. Inhibitory effects can be caused by pharmaceuticals such as irbesartan, citalopram, and cambisol. Stress effects can be caused by the pharmaceuticals in arenosol and cambisol haplic. Thus, dormancy effects in chernozem Celtic soils may result from prolonged pharmaceutical exposure, potentially leading to reduced microbial activity and delayed ecosystem recovery. The soil type, pharmaceutical compound, and time are the factors on which the microbial responses are highly dependent on (Frkova et al. 2020). Organic micropollutants impede several biogeochemical processes, decrease soil quality, reduce microbial diversity, and alter genes in microbial communities. They also influence human health as they move across food chains (Gosai et al. 2024). Four groups of pharmaceuticals, consisting nonsteroidal anti-inflammatory drugs (NSAIDs), antibiotics, β blockers, and blood lipid-lowering agents (BLLA), were assessed on soil microbiomes in a cultivated land; NSAIDs exhibited the least toxicity, whereas antibiotics showed the highest toxicity, however, β blockers and BLLA showed mid-level toxicity. The pharmaceuticals even at the level of 100 mg/L can affect microbiomes and alter the soil surroundings (Pino et al. 2017). They also reduces the diversity, alter metabolic activities, and change the metabolism of the carbon source. Antibiotics can alter the enzyme activity of soil microorganisms. Moreover, they can metabolize several carbon sources by altering not only microbial biomass but also the percentage composition of bacteria and fungi. Research based on nucleic acids analysis demonstrates that antibiotics can make changes in genes and enzymes (Cycoń et al. 2019). When oxytetracycline and sulfamethoxazole concentrations including 0, 1, 10, 25, 50, and 100 mg/kg were added to a sandy loam soil and incubated at 25°C for 21 days, the two antibiotics differently affected the biomass carbon and the microbial community (Molaei et al. 2017). These antibiotics are commonly used in poultry and livestock industries. When glyphosate (herbicide), aphacypermethrin (insecticide), and malathion (insecticide)

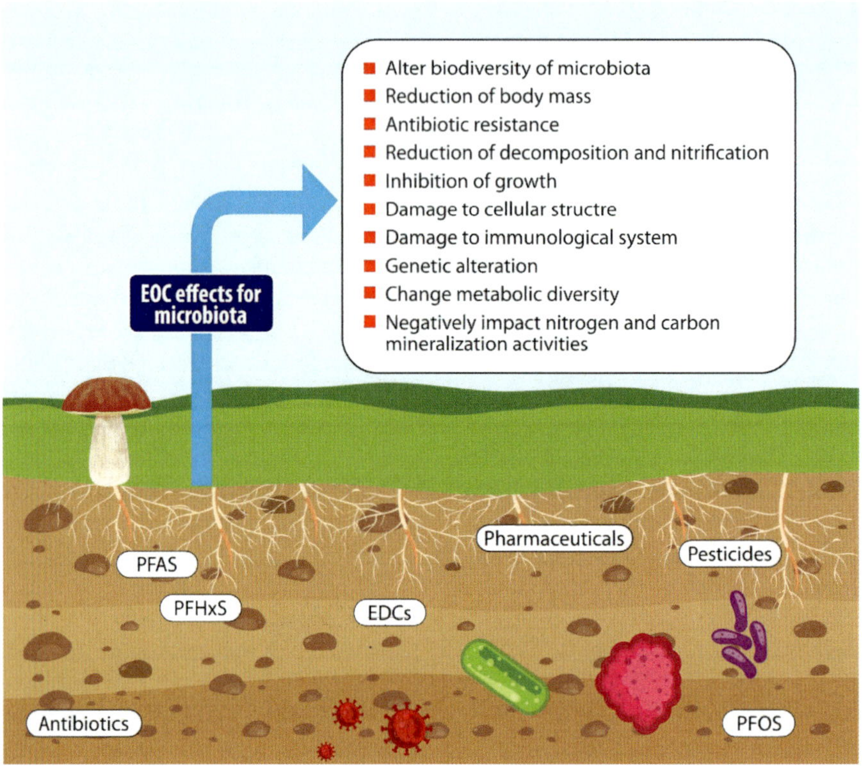

FIGURE 7.2 Effects of emerging organic contaminants on soil microbes.

were incorporated into soils individually at concentrations of 0, 50, 100, and 200 ppm, respectively, and placed under incubation at 30°C, their effects on the bacteria, fungi, actinomycetes counts, and CO_2 production were analyzed in the span of 7 weeks (every week). It was proven that supplementing the three above-stated pesticides has greatly reduced the population of soil bacteria, fungi, and actinomycetes (Al-Ani et al. 2019). The field application of glyphosate was shown to increase both microbial biomass carbon and nitrogen in nine soils by 17% and 76% respectively, 14 days after treatment (Arora and Sahni 2016). Experiments were performed in soil microcosms with imidacloprid, benomyl, and metribuzin pesticide formulations at single and decadic order application rates. The application of pesticides was followed by drastic changes in the fungal community structure at the alfa-diversity level (Streletskii et al. 2022) (Figure 7.2).

7.3.1 Impact of EOCs on Soil Invertebrates: Ecotoxicological Perspective

The soil invertebrate earthworm is commonly used to understand the toxicological effects on soil biota of different types of chemicals. As a detritivore, earthworms play a crucial role in absorbing and passing these chemicals into the organisms via food webs by the mechanisms such as bioaccumulation and biomagnification. (Carter

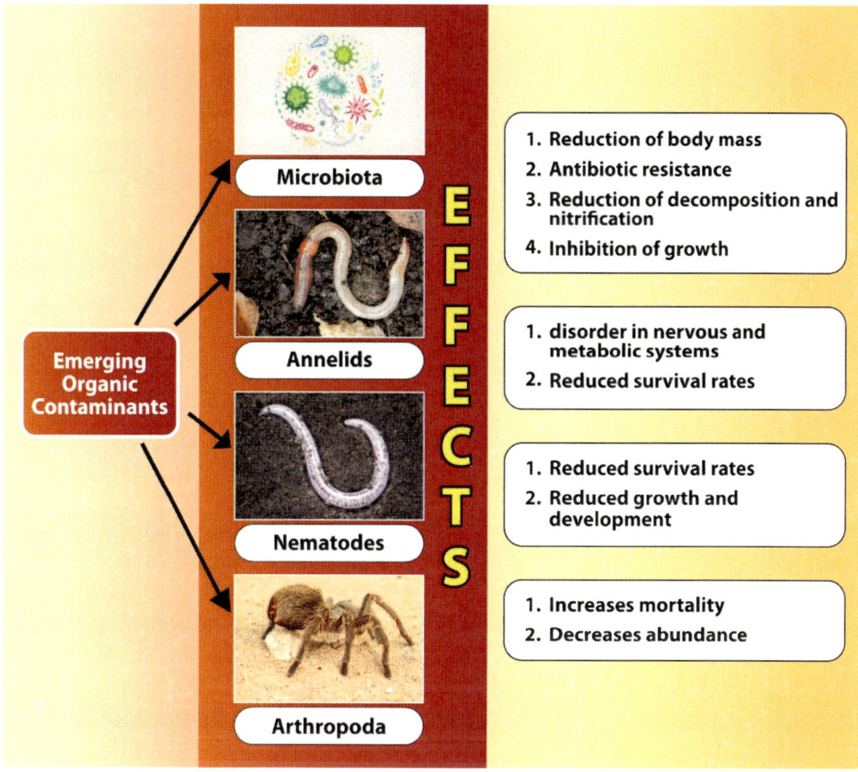

FIGURE 7.3 Eco risks of emerging organic contaminants in soil biota (Carter et al. 2016).

et al. 2016; Shore et al. 2014). The ecotoxicological effects of pesticides, pharmaceuticals and personal care products, endocrine disruptors, antibiotics, PFAS, and flame retardants on different soil biota are discussed in this section.

7.3.1.1 Pesticides

The accumulation of pesticide residues in soil can harm both soil health and biota. The intrinsic nature of pesticides enables them to affect biota both inside and outside of farming areas (Goritschnig et al. 2024). There are several studies that indicate the decrease in the body weight of treated worms exposed to different commercial pesticides (Yasmin and D'Souza 2010). An environmental risk assessment for pesticides on soil earthworm *E. foetida* showed that fungicides and herbicides were the primary cause of intermediate to high chronic toxicity, although insecticides had a lesser impact (Goritschnig et al. 2024). Certain insecticides including sulfoxaflor, tefluthrin, acetamiprid, and dazomet along with the fungicide mefentrifluconazole and plant growth regulator 1-methylcyclopropene showed highly acute earthworm toxicity (Goritschnig et al. 2024). Thus, soils in traditional crop cultivation contain higher levels of pesticide remainders and present a significant risk to soil invertebrates compared with organic fields. In a study by De Silva et al. (2010), earthworm toxicity tests were carried out in accordance with the Organization for Economic

Co-operation and Development (OECD) guideline 222 (OECD 2016), in an artificial soil, which consists of 10% organic matter, and pH ranged between 6.5 to 6.8. At the end of the exposure period (28 days), carbofuran showed the highest toxicity compared with mancozeb, both in their pure and formulated forms. It also reported that, overall, formulated versions were more toxic than the active ingredients alone. Nevertheless, variation between the LC_{50} and EC_{50} levels differ by less than a factor of 2.0. Accordingly, the study investigated the 28-day toxicity of abamectin and doramectin on soil invertebrates, including *Eisenia andrei*, *Folsomia candida*, and *Enchytraeus crypticus* (Kolar et al. 2008). In soil, *Eisenia andrei* earthworms were the most severely affected by the chemical, dying at a considerably lower concentration between 18 and 228 mg per kg of dry soil. *Folsomia candida*, *Porcellio scaber*, and *Enchytraeus crypticus* had a comparatively lower impact, requiring higher concentrations of the chemical between 67 and over 300 mg/kg of soil in the same type of harmful effect. The EC_{50} values for the reproductive effects on *Folsomia candida* and *Enchytraeus crypticus* were 13 and 38 mg/kg, respectively, for abamectin. For doramectin, the EC_{50} values were 42 mg/kg for springtails and 170 mg/kg for enchytraeids. For earthworms, no observed effect concentration for body weight changes was 10 mg/kg for abamectin on a species under field conditions (Solomon and Hanson 2014). However, the studies investigated on the toxicity and their potential threats to soil organisms are relatively limited. In 2020, a study investigated the occurrence and distribution of 15 pesticides that are widely utilized across 41 crop lands in Quzhou, North China Plain, during the cultivation period. Moreover, toxicity exposure ratios (TERs) and risk quotient (RQ) approaches were applied to investigate the pesticide ecological threat to soil organisms (earthworms, enchytraeids, springtails, mites) and the microbes involved in nitrogen mineralization. Based on RQs and pesticide detection rates, pesticide hazards were ranked using the Hasse diagram (Mu et al. 2023). The results revealed that chlorpyrifos, carbendazim, and imidacloprid posed a significant long-term exposure risks to *E. fetida*, *F. candida*, and *E. crypticus*, as their TERs exceeded the trigger value. In addition, pesticide mixtures tend to cause ecological threats to soil biota at more than two-thirds of the sites investigated. Almost half of the samples were categorized as high-risk due to their RQ exceeding the maximum value of 490. The Hasse diagram analysis identified abamectin, tebuconazole, chlorantraniliprole, and chlorpyrifos as the most hazardous pesticides to soil biota in the region. These findings highlight the urgent need to explore alternative pest management strategies to mitigate these risks.

7.3.1.2 Pharmaceuticals and Personal Care Products

The ecotoxicity of pharmaceuticals in soil invertebrates is often poorly documented in scientific literature, with many accessible data pertaining to earthworm species (Carter et al. 2020). One of a studies analyzed 26 pharmaceuticals and personal care products in soil samples (Gutierrez et al. 2016). Primarily, the results revealed that pollutants did not show significant impacts on the total number of nematodes. The overall nematode population remained intact, however, there were 21 specific genera present that were sensitive to pollutants. Further, ten specific genera were found to be particularly sensitive to pharmaceuticals and personal care products. The diversity and structure of the soil nematode community were significantly altered, but the long-term implications of these changes remain unclear (Gutiérrez et al. 2016). When considering the ecological risks of antibiotics on microbial communities, Du and Liu (2011), stated that apramycin inhibited soil bacteria growth significantly. Another study was conducted by Lin et al. (2012) to evaluate the toxic effects of chlortetracycline. It is an antibiotic introduced to soil, which was hazardous to *Eisenia fetida*, disrupting their reproduction, biochemistry, and genetics. EC_{50} values for juvenile and cocoon counts were at 96.1 mg/kg and 120.3 mg/kg of soil, respectively. There were observable decreases in juvenile and cocoon counts after exposure. Exposed earthworms also had severely altered levels of detoxication enzymes. These results highlight appreciable risks to soil health. Exposure to varying doses of chlortetracycline induced more DNA damage in coelomocytes, as a higher percentage of DNA in the tail measured using single-cell gel electrophoresis. The results showed a clear dose-dependent effect, with higher doses leading to greater DNA damage, concluding that chlortetracycline induces physiological responses and genotoxic effects in earthworms (Lin et al. 2012). Exposing earthworms to polystyrene microplastics (58 μm) at concentrations of 0%, 0.25%, 0.5%, 1%, and 2% w/w resulted in increased mortality (Cao et al. 2017; Khaldoon et al. 2022). Earthworms were exposed to low-density polyethene microplastics at different concentrations and sizes, with mortality rates ranging from 8% to 26% at 28% to 60% w/w and 20% to 30% at 1% w/w. Both polystyrene and low-density polyethene microplastics showed detrimental effects on earthworms. Polystyrene microplastics were found to be more hazardous and limit worm growth at concentrations of 1% and 2% w/w, respectively. Two investigations found that low-density polyethene microplastics had mortality rates of 28% to 60% w/w and 1% w/w (Huerta Lwanga et al. 2016; Lithner et al. 2011).

7.3.1.3 Endocrine Disruptors

Endocrine-disrupting chemicals have the capability to alter endocrine functions (Markman et al. 2007). In addition, xenobiotics like BPA, which are prevalent in biosolids used as organic fertilizers in farming. Their impact on soil biota remains underexplored, making it crucial to understand their impact on earthworms, which represent the primary animal biomass in this environment. A study examining the effects of BPA on gene expression in *Eisenia fetida* (Novo et al. 2018) found no significant changes in whole-body tissue during long-term exposure. However, the male reproductive organs showed altered expression of genes linked to endocrine function. One epigenetic-related gene was altered in relation to whole-body tissues and male reproductive organs in relation to the acute exposures. The molecular differences seen in the earthworms show that BPA exposure could cause negative impact on their populations (Novo et al. 2018).

7.3.1.4 Antibiotics

The antibiotic tetracycline (TC) has been largely reported in the environment due to its wide usage in farming. TCs are broad-spectrum antibiotics, and they are protein synthesis inhibitors. However, their adverse effects on nontarget organisms, although less understood, have also been observed. Vangheel et al. (2013), investigated the consequences of this widely used antibiotic on the nematode called *Caenorhabditis elegans* for reproduction, growth and population growth rate (PGR). The study found that all toxicological endpoints were negatively impacted. Exposure to TC concentrations as low as 5 mg/L was found to significantly impair growth and reproductive performance, while even lower concentration of 3 mg/L substantially diminished the PGR. However, antibiotics do not indicate a significant risk to nematodes inhabiting their natural setting (Vangheel et al. 2013). Another study that assessed the pathogenic and mutualistic phases of coryneforms, including *Staphylococcus*, *Pediococcus*, *Micrococcus*, and *Corynebacterium*, using the tobacco hornworm *Manduca sexta* (a model lepidopteran insect) showed that the midgut of *M. sexta* raised on an artificial diet contained mostly gram-positive cocci. Further, this study investigated the perturbation in the gut community on the addition of antibiotics to the diet (van der Hoeven et al. 2008). Additionally, this study observed that the reproduction of the nematode *Steinernema carpocapsae* was less efficient, and the colonization of the nematode by its symbiont *Xenorhabdus nematophila* decreased in insects that were fed a diet containing antibiotics.

7.4 ECOLOGICAL RISK ASSESSMENTS

Ecological risk is associated with the potential harm that human activities or environmental changes can cause to ecosystems, species, and biodiversity (Bartell 2008). The decision makers face a challenge of finding solutions to environmental issues caused by the growing pressure on natural resources. As a result, it is crucial to investigate ecological risks using an ecological risk assessment. Ecological risk assessment is the process of evaluating how environmental stressors affect plants, animals, and ecosystems scientifically. Prediction of potential risks, identifying vulnerable species or habitats, and guidance of decision making for environmental protection and management are crucial benefits of ecological risk assessment. Along with ecological risk assessments, this part addresses the ecological risks posed by emerging organic pollutants on soil biota. An ecological risk assessment of emerging contaminants on soil biota has become a globally concerned challange (Gomes et al. 2017). A kind of effective assessment can be done by identifying relevant tested parameters in each study. Measured variables assess the impact of a target contaminant on a specific organism by obtaining a defined outcome after exposure. The parameters are, mortality, abundance, biomass, behavioral responses (avoidance, digging behavior biomass degradation, nutrient intake, predator-prey interaction, fecal production, and locomotor functioning), reproduction, growth, richness and diversity, and structural changes (Gunstone et al. 2021). EOCs in soil pose significant toxicological threats to nature, public health, and ecological systems (Table 7.2).

7.5 THE RISK OF ANTIBIOTIC RESISTANCE

Antibiotic resistance genes in bacteria provide novel challenges for soil restoration, rehabilitation, and preservation (Li et al. 2023). Antibiotics have been widely utilized in human and livestock treatment for managing infectious diseases and as supplements for approximately five decades (Esiobu et al. 2002). Anthropogenic interactions are encouraging the occurrence of antibiotic-resistant bacteria in soils, and antibiotic-polluted soils are steadily rising globally (Xu et al. 2021). When

TABLE 7.2
Concentrations of Emerging Organic Contaminants in Soil Biota from Experimental Data

Type of Emerging Organic Contaminant	Chemical Name	Organism	Tested Concentration	End Point/Effect	References
Flame retardants	Tetrabromobisphenol A (TBBPA)	Enchytraeids	10 mg/kg	Inhibit reproduction	(Rothenbacher and Pecquet 2018; Sverdrup et al. 2006)
	Polychlorinated biphenyls (PCBs)	House cricket	150 mg/kg	Increase mortality	Paine et al. 1993
Poly-fluoroalkyl substances (PFAS)	Perfluorooctane sulfonic acid (PFOS)	Bacteria (*Acidobacteria*)		PFHxS exposure suppressed the richness and abundance of bacteria.	(Samarasinghe et al. 2023)
		Earthworms	100 mg/kg	Increase mortality	
	Perfluorooctanoic acid (PFOA)	Earthworms	100 mg/kg	Mortality and disruption in the nervous and metabolic systems of earthworms	(Li et al. 2023; Wang et al. 2022)
	Perfluorohexane sulfonic acid (PFHxS)	Earthworms	100 mg/kg	Disrupted energy balance and triggered inflammatory responses	
Endocrine distruptors	Bisphenol A (BPA)	Earthworms	2000 mg/kg	BPA does not significantly affect the survival	(Verdú et al. 2018)
		Bacteria		A change in the structural composition stimulated the multiplication of *Pseudomonas* sp. and slowed down the multiplication of fungi	
	Benzophenone-3 (oxybenzone)	Earthworm	364 mg/kg	Mortality increased	(Gautam et al. 2022)

antibiotic remnants infiltrate the soil, the primary processes revealing their durability are absorption to organic matter and subsequent breakdown or alteration. Half-life (DT50) values for antibiotic remnants in soils rely on various factors, e.g., composition of the residue, soil properties, and conditions of the nature (precipitation, temperature, and humidity). The use of pharmaceuticals and personal care products such as prophylactics contributes to initiating new antibiotic-resistant bacteria (Chaturvedi et al. 2021). In a 3-year experiment, the concentration of erythromycin (an antibiotic) in the soil varied between 0.83 and 76 μg/kg. Macrolide-lincosamide-streptogramin antibiotic resistance genes increased over time. The bacteria became more resistant to these classes of antibiotics, which is a concerning trend for antibiotic resistance development (Stanton et al. 2024).

The addition of antibiotic-resistant bacteria supplement increased the niche breadth and diversity, and changed the soil bacterial community structure and niche composition. TC contamination further enhanced these impacts probably via increasing the survival of foreign antibiotic-resistant bacteria in the soil. Antibiotic-resistant bacteria can induce compositional and functional changes in the bacterial community, potentially triggered by the drastic shifts in certain core taxa, including Proteobacteria, Bacteroidetes, Chloroflexi, and Patescibacteria (Xu et al. 2021). Manure application is one of the principal pathways of entry for antibiotic-resistant bacteria and antibiotic-resistant genes into the soil and leads to an accumulation of enormous quantities in the soil environment (Ondon et al. 2021). With respect to the impact of antibiotic resistance on environmental and human health, the transmission to humans is caused by contamination of food or water and by direct contact with contaminated animals, water, or soil. The resistant pathogens that originated from agricultural fields reach humans upon contacting livestock or by consuming bacteria from contaminated meat. Upon becoming infected or colonized by a resistant microorganism, transmission between humans becomes possible (Ondon et al. 2021). In a study (Symochko et al. 2023), 389 strains of leading bacterial species were isolated from the studied ecosystems, 57 of which were antibiotic-resistant, with the levels of resistance to antibiotics being greater than 70%. The soil microbiome in ancient forests was dominated by a limited proportion of antibiotic-resistant bacteria; only two species, *Bacillus cereus* and *Pantoea agglomerans*, were strongly resistant to antibiotics. In the soil of medicinal herbs in an agroecosystem, out of 106 bacterial strains, 13 species showed a high percentage of resistance to antibiotics. It was established that the largest number of antibiotic-resistant bacteria occur in the soil of agroecosystems contaminated with enrofloxacin. Among 190 assessed strains of bacteria, 42 (22%) had a high rate of antibiotic resistance. Nanoplastics (NPs) are also being viewed as a newly emerging environmental contaminant (Chan et al. 2023). NPs exhibit the capacity to enhance the colonization of microbial pathogens on their surface and adsorb toxic substances, such as heavy metals and leftover antibiotics. Soil animals, such as the bacterivorous nematode *Caenorhabditis elegans* (*C. elegans*), are highly mobile, and it is a likelihood for them to play an active role in propagating NPs and adsorbed contaminants. The research (Chan et al. 2023) illustrates that antibiotic-resistant bacteria would congregate with adsorbed NPs of antibiotics to form adsorbed NP-antibiotic resistant bacteria (ANP-ARB) aggregates, and they utilized polymyxins (colistin) as proof of concept. Colistin-resistant genes bear *Escherichia coli* from a mixed culture of resistant and susceptible bacteria. Selectively, colistin ANPs aggregate with a mixed population of resistant and susceptible bacteria. *C. elegans* consumed ANP-ARB aggregates in the soil microcosm, and the nematodes spread ANP-ARB rapidly throughout the soil at 40–60 cm/day. Root-knot nematodes are controlled widely with nematicide avermectin. The 5-year application of avermectin effects on ill tobacco plants' arable land rhizosphere microbiome and resistome was studied in this research (Zhao et al. 2023). Environmental risks of avermectin were investigated in this study using a metagenomic method to study antibiotic-resistant bacteria and avermectin-treated soil antibiotic resistance genes. The application of avermectin had an impact on the abundance and distribution of antibiotic-resistant bacteria and their antibiotic resistance genes. The antibiotic-resistant proteobacteria comprised the highest percentage (36%) in the rhizosphere soil and contained 530 antibiotic resistant genes. Opportunistic human pathogens carrying antibiotic resistance genes were enriched under avermectin-treated soil. Avermectin treatment increased the populations of many of the categories of antibiotic resistance genes.

7.6 RISK MINIMIZATION

Various emerging organic compounds are added to the soil due to the rapid growth in chemical industry across the globe and the breakthrough of new applications. Improved analytical techniques have recently revealed measurable amounts of EOCs in soil, which could be harmful in the future. This is because the unlimited capacity of the compartment to act as a chemical sink has hindered past observations. Meanwhile, global soils are facing grim threats that have not yet been included in successful policy formulation aimed at safeguarding the health and sustainable use of soils. Awareness of people, applying regulations, constant monitoring, bioremediation processes, and new water treatment technologies are some actions that can be taken to reduce the threats from the emerging organic pollutants. The development and spread of antibiotic-resistant bacteria, which are linked with the soil ecosystem, can cause potential threats to people. It is crucial to take relevant actions to control the soil microbiome and prevent the contamination of soil with antibiotics to minimize the threat of antibiotic resistance for human (Symochko et al. 2023). The implementation of waste management systems is also essential to decrease the antibiotic resistance. There are some practical approaches with other benefits of nutrient management to ensure the safeguard of the soil environment. Measures such as mitigating sediment erosion and the use of limited animal manure can be applied to reduce antibiotic resistance emerging from manure application. Apart from that, it is crucial to encourage the management of lagoon spills, surface runoff, and seepage. The surface runoff may be limited by enhancing the manure collection and expanding the storage capacity. The crop demand can be met with the application of manure in an appropriate method. Antibiotic residues can be treated by digested animal waste and composting. The breakdown of antibiotics, such as chlortetracycline, tylosin, and monensin, is escalated by processes such as turning of compost, aeration, and watering. In addition, complete antibiotic degradation can be induced by the simple storage of manure piles (Ondon et al. 2021).

REFERENCES

Ailijiang, N., N. Zhong, X. Zhou, et al. 2022. Levels, sources, and risk assessment of PAHs residues in soil and plants in urban parks of Northwest China. *Scientific Reports* 12:21448.

Al-Ani, M., R. M. Hmoshi, I. Kanaan, and A. Thanoon. 2019. Effect of pesticides on soil microorganisms. *Journal of Physics Conference Series* 1294:072007.

Armitage, J. M., and Gobas, F. 2007. A terrestrial food chain bioaccumulation model for POPs. *Environmental Science and Technology*, 41:4019–25.

Arora, S., and D. Sahni. 2016. Pesticides effect on soil microbial ecology and enzyme activity: An overview. *Journal of Applied and Natural Science* 8:1126–32.

Arrieta-Cortes, R., P. Farias, C. Hoyo-Vadillo, and M. Kleiche-Dray. 2017. Carcinogenic risk of emerging persistent organic pollutant perfluorooctane sulfonate (PFOS): A proposal of classification. *Regulatory Toxicology and Pharmacology* 83:66–80.

Bach, E. M., K. S. Ramirez, T. D. Fraser, and D. H. Wall. 2020. Soil biodiversity integrates solutions for a sustainable future. *Sustainability* 12:2662.

Barra Caracciolo, A., E. Topp, and P. Grenni. 2015. Pharmaceuticals in the environment: Biodegradation and effects on natural microbial communities: A review. *Journal of Pharmaceutical and Biomedical Analysis* 106:25–36.

Bartell, S. M. 2008. Ecological risk assessment. In *Encyclopedia of Ecology*, eds. S. E. Jørgensen, and B. D. Fath, pp. 1097–101. Academic Press, Oxford.

Blankenship, A., M. Zwiernik, K. Coady, et al. 2005. Differential accumulation of polychlorinated biphenyl congeners in the terrestrial food web of the Kalamazoo river superfund site, Michigan. *Environmental Science & Technology* 39:5954–63.

Burkhard, L. P., and L. K. Votava. 2023. Review of per- and polyfluoroalkyl substances (PFAS) bioaccumulation in earthworms. *Environmental Advances* 11:100335.

Cao, D., X. Wang, X. Luo, and G. Liu. 2017. Effects of polystyrene microplastics on the fitness of earthworms in an agricultural soil. *IOP Conference Series: Earth and Environmental Science* 61:012148.

Carter, L. J., C. D. Garman, J. Ryan, et al. 2014. Fate and uptake of pharmaceuticals in soil earthworm systems. *Environmental Science & Technology* 48:5955–63.

Carter, L. J., J. J. Ryan, and A. B. A. Boxall. 2016. Effects of soil properties on the uptake of pharmaceuticals into earthworms. *Environmental Pollution* 213:922–31.

Carter, L. J., M. Williams, and J. B. Sallach. 2020. Uptake and effects of pharmaceuticals in the soil-plant-earthworm system. In *The Handbook of Environmental Chemistry*. eds. S. P. Solsona, N. Montemurro, S. Chiron, and D. Barceló, pp. 175–220. Springer, Cham.

Chan, S. Y., S. Y. Liu, R. Wu, et al. 2023. Simultaneous dissemination of nanoplastics and antibiotic resistance by nematode couriers. *Environmental Science & Technology* 57:8719–27.

Chaturvedi, M., S. Joy, R. D. Gupta, S. Pandey, and S. Sharma. 2023. Endocrine disrupting chemicals (EDCs): Chemical fate, distribution, analytical methods and promising remediation strategies: A critical review. *Environmental Technology Reviews* 12:286–315.

Chaturvedi, P., P. Shukla, B. S. Giri, et al. 2021. Prevalence and hazardous impact of pharmaceutical and personal care products and antibiotics in environment: A review on emerging contaminants. *Environmental Research* 194:110664.

Chormare, R., and M. A. Kumar, 2022. Environmental health and risk assessment metrics with special mention to biotransfer, bioaccumulation and biomagnification of environmental pollutants. *Chemosphere* 302:134836.

Cycoń, M., A. Mrozik, and Z. Piotrowska-Seget. 2019. Antibiotics in the soil environment degradation and their impact on microbial activity and diversity. *Frontiers in Microbiology* 10:412419.

De Silva, P. M., A. Pathiratne, and C. Gestel. 2010. Toxicity of chlorpyrifos, carbofuran, mancozeb and their formulations to the tropical earthworm *Perionyx excavates*. *Applied Soil Ecology* 44:56–60.

Du, L., and W. Liu. 2011. Occurrence, fate, and ecotoxicity of antibiotics in agro-ecosystems: A review. *Agronomy for Sustainable Development* 32:309–327.

Ehsan, M. N., M. Riza, M. N. Pervez, et al. 2024. PFAS contamination in soil and sediment: Contribution of sources and environmental impacts on soil biota. *Case Studies in Chemical and Environmental Engineering* 9:100643.

Eltoukhy, A., Y. Jia, R. Nahurira, et al. 2020. Biodegradation of endocrine disruptor bisphenol A by Pseudomonas putida strain YC-AE1 isolated from polluted soil, Guangdong, China. *BMC Microbiology* 20:1–14.

Esiobu, N., L. Armenta, and J. Ike. 2002. Antibiotic resistance in soil and water environments. *International Journal of Environmental Health Research* 12:133–44.

Fabregat-Palau, J., M. Vidal, and A. Rigol. 2021. Modelling the sorption behaviour of perfluoroalkyl carboxylates and perfluoroalkane sulfonates in soils. *Science of the Total Environment* 801:149343.

Fremlin, K. M., J. E. Elliott, D. J. Green, et al. 2020. Trophic magnification of legacy persistent organic pollutants in an urban terrestrial food web. *Science of The Total Environment* 714:136746.

Frkova, Z., Y. Vystavna, A. Koubová, et al. 2020. Microbial responses to selected pharmaceuticals in agricultural soils: Microcosm study on the roles of soil, treatment and time. *Soil Biology and Biochemistry* 149:107924.

Furuno, S., S. Foss, E. Wild, et al. 2012. Mycelia promote active transport and spatial dispersion of polycyclic aromatic hydrocarbons. *Environmental Science & Technology* 46:5463–70.

García, J., M. J. García-Galán, J. W. Day, et al. 2020. A review of emerging organic contaminants (EOCs), antibiotic resistant bacteria (ARB), and antibiotic resistance genes (ARGs) in the environment: Increasing removal with wetlands and reducing environmental impacts. *Bioresource Technology* 307:123228.

Gautam, K., M. Seth, S. Dwivedi, et al. 2022. Soil degradation kinetics of oxybenzone (benzophenone-3) and toxicopathological assessment in the earthworm, *Eisenia fetida*. *Environmental Research* 213:113689.

Gobas, F. A., L. P. Burkhard, W. J. Doucette, et al. 2016. Review of existing terrestrial bioaccumulation models and terrestrial bioaccumulation modeling needs for organic chemicals. *Integrated Environmental Assessment and Management* 12:123–134.

Gomes, R., C. Justino, T. Rocha-Santos, et al. 2017. Review of the ecotoxicological effects of emerging contaminants on soil biota. *Journal of Environmental Science and Health, Part A* 52:1–16.

Goritschnig, L., H. Burtscher-Schaden, T. Durstberger, and J. G. Zaller. 2024. Ecotoxicity of pesticides approved for use in European conventional or organic agriculture for honeybees, birds, and earthworms. *Environments* 11:137.

Gosai, H. G., F. Jadeja, A. Sharma, and S. Jain. 2024. Occurrence and toxicity of organic microcontaminants in agricultural perspective: An overview. In *Organic Micropollutants in Aquatic and Terrestrial Environments*, eds. R. Bhadouria, S. Tripathi, P. Singh, R. Singh and H. P. Singh, pp. 107–26. Springer, Cham.

Gunstone, T., T. Cornelisse, K. Klein, A. Dubey, and N. Donley. 2021. Pesticides and soil invertebrates: A hazard assessment. *Frontiers in Environmental Science* 9:643847.

Gutiérrez, C., C. Fernández, M. Escuer, et al. 2016. Effect of soil properties, heavy metals and emerging contaminants in the soil nematodes diversity. *Environmental Pollution* 213:184–94.

Gworek, B., M. Kijeńska, J. Wrzosek, and M. Graniewska. 2021. Pharmaceuticals in the soil and plant environment: A review. *Water, Air, & Soil Pollution* 232:145.

Houida, S., L. Yakkou, M. Chelkha, et al. 2024. Exposure to emerging contaminants: Ecotoxicological effects on earthworms and the potential of gut-associated microorganisms in bioremediation. In *Earthworm Technology in Organic Waste Management*, eds. K. Huang, S. A. Bhat, F. Li, and V. Kumar, pp. 257–92. Elsevier.

Huerta Lwanga, E., H. Gertsen, H. Gooren, et al. 2016. Microplastics in the terrestrial ecosystem: Implications for *Lumbricus terrestris* (Oligochaeta, Lumbricidae). *Environmental Science & Technology* 50:2685–91.

Janssen, M. P. M., and R. Hogervorst. 1993. Metal accumulation in soil arthropods in relation to micro-nutrients. *Environmental Pollution* 79:181–189.

Karlsson, M. V., S. Marshall, T. Gouin, and A. Boxall. 2015. Routes of uptake of diclofenac, fluoxetine, and triclosan into sediment-dwelling worms: Uptake of pharmaceuticals and personal care products. *Environmental Toxicology and Chemistry* 35:836–42.

Kelly, B., M. Ikonomou, J. Blair, A. Morin, and F. Gobas. 2007. Food web specific biomagnification of persistent organic pollutants. *Science* 317: 236–239.

Khaldoon, S., J. Lalung, U. Maheer, et al. 2022. A review on the role of earthworms in plastics degradation: Issues and challenges. *Polymers* 14:4770.

Kim, H. 2016. A study on the utilization of the earthworms *Eisenia fetida* and *Eisenia andrei* for the disposal of polymers. *International Journal of Environmental Science and Development* 7:355–58.

Kinney, C., E. Furlong, D. Kolpin, et al. 2008. Bioaccumulation of pharmaceuticals and other anthropogenic waste indicators in earthworms from agricultural soil amended with biosolid or swine manure. *Environmental Science & Technology* 42:1863–70.

Kolar, L., N. K. Eržen, L. Hogerwerf, and C. Gestel. 2008. Toxicity of abamectin and doramectin to soil invertebrates. *Environmental Pollution* 151:182–89.

Li, H., and P. Koosaletse-Mswela. 2023. Occurrence, fate, and remediation of per- and polyfluoroalkyl substances in soils: A review. *Current Opinion in Environmental Science & Health* 34:100487.

Li, S., B. S. Ondon, S. H. Ho, and F. Li. 2023. Emerging soil contamination of antibiotics resistance bacteria (ARB) carrying genes (ARGs): New challenges for soil remediation and conservation. *Environmental Research* 219:115132.

Liang, X., S. Zhu, P. Chen, and L. Zhu. 2010. Corrigendum to bioaccumulation and bioavailability of polybrominated diphenyl ethers (PBDEs) in soil. *Environmental Pollution* 158:3011.

Lin, D., Q. Zhou, Y. Xu, C. Chen, and Y. Li. 2012. Physiological and molecular responses of earthworm (*Eisenia fetida*) to soil chlortetracycline contamination. *Environmental Pollution* 171:46–51.

Lithner, D., A. Larsson, and G. Dave. 2011. Environmental and health hazard ranking and assessment of plastic polymers based on chemical composition. *Science of the Total Environment* 409:3309–24.

Mali, H., C. Shah, B. H. Raghunandan, et al. 2023. Organophosphate pesticides an emerging environmental contaminant: Pollution, toxicity, bioremediation progress, and remaining challenges. *Journal of Environmental Sciences* 127:234–50.

Marchi, E. 2015. Impact of wildfire on the edaphic microarthropod community in a *Pinus pinaster* forest in central Italy. *iForest: Biogeosciences and Forestry* 8:874–83.

Markman, S., I. A. Guschina, S. Barnsley, et al. 2007. Endocrine disrupting chemicals accumulate in earthworms exposed to sewage effluent. *Chemosphere* 70:119–25.

Mortensen, L. H., R. Rønn, and M. Vestergård. 2018. Bioaccumulation of cadmium in soil organisms – with focus on wood ash application. *Ecotoxicology and Environmental Safety* 156:452–462.

Molaei, A., A. Lakzian, G. Haghnia, et al. 2017. Assessment of some cultural experimental methods to study the effects of antibiotics on microbial activities in a soil: An incubation study. *PLoS One* 12:e0180663.

Mu, H., X. Yang, K. Wang, et al. 2023. Ecological risk assessment of pesticides on soil biota: An integrated field-modelling approach. *Chemosphere* 326:138428.

Navarro, I., A. de la Torre, P. Sanz, et al. 2016. Bioaccumulation of emerging organic compounds (perfluoroalkyl substances and halogenated flame retardants) by earthworm in biosolid amended soils. *Environmental Research* 149:32–39.

Novo, M., I. Verdú, D. Trigo, and J. L. Martínez-Guitarte. 2018. Endocrine disruptors in soil: Effects of bisphenol A on gene expression of the earthworm *Eisenia fetida*. *Ecotoxicology and Environmental Safety* 150:159–67.

OECD. 2016. *Test No. 222: Earthworm Reproduction Test* (*Eisenia fetida/Eisenia andrei*), OECD Guidelines for the Testing of Chemicals, Section 2, OECD Publishing, Paris.

Ojha, R. B., and D. Devkota. 2014. Earthworms: Soil and ecosystem engineers: A review. *World Journal of Agricultural Research* 2:257–60.

Ondon, B. S., S. Li, Q. Zhou, and F. J. Li. 2021. Sources of antibiotic resistant bacteria (ARB) and antibiotic resistance genes (ARGs) in the soil: A review of the spreading mechanism and human health risks. *Reviews of Environmental Contamination and Toxicology* 256:121–53.

Osuoha, J. O., B. O. Anyanwu, and C. Ejileugha. 2023. Pharmaceuticals and personal care products as emerging contaminants: Need for combined treatment strategy. *Journal of Hazardous Materials Advances* 9:100206.

Paine, J. M., M. J. McKee, and M. E. Ryan. 1993. Toxicity and bioaccumulation of soil PCBs in crickets: Comparison of laboratory and field studies. *Environmental Toxicology and Chemisrry* 12:2097–103.

Pashaei, R., P. Zahedipour-Sheshglani, R. Dzingelevičienė, S. Abbasi, and R. M. Rees. 2022. Effects of pharmaceuticals on the nitrogen cycle in water and soil: A review. *Environmental Monitoring and Assessment* 194:105.

Peijnenburg, W., E. Capri, C. Kula, et al. 2012. Evaluation of exposure metrics for effect assessment of soil invertebrates. *Critical Reviews in Environmental Science and Technology* 42:1862–93.

Pino, M. R., J. Val, A. M. Mainar, et al. 2015. Acute toxicological effects on the earthworm *Eisenia fetida* of 18 common pharmaceuticals in artificial soil. *Science of the Total Environment* 518–519:225–37.

Pino, M. R., S. Muñiz, J. Val, and E. Navarro. 2017. Effects of 18 Pharmaceuticals on the Physiological Diversity of Edaphic Microorganisms. *Science of the Total Environment* 595: 441–450.

Pinto, A. G., T. Kos, J. Puškarić, et al. 2024. Soil ecosystem functioning through interactions of nematodes and fungi *Trichoderma sp. Sustainability* 16:2885.

Rothenbacher, K. P., and A. M. Pecquet. 2018. Summary of historical terrestrial toxicity data for the brominated flame retardant tetrabromobisphenol A (TBBPA): Effects on soil microorganisms, earthworms, and seedling emergence. *Environmental Science and Pollution Research* 25:17268–77.

Saha, S., N. Narayanan, N. Singh, and S. Gupta. 2022. Occurrence of endocrine disrupting chemicals (EDCs) in river water, ground water, and agricultural soils of India. *International Journal of Environmental Science and Technology* 19:11459–74.

Samarasinghe, S. V. A. C., M. M. Bahar, F. Qi, et al. 2023. Evaluating PFHxS toxicity to invertebrates and microbial processes in soil. *Environmental Chemistry and Ecotoxicology* 5:120–28.

Sandeep, S. Sharma, A. Sharma, et al. 2024. Pesticide biology in soil: Sorption, leaching, and accumulation. In *Pesticides in the Environment*, eds. A. Sharma, V. Kumar, and B. Zheng, pp. 49–66. Elsevier, Amsterdam.

Shore, R., M. Taggart, J. Smits, et al. 2014. Detection and drivers of exposure and effects of pharmaceuticals in higher vertebrates. *Philosophical Transactions of the Royal Society B Biological Sciences* 369:20130570.

Solomon, K. R., and M. Hanson. 2014. Mesocosms and microcosms (aquatic). In *Encyclopedia of Toxicology* (Third Edition), ed. P. Wexler, pp. 223–26. Academic Press, Oxford.

Stanton, I. C., H. J. Tipper, K. Chau, et al. 2024. Does environmental exposure to pharmaceutical and personal care product residues result in the selection of antimicrobial-resistant microorganisms, and is this important in terms of human health outcomes? *Environmental Toxicology and Chemistry* 43:623–36.

Stenrød, M., S. S. Klemsdal, H. R. Norli, and O. M. Eklo. 2013. Effects of picoxystrobin and 4-n-nonylphenol on soil microbial community structure and respiration activity. *PLOS One* 8:e66989.

Streletskii, R., A. Astaykina, G. Krasnov, and V. Gorbatov. 2022. Changes in bacterial and fungal community of soil under treatment of pesticides. *Agronomy* 12:124.

Sverdrup, L. E., T. Hartnik, E. Mariussen, and J. Jensen. 2006. Toxicity of three halogenated flame retardants to nitrifying bacteria, red clover (*Trifolium pratense*), and a soil invertebrate (*Enchytraeus crypticus*). *Chemosphere* 64:96–103.

Svobodová, M., K. Šmídová, M. Hvězdová, and J. Hofman. 2018. Uptake kinetics of pesticides chlorpyrifos and tebuconazole in the earthworm *Eisenia andrei* in two different soils. *Environmental Pollution* 236:257–64.

Symochko, L., O. Demyanyuk, V. Symochko, et al. 2023. The spreading of antibiotic-resistant bacteria in terrestrial ecosystems and the formation of soil resistome. *Land* 12:769.

Tian, Y. X., H. Y. Chen, J. Ma, et al. 2023. A critical review on sources and environmental behavior of organophosphorus flame retardants in the soil: Current knowledge and future perspectives. *Journal of Hazardous Materials* 452:131161.

Uddin, M., F. Desai, and E. Asmatulu. 2021. Review of bioaccumulation, biomagnification, and biotransformation of engineered nanomaterials. *Environmental Chemistry for a Sustainable World* 2:133–164.

Van Den Hoogen, J., S. Geisen, D. Routh, et al. 2019. Soil nematode abundance and functional group composition at a global scale. *Nature* 572:194–98.

Van Der Hoeven, R., G. Betrabet, and S. Forst. 2008. Characterization of the gut bacterial community in *Manduca sexta* and effect of antibiotics on bacterial diversity and nematode reproduction. *FEMS Microbiology Letters* 286:249–56.

Vangheel, M., W. Traunspurger, and N. Spann. 2013. Effects of the antibiotic tetracycline on the reproduction, growth and population growth rate of the nematode *Caenorhabditis elegans*. *Nematology* 16:19–29.

Verdú, I., D. Trigo, J. L. Martínez-Guitarte, and M. Novo. 2018. Bisphenol A in artificial soil: Effects on growth, reproduction and immunity in earthworms. *Chemosphere* 190:287–95.

Wang, J., T. Liu, J. Zhao, et al. 2024. Energy flows through nematode food webs depending on the soil carbon and nitrogen contents after forest conversion. *Science of the Total Environment* 935:173322.

Wang, Z., F. Qi, Y. Shi, et al. 2022. Evaluation of single and joint toxicity of perfluorooctanoic acid and arsenite to earthworm (*Eisenia fetida*): A multi-biomarker approach. *Chemosphere* 291:132942.

Xu, H., Z. Chen, X. Wu, et al. 2021. Antibiotic contamination amplifies the impact of foreign antibiotic-resistant bacteria on soil bacterial community. *Science of the Total Environment* 758:143693.

Yang, Y., P. Liu, and M. Li. 2020. Tri-n-butyl phosphate induced earthworm intestinal damage by influencing nutrient absorption and energy homeostasis of intestinal epithelial cells. *Journal of Hazardous Materials* 398:122850.

Yasmin, S., and D. D'Souza. 2010. Effects of pesticides on the growth and reproduction of earthworm: A review. *Applied and Environmental Soil Science* 2010:678360.

Zhao, X., G. Qi, Y. Feng, and C. Du. 2023. Application of nematicide avermectin enriched antibiotic-resistant bacteria and antibiotic resistance genes in farmland soil. *Environmental Research* 227:115802.

Zhao, Y., G. Li, D. Qi, et al. 2017. Biomarker responses of earthworms (*Eisenia fetida*) to soils contaminated with perfluorooctanoic acid. *Environmental Science and Pollution Research* 24:22073–81.

8 Global Regulations on Emerging Contaminants

Fang Tao

8.1 INTRODUCTION

The international community has begun to pay attention to the issue of emerging contaminants and has implemented corresponding laws and regulations to control them. Among these, the Stockholm Convention, as one of the international environmental treaties, has made detailed provisions for the management of persistent organic pollutants (POPs). In addition, the EU's REACH regulation (European Chemicals Agency, 2021) and the United States Environmental Protection Agency (EPA) regulations (2005) are also advancing the governance of emerging contaminants. China is also gradually improving its environmental protection legal system to manage the environmental risks from these new pollutants (Chinese National Standard, 2002; Chinese National Standard, 2018a; Chinese National Standard, 2022).

This chapter aims to systematically introduce and analyze the laws and regulations concerning emerging contaminants, with a focus on the limit values and remediation standards for these pollutants in soil and water environments. By reviewing and comparing international and domestic regulations, we will identify the shortcomings of the current legal framework and suggest directions for improvement, providing scientific basis and policy recommendations for environmental management.

In this chapter, we first outline the international and domestic management regulations for emerging contaminants. We then delve into the limit standards for different types of emerging contaminants, especially focusing on the limit values and remediation standards for POPs, antibiotics, microplastics, and endocrine disruptors in soil and water environments. Next, we analyze the current status and challenges of monitoring and enforcement, using case studies to illustrate practical problems and solutions. Finally, we look forward to future trends in the management of emerging contaminants and propose corresponding policy recommendations.

Through the research presented in this chapter, we hope to provide valuable reference materials for environmental protection workers, policymakers, and the general public, jointly promoting the in-depth development of emerging contaminant management, and protecting our ecological environment and human health.

DOI: 10.1201/9781003509523-8

8.2 INTERNATIONAL REGULATORY FRAMEWORK

8.2.1 The Stockholm Convention

The Stockholm Convention, an international environmental treaty adopted in 2001, aims to eliminate or reduce the production and release of POPs (UNEP, n.d.). POPs are chemical substance that persist in the environment for extended periods, distributed widely across regions, accumulated in the fatty tissue of organisms, and pose toxicity risks to both humans and wildlife. According to the Stockholm Convention (UNEP, 2001), The chemicals targeted by the Stockholm Convention are listed in three annexes of the convention text: Parties must take measures to eliminate the production and use of the chemicals listed under Annex A, it mainly consists of some pesticides; Parties must take measures to restrict the production and use of the chemicals listed under Annex B; Parties must take measures to reduce the unintentional releases of chemicals listed under Annex C (ECHA, 2025a). The initially 12 POPs, often referred to as the "dirty dozen," are aldrin, chlordane, dichlorodiphenyltrichloroethane (DDT), dieldrin, endrin, heptachlor, hexachlorobenzene (HCB), mirex, toxaphene, polychlorinated biphenyls (PCBs), polychlorinated dibenzo-*p*-dioxins (PCDDs), and polychlorinated dibenzofurans (PCDFs) (UNEP, 2001). Over time, the list has expanded to include more chemicals as scientific evidence of their harmful effects has emerged (Table 8.1).

The Stockholm Convention aims to eliminate or restrict the production and use of these hazardous chemicals. Specific control measures include (UNEP, 2023):

1. **Reduce or Eliminate from Intentional Production and Use:**
    - States shall prohibit or strictly regulate the commercial production and application of POPs, with specific exceptions (e.g., DDT for disease vector control).
2. **Reduction or Elimination of Unintentional Releases:**
    - Countries must apply measures to minimize or eliminate releases of by-products including dioxins and furans, using optimal feasible technologies and sustainable management approaches.
3. **Reduce or Eliminate Releases from Stockpiles and Wastes:**
    - POP stockpiles and wastes must be controlled and disposed of in an eco-friendly manner to avoid recovery, recycling, reclamation, immediate reuse, or alternative functions
4. **Information Exchange:**
    - Parties must monitor POPs in the environment and report on the actions taken to execute the Convention, submitting regular national reports to the Convention's Secretariat.
5. **Public Information, Awareness and Education:**
    - Promoting general information, consciousness, and education regarding the risks and impacts of POPs, and encouraging public participation in addressing POP issues.
6. **Research, Monitoring and Development:**
    - Encouraging research and development of safer alternatives to POPs and supporting studies on the environmental and health effects of POPs.

7. **Technical Assistance:**
   - Providing technical assistance to developing countries and countries with transitional economies to help them operationalize the Convention.
8. **Financial Resources and Mechanisms:**
   - Parties need to provide financial support and appropriate guidance of mechanism to countries that fulfill the strategic goals.

Through these comprehensive control measures, the Stockholm Convention aims to minimize adverse effects on public health and ecological systems from POPs, promoting global efforts to remove or reduce the release of these health-threatening agents.

8.2.2 EU REACH Regulation

8.2.2.1 Overview of REACH

The Registration, Evaluation, Authorization, and Restriction of Chemicals (REACH) regulation is a European Union (EU) legislative framework enacted in 2006. It aims to improve the protection of human health and the environment from the risks posed by chemicals, while enhancing the competitiveness of the EU chemical industries (European Union, 2006). REACH imposes the obligation on manufactures to ensure the safety of their productions.

8.2.2.2 Key Components of REACH

The REACH regulation encompasses several key components: registration, evaluation, authorization, and restriction, each designed to create a comprehensive framework for chemical safety.

In terms of registration, all manufacturers and importers handling chemical substances in annual volumes exceeding 1 ton are obligated to register these compounds with the European Chemicals Agency (ECHA). This process demands the submission of extensive information regarding the characteristics, applications, and safe management of the chemicals (ECHA, 2025a). For chemicals manufactured or imported annual volumes exceeding ten tons, a comprehensive chemical safety management report is also required. A comprehensive analysis of chemical safety measures is presented in the report, ensuring that companies take responsibility for the safety of their products.

Evaluation is another critical aspect of REACH. The ECHA evaluates the information provided in registration dossiers to ensure compliance with REACH requirements. This evaluation involves examining testing proposals and the adequacy of the data submitted. Additionally, the ECHA or member states may conduct further evaluations of compounds that could present a risk to public health or ecosystems. Such evaluations can lead to additional testing or regulatory actions to mitigate these risks.

Authorization under REACH targets substances of very high concern (SVHCs) (ECHA, 2025b). These include carcinogens, mutagens, reproductive toxicants, persistent, bioaccumulative and toxic substances, and endocrine disruptors (ECHA, 2025b). Companies must apply for authorization to continue using SVHCs, demonstrating either that the hazards are sufficiently mitigated or the collective societal

TABLE 8.1
Types of Pollutants under the Stockholm Convention and Their Control Measures (UNEP, 2023)

Pollutant Types	Pollutants	Control Measures
Pesticides	Aldrin	Elimination of production and use; environmentally sound disposal of existing stockpiles.
	Chlordane	Elimination of production and use; exceptions for specific uses like termite control in certain countries; safe disposal of stockpiles.
	DDT	Restricted use for disease vector control in line with WHO guidelines; encouragement of safer substitutes; proper handling and disposal of stockpiles.
	Dieldrin	Phase-out of production and use; safe disposal of existing stockpiles.
	Endrin	Phase-out of production and use; environmentally sound disposal of stockpiles.
	Heptachlor	Elimination of production and use; exceptions for specific uses like termite control in certain countries; environmentally sound disposal of stockpiles.
	HCB	Phase-out of production and use as a pesticide and industrial chemical; proper management and disposal of stockpiles; reduction of unintentional releases.
	Mirex	Phase-out of production and use; safe disposal of existing stockpiles.
	Toxaphene	Elimination of production and use; safe disposal of existing stockpiles.
Industrial chemicals	PCBs	Elimination of production; phased elimination of use in existing equipment by 2025; safe and environmentally sound disposal of PCB-containing equipment and wastes by 2028.
	HCB	Elimination of production and use; proper disposal of stockpiles and wastes; reduction of unintentional releases.
By-products	HCB	Reduction and eventual elimination of unintentional releases.
	PCDD and PCDF	Reduction and eventual elimination of unintentional releases; adoption of optimal technical methods and leading sustainability practices, coupled with the creation of systematic registries and strategic operational frameworks.
	PCBs	Reducing and ultimately eliminating unintentional releases.

(Continued)

TABLE 8.1 (*Continued*)
Types of Pollutants under the Stockholm Convention and Their Control Measures (UNEP, 2023)

Pollutant Types	Pollutants	Control Measures
Newly added POPs	PFOS, its derivatives and PFOSF	Restriction on production and use with specific exemptions; development and promotion of safer alternatives; safe disposal of PFOS-containing products and wastes.
	PCNs	Phase-out of production and use; proper disposal of stockpiles and wastes; reduction of unintentional releases from industrial processes.
	HCBD	Phase-out of production and use; safe disposal of stockpiles and wastes; reduction of unintentional releases as a by-product.
	PeCB	Phase-out of production and use; safe disposal of stockpiles and wastes; reduction of unintentional releases from industrial processes.
	PBDEs	Elimination of production and use; management and disposal of PBDE-containing products in an eco-friendly manner; promotion of alternatives and recycling practices that prevent environmental release.
	α-HCH	Elimination of production and use; reduction of unintentional releases as a by-product.
	β-HCH	Elimination of production and use; reduction of unintentional releases as a by-product.
	Chlordecone	Elimination of production and use; safe disposal of stockpiles and wastes; development and promotion of safer alternatives.
	c-decaBDE (commercial mixture)	Elimination of production and use, except for specific exemptions; reduction of unintentional releases from industrial processes.
	Dechlorane Plus	Elimination of production and use, except for specific exemptions; development and promotion of safer alternatives.
	Dicofol	Elimination of production and use;
	HBB	Elimination of production and use; safe disposal of stockpiles and wastes.
	HBCD	Elimination of production and use; safe disposal of stockpiles and wastes; development and promotion of safer alternatives.
	hexaBDE and HpBDE (commercial octabromodiphenyl ether)	Elimination of production; development and promotion of safer alternatives.

(Continued)

TABLE 8.1 (*Continued*)
Types of Pollutants under the Stockholm Convention and Their Control Measures (UNEP, 2023)

Pollutant Types	Pollutants	Control Measures
	HCB	Elimination of production and use; reduction of unintentional releases as a by-product.
	Lindane	Elimination of production; reduction of unintentional releases from industrial processes.
	Methoxychlor	Elimination of production and use; development and promotion of safer alternatives.
	PCP and its salts and esters	Elimination of production and use; reduction of unintentional releases as a by-product.
	PFHxS, its salts and PFHxS-related compounds Draft indicative list of substances covered by the listing of PFHxS, its salts and PFHxS-related compound	Elimination of production and use; development and promotion of safer alternatives.
	PFOA, its salts and PFOA-related compounds Indicative list of substances covered by the listing of PFOA, its salts and PFOA-related compounds	Restriction on production and use with specific exemptions; development and promotion of safer alternatives; safe disposal of stockpiles and wastes.
	SCCPs	Elimination of production and use, except for specific exemptions; development and promotion of safer alternatives.
	Technical endosulfan and its related isomers	Elimination of production and use; development and promotion of safer alternatives.
	BDE-47 and BDE-99 (commercial pentabromodiphenyl ether)	Elimination of production and use; development and promotion of safer alternatives.
	UV-328	Elimination of production and use; development and promotion of safer alternatives.

BDE-47, Tetrabromodiphenyl ether; BDE-99, Pentabromodiphenyl ether; c-decaBDE, Decabromodiphenyl ether; DDT, Dichlorodiphenyltrichloroethane; HBB, Hexabromobiphenyl; HBCD, Hexabromocyclododecane; HCB, Hexachlorobenzene; HCBD, Hexachlorobutadiene; HexaBDE, Hexabromodiphenyl ether; HpBDE, Heptabromodiphenyl ether; PBDEs, Polybrominated diphenyl ethers; PCB, Polychlorinated biphenyls; PCDD, Polychlorinated dibenzo-p-dioxins; PCDF, Polychlorinated dibenzofurans; PCNs, Polychlorinated naphthalenes; PCP, Pentachlorophenol; PeCB, Pentachlorobenzene; PFH_XS, Perfluorohexane sulfonic acid; PFOA, Perfluorooctanoic acid; PFOS, Perfluorooctane sulfonic acid; PFOSF, Perfluorooctane sulfonyl fluoride; $SCCP_S$, Short-chain chlorinated paraffins; α-HCH, Alpha hexachlorocyclohexane; β-HCH, Beta hexachlorocyclohexane.

and economic gains justify the associated risks. Following a specified sunset date, the application of SVHCs is strictly banned unless authorization is granted, ensuring that only those uses that can be justified are allowed to continue.

The restriction component of REACH focuses on limiting the manufacture, placing on the market, or use of hazardous substances that pose unacceptable risks to health or the environment (ECHA, 2025b). Restrictions can apply to specific uses or to all uses of a substance. The process of imposing restrictions involves a proposal, evaluation by the ECHA, and a decision-making process involving the European Commission and member states. This mechanism ensures that the EU can respond effectively to emerging risks associated with chemical substances.

8.2.2.3 Impact of REACH on Industry and Environment

The implementation of REACH has significant implications for both industry and the environment. For the industry, REACH requires companies to share data, thus avoiding unnecessary testing, particularly animal testing, and reducing costs. This data-sharing fosters transparency and collaboration among companies. Additionally, REACH promotes innovation and the development of safer alternatives to hazardous substances, driving the industry toward more sustainable practices.

From an environmental and health perspective, REACH aims to reduce the risks associated with chemical exposure. By demanding detailed information and assessments of chemicals, it ensures that potential hazards are identified and managed effectively. This contributes to the sustainable management of chemicals, aligning with the EU's broader environmental and health policies.

Compliance and enforcement of REACH are overseen by national authorities in each EU member state. These authorities are responsible for ensuring that companies adhere to REACH requirements through inspections and enforcement actions. Companies that fail to comply with REACH face penalties, including fines and restrictions on their ability to market their products within the EU.

The REACH regulation marks a significant advancement in the regulatory governance of hazardous substances within the EU (Fisher, 2008). By requiring companies to demonstrate the safety of their products, REACH promotes transparency, innovation, and sustainable practices in the chemicals sector. As the regulatory landscape continues to evolve, REACH will serve as a cornerstone in determining future development trajectories of chemical safety and environmental protection, both within the EU and globally (ECHA, 2025a).

8.2.3 EPA Regulations

8.2.3.1 Overview of EPA Regulations

EPA is the primary federal agency charged with supervising chemical usage for the preservation of human well-being and natural ecosystems (U.S.EPA, 2024a). Established in 1970, the EPA administers several key pieces of legislation that govern chemical safety, including the Toxic Substances Control Act (TSCA), the Clean Water Act (CWA), and the Clean Air Act (CAA) (U.S.EPA, 2025a). These laws establish an integrated comprehensive framework governing the manufacturing, use, and chemical waste management in the United States.

8.2.3.2 Key Legislation Administered by the EPA

One of the cornerstones of the EPA's regulatory framework is the TSCA, enacted in 1976 (U.S.EPA, 2016). TSCA authorizes the EPA the jurisdictional mandate to regulate both novel and current chemicals, ensuring they do not generate excessive threats to health or the ecosystems. Central to TSCA is the requirement for companies to notify the EPA before manufacturing or importing new chemicals. This premanufacture notification includes extensive records on the chemical's characteristics, uses, and potential health and environmental effects. The EPA evaluates this information to determine whether the chemical can be safely used or if restrictions are necessary.

Another significant element of TSCA is the chemical inventory maintained by the EPA, which includes all chemicals currently in commerce. For existing chemicals, the EPA conducts risk evaluations to assess whether they pose an unreasonable risk under their conditions of use. If a chemical is deemed risky, the EPA can impose a range of risk management measures, from restrictions and labeling requirements to outright bans. The Lautenberg Chemical Safety Act, a 2016 amendment to TSCA, strengthened the EPA's ability to evaluate and manage chemicals by mandating risk evaluations for all active chemicals in commerce.

The CWA is another critical piece of legislation managed by the EPA. Originally enacted in 1948 and significantly amended in 1972, the CWA aims to rehabilitate and safeguard the ecological health of domestic water systems (U.S.EPA, 2024b). The National Pollutant Discharge Elimination System permit regime, a critical mechanism under the CWA, manages the emission of pollutants into federally protected water bodies (U.S.EPA, 2025b). Enterprises are mandated to obtain permits that define permissible contaminant types and discharge volumes, thereby aligning with statutory requirements for aquatic ecosystem protection (U.S.EPA, 2024c). These standards are established by the EPA and must be met or exceeded by state standards. For impaired waters, EPA defines Total Maximum Daily Loads as pollutant thresholds ensuring standard attainment (U.S.EPA, 2025c).

The CAA, legislated in 1963 and strengthened via critical updates in 1970 and 1990, targets the maintenance of national air purity and mitigation of stratospheric ozone depletion (U.S.EPA, 2024d). Under the CAA, the EPA defines National Ambient Air Quality Standards to regulate substances posing risks to human health and ecological stability (U.S.EPA, 2024d). States are incentivized to formulate implementation frameworks, ensuring continuous adherence to air quality attainment goals. Additionally, the EPA regulates emissions of hazardous air pollutants through technology-based standards and risk-based standards, aiming to minimize exposure to toxic atmospheric contaminants. The New Source Performance Standards establish emission standards for new and modified sources of air pollution, ensuring the adoption of the best available control technologies (U.S.EPA, 2021). Major emitters of atmospheric contaminants are required to secure CAA-mandated permits, integrating all applicable emission regulations into a unified compliance framework.

8.2.3.3 Impact of EPA Regulations on Industry and Environment

The regulatory framework administered by the EPA has profound implications for both industry and the environment. For the industry, the requirement to assess and address potential risks from hazardous materials promotes safer practices and innovation in the development of less hazardous substances. This regulatory oversight ensures that industries operate eco-conscious protocols to mitigate public health risks, reducing the incidence of pollution-related illnesses and environmental degradation.

The benefits of EPA regulations extend beyond industry compliance. By requiring detailed information and assessments of chemicals, these regulations help to identify and mitigate potential hazards effectively. This proactive approach contributes to the sustainable management of chemicals, aligning with the EPA's broader environmental and health policies. The regulatory measures ensure that the release of harmful substances into the air, water, and soil is minimized, thereby protecting ecosystems and human communities.

The EPA enforces compliance with its regulations through a robust system of inspections, monitoring, and penalties for violations. This enforcement ensures that companies adhere to established standards and regulations, fostering a culture of compliance and accountability. Noncompliant entities face significant penalties, including fines and operational restrictions, reinforcing the importance of adhering to environmental laws. This rigorous enforcement regime underscores the EPA's commitment to maintaining high standards of chemical safety and environmental protection.

8.2.3.4 Conclusion

The EPA's regulations form a comprehensive framework for managing chemical safety in the United States. Through key legislation such as TSCA, CWA, and CAA, the EPA ensures that chemicals are evaluated, regulated, and managed to protect human well-being and the natural resources. As the regulatory landscape continues to evolve, the EPA's commitment to rigorous oversight and innovation will have a profound impact on the future of hazardous material regulation and pollution prevention, both domestically and internationally.

8.2.4 Other International Regulations and Standards

8.2.4.1 Overview of International Chemical Regulations

In addition to the Stockholm Convention, various other international agreements and standards play crucial roles in managing chemicals and mitigating their threats to human health and ecosystems. These regulations and standards facilitate global cooperation and establish common frameworks for handling hazardous substances. Key international instruments include the Basel Convention, the Rotterdam Convention, the Minamata Convention, and the Strategic Approach to International Chemicals Management (SAICM) (Basel Convention, 2025; Rotterdam Convention, 2025a; SAICM, 2025a).

8.2.4.2 The Basel Convention

The Basel Convention on the Control of Transboundary Movements of Hazardous Wastes and Their Disposal, adopted in 1989, aims to protect human health and the environment against the adverse effects of hazardous wastes (Basel Convention, 1989). This convention regulates the movement of hazardous wastes across international borders and ensures that such wastes are managed and disposed off in an environmentally sound manner (Basel Convention, 1989).

8.2.4.2.1 Key Provisions

- **Transboundary Movement Control:** The Basel Convention requires prior informed consent from the receiving country before hazardous waste can be shipped across borders (Basel Convention, 1989). This ensures that the receiving country is aware of and consents to the importation of dangerous waste.
- **Hazardous Waste Management:** The convention promotes minimizing the production of hazardous waste and encourages the treatment and decentralized management of wastes at or near their origin.
- **Prohibition of Exports:** Exports of hazardous waste to countries lacking the capacity to manage it in an environmentally sound manner are prohibited (Basel Convention, 1989).
- **Technical Assistance and Information Exchange:** The agreement promotes knowledge sharing regarding toxic material regulation and delivers expert support to emerging economies (Basel Convention, 2025).

8.2.4.3 The Rotterdam Convention

Formulated in 1998, the Rotterdam Convention establishes a Prior Informed Consent mechanism to enhance joint accountability in the global trade of toxic substances and regulated agrochemicals (Rotterdam Convention, 2025b). The agreement strengthens international data sharing on dangerous chemical substances and establishes a standardized procedure to collect and communicate importing nations' regulatory approvals for these substances.

8.2.4.3.1 Key Provisions

- **Prior Informed Consent (PIC) Procedure:** Exporters are obligated by the Rotterdam Convention to acquire explicit importing country clearance for specific high-risk chemical consignments prior to cross-border dispatch (Rotterdam Convention, 2025c). This mechanism enables governments to acquire validated documentation necessary for compliant chemical import governance.
- **Information Exchange:** The convention facilitates the sharing of knowledge about the characteristics and risks of hazardous chemicals, helping countries to improve their chemical management practices.
- **National Decision-Making:** Each party must designate a national authority responsible for implementing the PIC procedure and making import decisions for hazardous chemicals.

- **Chemicals Covered:** The protocol identifies synthetic chemical products and pest control substances falling under the PIC framework, ensuring that their trade is carefully monitored and controlled.

8.2.4.4 Strategic Approach to International Chemicals Management (SAICM)

Established in 2006, the SAICM functions as a governance system designed to enhance safety protocols for chemical substances across all stages of production, use, and disposal (SAICM, 2025b).

SAICM addresses chemical safety in a comprehensive and multi-sectoral manner, involving state institutions, global organizations, the industrial sector, and community groups.

8.2.4.4.1 Key Objectives

- **Risk Reduction:** SAICM promotes the adoption of sound management practices and policies to minimize the risks associated with the production, application, and disposal of substances (SAICM, 2006).
- **Knowledge and Information:** The framework encourages the sharing of scientific information and knowledge about chemical hazards and risks, enhancing global understanding and capacity for chemical management.
- **Governance:** SAICM supports the creation and execution of country-specific and worldwide frameworks and regulations that ensure the safe regulation of chemicals (SAICM, 2006).
- **Capacity Building:** The framework provides technical support and financial resources to emerging economies and transitional market nations to build their capacity for sound chemical management (SAICM, 2006).
- **Public Health and Environment:** SAICM aims to reduce the detrimental effects of chemical substances on public health and ecological systems, contributing to sustainable development goals (SAICM, 2006).

8.2.4.4.2 Regional Agreements and Standards

Beyond worldwide accords, area-specific arrangements and standards also play significant roles in chemical management. For example, the EU's REACH regulation serves as a model for comprehensive chemical safety regulation, influencing policies beyond Europe. Similarly, the Asia-Pacific region has developed frameworks such as the ASEAN Hazardous Chemicals Management Scheme to address chemical safety within member states.

The international regulatory landscape for chemical management is shaped by a network of conventions and agreements that establish common standards and promote cooperation among nations. These instruments provide a robust framework for addressing the risks associated with hazardous chemicals, ensuring that operational practices uphold health standards and ecological preservation. By fostering information exchange, capacity building, and collaborative action, these international regulations and standards contribute to a safer and more sustainable world.

8.3 CHINA'S CONTROL MEASURES FOR EMERGING CONTAMINANTS

8.3.1 Overview of China's Approach to Emerging Contaminants

China has recognized the significant environmental and health risks posed by emerging contaminants. In response, the Chinese government has implemented a series of regulatory measures aimed at managing and mitigating these risks. The government's commitment to addressing these issues is reflected in several key policy documents and regulatory frameworks.

8.3.2 National Strategy and Policy Initiatives

The Chinese government has placed significant emphasis on the control and management of emerging contaminants. A pivotal move in this regard was the issuance of the "Action Plan for Controlling Emerging Contaminants" by the Department of the State Council. This document delineates Chinese government's strategic approach, establishing clear objectives and measures for the governance of these pollutants.

The action plan outlines specific goals aimed at reducing the presence and impact of emerging contaminants in the environment. It stresses the necessity for comprehensive monitoring, risk assessment, and the advancement of treatment technologies. Key measures within the plan include boosting research and development, fortifying monitoring and evaluation systems, advocating for the use of safer alternatives, and enhancing public awareness and education about the risks associated with emerging contaminants.

Beyond strategic policy documents, China has instituted specific regulations to control and manage these pollutants. A noteworthy regulatory development is the "List of Key Controlled Emerging Contaminants (2023 Edition)," which came into effect on March 1, 2023. This list identifies particular emerging contaminants (shown in Table 8.2) that are under strict regulatory controls due to their **considerable threats** to human health and the environment. This list also specifies control measures, such as restrictions on production, use, and disposal, along with requirements for monitoring and reporting (shown in Table 8.2). Entities involved with these contaminants must adhere to detailed regulatory requirements, including obtaining necessary permits, conducting regular environmental impact assessments, and implementing the best available technologies for pollution control.

The implementation of these regulatory measures involves multiple components and stakeholders to ensure a comprehensive approach to managing emerging contaminants. Enhanced monitoring systems have been established to track the presence and levels of these pollutants in water, soil, and air. Industries and facilities handling the listed contaminants are obligated to regularly report their emissions and waste management practices to regulatory authorities.

Significant investments are being directed toward research and development to advance treatment technologies for emerging contaminants. This includes creating more efficient methods for detecting, removing, and degrading these substances, as well as promoting the use of safer alternatives to hazardous contaminants. Public

TABLE 8.2
Key Controlled Emerging Pollutants List and Control Measures (Ministry of Ecology and Environment et al. 2022)

No.	Name of Emerging Pollutant	CAS Number[a]	Main Environmental Risk Control Measures[bc]
I	Perfluorooctanesulfonic acid and its salts and perfluorooctanesulfonyl fluoride (PFOS class)	e.g. 1763-23-1 307-35-7 2795-39-3 29457-72-5 29081-56-9 70225-14-8 56773-42-3 251099-16-8	1. Prohibition of production. 2. Prohibition of processing and use (except for listed uses)[d]. 1. For the production of fire-fighting foam agents (the exemption period for this use ends on December 31, 2023). 3. Enterprises using PFOS class for the production of fire-fighting foam agents shall implement mandatory clean production audits according to law. 4. Import or export of PFOS, should handle toxic chemicals in (out) environmental management release notice. Import and export will be banned from January 1, 2024. 5. The use of PFOS that has been prohibited, declared abandoned by the owner, or collected/received by relevant departments according to law and requiring destruction, shall be managed as hazardous waste in accordance with the national Hazardous Waste List or the identification standard of hazardous waste. 6. Enterprises involved in the production or use of PFOS class as key soil pollution supervision units shall establish a soil pollution hazard investigation system according to law, ensuring continuous and effective prevention of the seepage, loss, and dispersion of toxic and harmful substances.

(Continued)

TABLE 8.2 (*Continued*)
Key Controlled Emerging Pollutants List and Control Measures (Ministry of Ecology and Environment et al. 2022)

No.	Name of Emerging Pollutant	CAS Number[a]	Main Environmental Risk Control Measures[bc]
II	Perfluorooctanoic acid and its salts and related compounds[e] (PFOA class)	—	1. Prohibition of the construction of new perfluorooctanoic acid production facilities. 2. Prohibition of production, processing, and use (except for the following uses). 1. Photoengraving or etching processes in semiconductor manufacturing; 2. Photographic coatings for film; 3. Oil and water-repellent textiles to protect workers from health and safety risks caused by hazardous liquids; 4. Invasive and implantable medical devices; 5. Production of perfluorobutane sulfonic acid from perfluorooctane sulfonic acid for pharmaceutical production purposes; 6. Producing polytetrafluoroethylene (PTFE) and polyvinylidene fluoride (PVDF) for corrosion-resistant membranes, heat exchangers, and sealants to block VOCs and PM2.5; 7. Manufacturing polychlorotrifluoroethylene (FEP) for the production of high-voltage power transmission cable. 3. Enterprises that use PFOA for production for the above purposes shall implement mandatory cleaner production audit in accordance with the law. 4. Importers and exporters of PFOA listed in China's restricted toxic chemicals must obtain the environmental management release notice for toxic chemicals' entry or exit. 5. The use of PFOA that has been prohibited, or that the owner has declared to be abandoned, or that the relevant departments have collected or received according to law and need to be destroyed, and that is determined to be hazardous waste according to the national list of dangerous waste or the identification standard of dangerous waste, shall be implemented as dangerous waste environmental regulation. 6. Enterprises involved in the production or use of PFOA in key soil pollution supervision units shall establish a soil pollution hidden danger investigation system according to law to ensure that toxic and harmful substances are continuously and effectively prevented from leaking, losing, and spreading.

(*Continued*)

TABLE 8.2 (*Continued*)

Key Controlled Emerging Pollutants List and Control Measures (Ministry of Ecology and Environment et al. 2022)

No.	Name of Emerging Pollutant	CAS Number[a]	Main Environmental Risk Control Measures[bc]
III	Decabromodiphenyl ether	1163-19-5	1. Prohibition of production and processing (except for listed uses). 1. Textile products that require flame retardancy (excluding clothing and toys); 2. Additives for plastic parts in household appliances (heaters, irons, fans) that contact electrical components or need flame retardancy, with less than 10% by weight; 3. Polyurethane foam for architectural insulation; 4. The exemption for these uses will end on December 31, 2023. 2. Enterprises using decabromodiphenyl ether for the production of the above uses shall implement mandatory clean production audits according to law. 3. Import or export of decabromodiphenyl ether, included in the list of toxic chemicals strictly restricted by China, shall be subject to the management notice for the import and export of toxic chemicals. From January 1, 2024, import and export are prohibited. 4. Decabromodiphenyl ether shall be classified and managed as hazardous waste if it meets any of the following criteria: 1. Banned by applicable regulations; 2. Declared as waste by the owner; 3. Confiscated or received by authorities and requiring destruction; and is confirmed as hazardous waste through its inclusion in the National Hazardous Waste Inventory or by meeting the Hazardous Waste Identification Standards. 5. Enterprises involved in the production or use of decabromodiphenyl ether as key soil pollution supervision units shall establish a soil pollution hazard investigation system according to law, ensuring continuous and effective prevention of the seepage, loss, and dispersion of toxic and harmful substances.

(*Continued*)

TABLE 8.2 (*Continued*)
Key Controlled Emerging Pollutants List and Control Measures (Ministry of Ecology and Environment et al. 2022)

No.	Name of Emerging Pollutant	CAS Number[a]	Main Environmental Risk Control Measures[bc]
IV	Short-chain chlorinated paraffins[f]	e.g. 85535-84-8 68920-70-7 71011-12-6 85536-22-7 85681-73-8 108171-26-2	1. Prohibition of production and processing (except for listed uses). 1. Additives for conveyor belt production in rubber industry; 2. Spare parts for rubber conveyor belts used in the mining and forestry industries; 3. Leather sector, particularly for leather fattening; 4. Lubricant additives, especially for engines used in automobiles, generators, and wind energy facilities, oil and gas exploration drilling, and production diesel refineries; 5. Outdoor decorative lamp tubes; 6. Waterproof and flame-retardant paints; 7. Adhesives; 8. Metalworking; 9. Secondary plasticizer for flexible polyvinyl chloride (not for processing and use in toys and children's products); 10. The exemption for these uses ends on December 31, 2023. 2. Enterprises using short-chain chlorinated paraffins for the production of the above uses shall implement mandatory clean production audits according to law. 3. Import or export of short-chain chlorinated paraffins shall be subject to the management notice for the import and export of toxic chemicals. From January 1, 2024, import and export are prohibited. 4. Short-chain chlorinated paraffins that have been banned, or declared as waste by the owner, or legally confiscated or received and needs to be destroyed by relevant departments, if determined as dangerous waste according to the national dangerous waste list or dangerous waste identification standards, shall be treated as dangerous waste. 5. Enterprises involved in the production or use of short-chain chlorinated paraffins as key soil pollution supervision units shall establish a soil pollution hazard investigation system according to law, ensuring continuous and effective prevention of the seepage, loss, and dispersion of toxic and harmful substances.

(*Continued*)

TABLE 8.2 (*Continued*)
Key Controlled Emerging Pollutants List and Control Measures (Ministry of Ecology and Environment et al. 2022)

No.	Name of Emerging Pollutant	CAS Number[a]	Main Environmental Risk Control Measures[bc]
V	Hexachlorobutadiene	87-68-3	1. Prohibition of production, processing, and import and export. 2. Enterprises involving hexachlorobutadiene shall implement emission standards in accordance with the "Petroleum Chemical Industry Pollutant Emission Standards" (GB 31571). 3. Hexachlorobutadiene that has been banned, or declared as waste by the owner, or legally confiscated or received and needs to be destroyed by relevant departments, if determined as dangerous waste according to the national dangerous waste list or dangerous waste identification standards, shall be treated as dangerous waste. Strict implementation of dangerous waste management requirements for heavy fractions and high-boiling point residue bottoms containing hexachlorobutadiene in chemical production processes. 4. Enterprises involved in the production or use of hexachlorobutadiene as key soil pollution supervision units shall establish a soil pollution hazard investigation system according to law, ensuring continuous and effective prevention of the seepage, loss, and dispersion of toxic and harmful substances.
VI	Pentachlorophenol and its Salts and Esters	87-86-5 131-52-2 27735-64-4 3772-94-9 1825-21-4	1. Prohibition of production, processing, and import and export. 2. Pentachlorophenol and its salts and esters that have been prohibited, or declared as waste by the owner, or legally confiscated or received and needs to be destroyed by relevant departments, shall be managed as dangerous waste according to the national dangerous waste list or dangerous waste identification standards. 3. Enterprises involved in the production or use of Pentachlorophenol and its salts and esters as key soil pollution supervision units shall establish a soil pollution hidden danger investigation system according to law to ensure continuous and effective prevention of the seepage, loss, and dispersion of toxic and harmful substances.

(*Continued*)

TABLE 8.2 (*Continued*)
Key Controlled Emerging Pollutants List and Control Measures (Ministry of Ecology and Environment et al. 2022)

No.	Name of Emerging Pollutant	CAS Number[a]	Main Environmental Risk Control Measures[bc]
VII	Dicofol	115-32-2 10606-46-9	1. Prohibition of production, processing, and import and export. 2. Trichlorfon that has been prohibited, or declared as waste by the owner, or legally confiscated or received and needs to be destroyed by relevant departments, shall be managed asdangerous waste according to the national dangerous waste list or dangerous waste identification standards.
VIII	Perfluorohexanesulfonic acid and its salts and related compounds[g] (PFHxS Class)	—	1. Prohibition of production, processing, and import and export. 2. PFHxS class that has been prohibited, or declared as waste by the owner, or legally confiscated or received and needs to be destroyed by relevant departments, shall be managed as dangerous waste according to the national dangerous waste list or dangerous waste identification standards.
IX	Dioxins and their Cis and Trans Isomers	13560-89-9 135821-03-3 135821-74-8	1. From January 1, 2024, prohibits on manufacturing, processing, and trade activities. 2. Dioxins and their cis and trans isomers that have been prohibited, or declared as waste by the owner, or legally confiscated or received and needs to be destroyed by relevant departments, shall be managed as toxic materials according to the national dangerous substances list or hazardous waste identification standards.

(*Continued*)

TABLE 8.2 (*Continued*)
Key Controlled Emerging Pollutants List and Control Measures (Ministry of Ecology and Environment et al. 2022)

No.	Name of Emerging Pollutant	CAS Number[a]	Main Environmental Risk Control Measures[bc]
X	Dichloromethane	75-09-2	1. Prohibition of producing paint removers containing dichloromethane. 2. According to the Cosmetic Safety Technical Specifications, dichloromethane is prohibited as a cosmetic ingredient. 3. According to the "Cleaning Agent Volatile Organic Compound Content Limit" (GB 38508), the total content of dichloromethane, chloroform, trichloroethylene, and tetrachloroethylene in water-based cleaners, semi-water-based cleaners, and organic solvent cleaners shall not exceed 0.5%, 2%, 20% respectively. 4. According to the emission control requirements for dichloromethane in the "Petroleum Chemical Industry Pollutant Emission Standard" (GB 31571), "Synthetic Resin Industry Pollutant Emission Standard" (GB 31572), and "Chemical Synthesis Pharmaceutical Industry Water Pollutant Emission Standard" (GB 21904), implement emission standards. 5. According to the "Air Pollution Prevention and Control Law of the People's Republic of China," enterprises and institutions shall build an environmental risk early warning system according to national regulations, regularly monitor the discharge outlet and surrounding environment, assess environmental risks, investigate environmental safety hidden dangers, and implement proactive strategies to mitigate ecological hazards. 6. According to the "Water Pollution Prevention and Control Law of the People's Republic of China," enterprises and institutions shall observe the discharge outlet and surrounding environment, assess environmental risks, investigate environmental safety hidden dangers, and publicize data regarding hazardous water contaminants while implementing risk control protocols to avoid ecological hazards. 7. Companies involved in the production or use of dichloromethane as key soil pollution supervision units shall establish a soil pollution hidden danger investigation system according to law to ensure continuous and effective prevention of the seepage, loss, and dispersion of toxic and harmful substances. 8. Strict implementation of soil pollution risk control standards to identify and control related soil environmental risks.

(*Continued*)

TABLE 8.2 (*Continued*)
Key Controlled Emerging Pollutants List and Control Measures (Ministry of Ecology and Environment et al. 2022)

No.	Name of Emerging Pollutant	CAS Number[a]	Main Environmental Risk Control Measures[bc]
XI	Chloroform	67-66-3	1. Prohibition of producing paint removers containing chloroform. 2. According to the "Cleaning Agent Volatile Organic Compound Content Limit" (GB 38508), the total content of dichloromethane, chloroform, trichloroethylene, and tetrachloroethylene in water-based cleaners, semi-water-based cleaners, and organic solvent cleaners shall not exceed 0.5%, 2%, and 20%, respectively. 3. According to the emission control requirements for chloroform in the "Petroleum Chemical Industry Pollutant Emission Standard" (GB 31571), implement emission standards. 4. According to the "Air Pollution Prevention and Control Law of the People's Republic of China," enterprises and institutions shall build an environmental risk early warning system according to national regulations, regularly monitor the discharge outlet and surrounding environment, assess environmental risks, investigate environmental safety hidden dangers, and implement proactive strategies to mitigate ecological hazards. 5. According to the "Water Pollution Prevention and Control Law of the People's Republic of China," enterprises and institutions shall observe the discharge outlet and surrounding environment, assess environmental risks, investigate environmental safety hidden dangers, and publicize data regarding hazardous water contaminants while implementing risk control protocols to avoid ecological hazards. 6. Enterprises involved in the production or use of chloroform as key soil pollution supervision units shall establish a soil pollution hidden danger investigation system according to law to ensure continuous and effective prevention of the seepage, loss, and dispersion of toxic and harmful substances.
XII	Nonylphenol	25154-52-3 84852-15-3	1. Prohibition of using nonylphenol as an auxiliary agent in the production of pesticide products. 2. Prohibition of using nonylphenol to produce nonylphenol ethoxylates. 3. According to the Cosmetic Safety Technical Specifications, nonylphenol is prohibited as a cosmetic ingredient.

(*Continued*)

TABLE 8.2 (*Continued*)
Key Controlled Emerging Pollutants List and Control Measures (Ministry of Ecology and Environment et al. 2022)

No.	Name of Emerging Pollutant	CAS Number[a]	Main Environmental Risk Control Measures[bc]
XIII	Antibiotics	—	1. Strict implementation of retail pharmacies selling prescription antibiotic drugs on a prescription basis, and the implementation of veterinary prescription sales and use of veterinary antibiotics. 2. Antibiotic residue produced in the production process of antibiotics, according to the national toxic material list or hazardous waste identification standards, shall be managed as harmful environmental pollutants if determined to be dangerous substances. 3. Strict implementation of the "Fermentation Pharmaceutical Industry Water Pollutant Emission Standard" (GB 21903) and the "Chemical Synthesis Pharmaceutical Industry Water Pollutant Emission Standard" (GB 21904) related emission control requirements.

(Continued)

TABLE 8.2 (*Continued*)
Key Controlled Emerging Pollutants List and Control Measures (Ministry of Ecology and Environment et al. 2022)

No.		Name of Emerging Pollutant	CAS Number[a]	Main Environmental Risk Control Measures[bc]
XIV	Obsolete class[h]	Hexabromocyclododecane	25637-99-4 3194-55-6 134237-50-6 134237-51-7 134237-52-8	1. Prohibition of production, processing, use, and import and export. 2. For those that have been banned, or declared as waste by the owner, or legally confiscated or received by relevant departments and need to be destroyed, if they are determined to be hazardous waste according to the national toxic material list or harmful environmental identification standards, processing shall follow national safety protocols for dangerous waste disposals. 3. For those already included in the soil pollutant regulatory thresholds, strictly implement the soil contamination hazard control standards to identify and manage the related soil environmental risks.
		Chlordane	57-74-9	
		Mirex	2385-85-5	
		Hexachlorobenzene	118-74-1	
		DDT (dichloro-diphenyl-trichloroethane)	50-29-3	
		α-Hexachlorocyclohexane	319-84-6	
		β-Hexachlorocyclohexane	319-85-7	
		Lindane	58-89-9	
		Endosulfan and its related isomers	115-29-7 959-98-8 33213-65-9 1031-07-8	
		Polychlorinated biphenyls	—	

TABLE 8.2 (*Continued*)
Key Controlled Emerging Pollutants List and Control Measures (Ministry of Ecology and Environment et al. 2022)

Notes:

[a] CAS number, i.e., the registration number of the Chemical Abstracts Service (Chemical Abstracts Service, abbreviated as CAS).

[b] Chemical compounds employed in small-scale experimental studies or serving as calibration benchmarks are exempt from the specified bans on manufacturing, handling, or international trade activities. Unless otherwise specified, inadvertent chemical traces below defined concentration thresholds within product matrices are not subject to restricted substance listings.

[c] Entries without a specified period are those that the state has already ordered to be implemented or to be implemented immediately. If there are other requirements in the main environmental risk control measures, follow the regulations.

[d] Processing and use refer to business operations and other activities using chemical substances, and do not include activities such as commercial behavior, stockpiling, conveyance, and the application of items containing chemical materials.

[e] PFOA class refers to: (i) Perfluorooctanoic acid (335-67-1), including any branched isomers; (ii) Perfluorooctanoic acid salts; (iii) Perfluorooctanoic acid-related compounds, i.e., any substance that degrades to perfluorooctanoic acid, including any substance containing a straight or branched perfluorinated carbon chain with the $(C_7F_{15})C$ part as one of the structural elements (including salts and polymers). The following compounds are not considered perfluorooctanoic acid-related compounds: (i) C_8F_{17}-X, where X = F, Cl, Br; (ii) $CF_3[CF_2]n$-R′ covering fluoropolymers, where R′ = any group, $n > 16$; (iii) Perfluoroalkyl carboxylic acids and phosphonic acids with ≥8 perfluorinated carbon atoms (including their salts, esters, halides, and anhydrides); (iv) Perfluoroalkylsulfonic acids with ≥9 perfluorinated carbon atoms (including their salts, esters, halides, and anhydrides); (v) Perfluorooctanesulfonic acid and its salts and perfluorooctanesulfonyl fluoride.

[f] Short-chain chlorinated paraffins refer to straight-chain chlorinated hydrocarbons with a chain length of C_{10} to C_{13}, with a chlorine content by weight exceeding 48%, and a concentration by weight of greater than or equal to 1% in the mixture.

[g] PFHxS class refers to: (i) Perfluorohexanesulfonic acid (355-46-4), including branched isomers; (ii) Perfluorohexanesulfonic acid salts; (iii) Perfluorohexanesulfonic acid congeners and their analogues, that are all materials with a structural component containing $C_6F_{13}SO_2$- and may degrade to perfluorohexanesulfonic acid.

[h] The definition range of the eliminated class of emerging pollutants is consistent with the definition range of the corresponding chemical substances in the Stockholm Convention on Persistent Organic Pollutants.

awareness campaigns are also being conducted to educate citizens about the risks and importance of proper disposal and waste management practices. Additionally, efforts are made to engage various stakeholders, including industry representatives, academic institutions, and civil society organizations, in the formulation and implementation of regulatory measures.

China's proactive approach to controlling emerging contaminants demonstrates its commitment to protecting public health and the ecosystem. Through strategic policies such as the "Action Plan for Controlling Emerging Contaminants" and specific regulatory measures such as the "List of Key Controlled Emerging Contaminants (2023 Edition)," China has established a robust framework for managing these hazardous substances. As the country continues to advance its regulatory and technological capabilities, it sets a strong example for global efforts to address the challenges posed by emerging contaminants.

8.4 LIMITS OF EMERGING POLLUTANTS

8.4.1 Persistent Organic Pollutants (POPs)

8.4.1.1 Limits in China

China's control standards for POPs are reflected in several national standards and regulations. As per the Standard for Drinking Water Quality (GB 5749-2022), the allowable limits for perfluorooctane sulfonate (PFOS) and perfluorooctanoic acid (PFOA) in water are set at 80 and 40 ng/L, respectively. Currently, no specific limits are set for other perfluorinated compounds such as perfluorohexane sulfonate (PFHxS).

Regarding flame retardants, China has not yet set specific limits for decabromodiphenyl ether, hexabromodiphenyl ether, heptabromodiphenyl ether, tetrabromodiphenyl ether, pentabromodiphenyl ether, hexabromocyclododecane, short-chain chlorinated paraffins, and Dechlorane Plus and its isomers.

For PCBs, the Environmental Quality Standards for Surface Water (GB 3838-2002) specifies a water limit of 20 ng/L, while the Soil Environmental Quality Standards (GB 36600-2018) sets a soil limit of 1.4 mg/kg.

Dioxins, encompassing PCDDs and PCDFs, have water and soil limits of 30 pg/L and 0.1 TEQ μg/kg, respectively. These limits are specified in the Standard for Drinking Water Quality (GB 5749-2022) and the Soil Environmental Quality Standards (GB 36600-2018).

China's regulations for organochlorine pesticides are stringent. The Identification Standards of Soil Quality Assessment for Exhibition Sites (HJ 350–2007) stipulates a soil limit of 0.04 ug/kg for aldrin, while the Soil Environmental Quality Standards (GB 36600-2018) sets a soil limit of 20 mg/kg for chlordane. The soil limits for dieldrin and endrin are 0.04 and 2.3 ug/kg, respectively, according to HJ 350–2007. The water limit for heptachlor is 0.2 μg/L (GB 3838-2002), and the soil limit is 1.3 mg/kg (HJ 350–2007). HCB has water and soil limits of 0.05 mg/L (GB 3838-2002) and 3.3 mg/kg (GB 36600-2018), respectively. Hexachlorobutadiene has water and soil limits of 0.6 μg/L (GB 3838-2002) and 1 mg/kg (HJ 350–2007), respectively. Alpha-and beta-hexachlorocyclohexane has water and soil limits of 5 μg/L (GB 5749-2022) and 0.1 mg/kg (GB 15618-2018), respectively. The water limit for lindane is 2 µg/L (GB 3838-2002). Endosulfan's soil limit is 470 mg/kg (GB 36600-2018). The water limit for DDT is 1 µg/L (GB 3838-2002) and the soil limit is 0.1 mg/kg (GB 15618-2018). The water limit for pentachlorophenol and its salts and esters is 9 µg/L (GB 3838-2002) and the soil limit is 12 mg/kg (GB 36600-2018).

8.4.1.2 Limits in the US

The U.S. EPA has detailed standards for POPs. According to EPA-HQ-OW-2022–0114, the water limits for PFOS, PFOA, PFHxS, PFNA, and Hexafluoropropylene oxide dimer acid (HFPO-DA) are 4, 4, 10, 10, and 10 ng/L, respectively. Additionally, the mixture of PFHxS, PFNA, HFPO-DA, and Perfluorobutane sulfonate (PFBS) has a hazard index limit of 1 (U.S.EPA, 2024e).

For flame retardants, decabromodiphenyl ether must not be detectable in water, and its soil limit is 430 mg/kg, as per EPA TSCA and the EPA Soil Screening Guidance. The water limit for pentabromodiphenyl ether is 73 µg/L, and the soil limit is 12 mg/kg (U.S.EPA, 1996).

According to EPA-HQ-OW-2022–0114, the water limit for PCBs and PCDDs advised to be 0.5 µg/L, whereas that for PCDFs is 30 pg/L.

Stringent limits are also set for organochlorine pesticides. Chlordane's water and soil limits are 2 µg/L (U.S.EPA, 2024e) and 1.6 mg/kg (U.S.EPA, 1996), respectively. Dieldrin has water and soil limits of 4.2 pg/L and 0.03 mg/kg, respectively (U.S.EPA, 1996). Endrin's water limit is 2 µg/L (U.S.EPA, 2024e) and the soil limit is 18 mg/kg (U.S.EPA, 1996). Heptachlor has water and soil limits of 0.4 µg/L (U.S.EPA, 2024e) and 0.11 mg/kg (U.S.EPA, 1996), respectively. HCB has water and soil limits of 1 µg/L (U.S.EPA, 2024e) and 0.3 mg/kg, respectively. Hexachlorobutadiene has water and soil limits of 0.86 µg/L and 6.2 mg/kg, respectively (U.S.EPA, 1996). Alpha-hexachlorocyclohexane has water and soil limits of 0.011 µg/L and 0.077 mg/kg, respectively (U.S.EPA, 1996). Beta-hexachlorocyclohexane has water and soil limits of 0.037 µg/L and 0.27 mg/kg, respectively (U.S.EPA, 1996). The water limit for lindane is 0.2 µg/L (U.S.EPA, 2024e) and the soil limit is 0.52 mg/kg (U.S.EPA, 1996). The water limit for pentachlorobenzene is 29 µg/L (U.S.EPA, 1996) and the soil limit is 49 mg/kg (U.S.EPA, 1996). Endosulfan has water and soil limits of 3 µg/L (U.S.EPA, 1996) and 370 mg/kg (U.S.EPA, 1996), respectively. The water limit for DDT is 0.2 µg/L (U.S.EPA, 1996) and the soil limit is 1.7 mg/kg (U.S.EPA, 1996). The water limit for pentachlorophenol and its salts and esters is 3 and 1 µg/L, respectively, (U.S.EPA, 2024e) and the soil limit is 3 mg/kg (U.S.EPA, 1996).

8.4.1.3 Limits in the EU

The EU's limits for POPs are mainly specified in (EU) 2019/1021. For perfluorinated compounds, PFOS limits are 10 ppm for substances and mixtures, 1000 ppm for articles, and 1 µg/m^2 for textiles or coated materials. PFOA and PFHxS limits are set at 0.025 mg/kg for substances, mixtures, or articles containing the respective compounds and their salts, and at 1 mg/kg for related substances (EU, 2019). In firefighting foam mixtures, the limit is 0.1 mg/kg, and this exemption will be reviewed by August 28, 2026 (EU, 2019).

For flame retardants, the limit for decabromodiphenyl ether is 10 ppm for single items and 500 ppm for the sum of five items in mixtures and articles. Hexabromocyclododecane has a limit of 100 mg/kg for substances, mixtures, articles, and the flame-retardant part of articles. The limit for short-chain chlorinated paraffins is 1% for substances/mixtures and 0.15% for articles. PCBs and polychlorinated naphthalenes are prohibited (EU, 2019).

The limit for dioxins, including PCDDs and PCDFs, in substances and mixtures is set at 5 μg/kg (EU, 2019).

For organochlorine pesticides, aldrin, chlordane, dieldrin, endrin, heptachlor, hexachlorobutadiene, lindane, mirex, pentachlorobenzene, endosulfan, DDT, and toxaphene are prohibited. The limit for HCB is 10 mg/kg. The limit for pentachlorophenol, including its salts and esters, is 5 mg/kg (EU, 2019).

8.4.2 Antibiotics

8.4.2.1 Limits in China

In China, there are detailed regulations governing the discharge of pollutants from the antibiotic manufacturing industry. The *Discharge Standard of Water Pollutants for Antibiotic Manufacture* (GB 21903-2008) specifies limits for penicillin, cephalosporins, and other antibiotics. The limit for penicillin is 1000 m^3/t, for cephalosporins is 1900 m^3/t, and for other antibiotics is 1200 m^3/t. The limits for oxytetracycline and tetracycline are both 750 m^3/t; the limit for demethylchlortetracycline is 1200 m^3/t; the limits for chlortetracycline and other antibiotics are both 500 m^3/t. The limits for streptomycin and dihydrostreptomycin are 1450 m^3/t; the limit for gentamicin is 6500 m^3/t; the limit for spectinomycin is 1500 m^3/t; the limit for other aminoglycoside antibiotics is 3000 m^3/t. The limit for erythromycin is 850 m^3/t; the limit for spiramycins is 750 m^3/t; the limit for other macrolide antibiotics is 850 m^3/t. The limit for bacitracin is 6500 m^3/t; the limit for desmycosin is 5000 m^3/t; and the limit for other polypeptide antibiotics is 5000 m^3/t. The limit for hygromycin, adriamycin, rifamycins, and other similar antibiotics is 6000 m^3/t (Chinese National Standard, 2008c).

8.4.2.2 Limits in the US

The U.S. requires further supplementation and confirmation for its limits on antibiotics. Currently, no specific data are available to support these limits.

8.4.2.3 Limits in the EU

The EU also needs further supplementation and confirmation regarding its limits for antibiotics. At present, no specific data are available to support these limits.

8.4.3 Microplastics

8.4.3.1 Limits in China

China has not yet established specific standards for microplastics. Currently, there are no relevant regulations or data.

8.4.3.2 Limits in the US

The U.S. has not yet established specific standards for microplastics. Currently, there are no relevant regulations or data.

8.4.3.3 Limits in the EU

The EU has codified limits for microplastics in REACH Annex XVII. According to these regulations, the limit for intentionally added microplastic is less than 0.01% w/w in products placed on the market after October 15, 2023 (European Chemicals Agency, 2021).

8.4.4 Endocrine Disrupting Chemicals (EDCs)

8.4.4.1 Limits in China

China has not yet established specific standards for endocrine disrupting chemicals (EDCs). Currently, there are no relevant regulations or data.

8.4.4.2 Limits in the US

The U.S. EPA has set limits for EDCs. For example, the water limit for nonylphenol is 1.7 μg/L (U.S.EPA, 2005).

8.4.4.3 Limits in the EU

The EU's limits for EDCs are specified in Directive 2003/53/EC. According to this directive, the limit for nonylphenol is 100 mg/kg (European Union. 2003).

8.5 CONCLUSION

The pollution limits set by different countries reflect varying environmental protection strategies and technological levels. These standards reflect the importance that each country places on environmental protection by considering the impacts of pollutants on the environment and human health. Through the implementation of these regulations, countries can better protect the ecological environment, ensuring human health and sustainable development.

However, there are notable gaps in the standards and regulations for emerging pollutants, particularly for antibiotics and microplastics. Neither China nor the United States has specific standards for microplastics, and both the United States and the EU require further supplementation and confirmation of limits for antibiotics.

The absence of comprehensive regulations for these pollutants highlights the need for future legislative and regulatory efforts. Establishing clear and enforceable limits for emerging pollutants is crucial for addressing the growing concerns over environmental contamination and public health risks. As scientific understanding and technological capabilities advance, it is imperative for governments to update and expand their environmental regulations to ensure they effectively mitigate the risks posed by these contaminants.

The proactive development of standards and regulations for antibiotics, microplastics, and other emerging pollutants will play a vital role in protecting ecosystems and human health, ensuring a sustainable and safe environment for future generations.

REFERENCES

Basel Convention. (1989). Basel Convention on the Control of Transboundary Movements of Hazardous Wastes and Their Disposal. Accessed February 15, 2025. https://www.basel.int/

Basel Convention. (2025). Basel Convention Article 10. Accessed February 15,https://www.basel.int/TheConvention/Overview/tabid/1271/Default.aspx

Chinese National Standard. (2002). *Environmental Quality Standards for Surface Water (GB 3838-2002).* Beijing: Standardization Administration of China.

Chinese National Standard. (2008c). *Discharge Standard of Water Pollutants for Antibiotic Manufacture (GB 21903-2008).* Beijing: Ministry of Ecology and Environment.

Chinese National Standard. (2018a). *Soil Environmental Quality Standards (GB 15618-2018).* Beijing: Standardization Administration of China.

Chinese National Standard. (2018b). *Soil Environmental Quality Standards (GB 36600-2018).* Beijing: Standardization Administration of China.

Chinese National Standard. (2022). *Standard for Drinking Water Quality (GB 5749-2022).* Beijing: Standardization Administration of China.

ECHA. (2025a). Legislation. Accessed February 15 2025, https://echa.europa.eu/legislation

ECHA. (2025b). Proposals to Identify Substances of Very High Concern Previous Consultation. Accessed February 15, https://echa.europa.eu/proposals-to-identify-substances-of-very-high-concern-previous-consultations

EU. (2019). Union, The Journal, Official Union, European (Recast) (Text with EEA Relevance). https://data.europa.eu/eli/reg/2024/996/oj

European Chemicals Agency. (2021). REACH Annex XVII.

European Union. (2003). Directive 2003/53/EC.

European Union. (2006). REACH Regulation (EC) No. 1907/2006 concerning the registration, evaluation, authorisation and restriction of chemicals. *Official Journal of the European Union*. Accessed February 14, 2025. https://echa.europa.eu/regulations/reach/understanding-reach

Fisher, Elizabeth. (2008). The 'perfect storm' of REACH: Charting regulatory controversy in the age of information, sustainable development, and globalization. *Journal of Risk Research* 11 (4): 541–63. doi:10.1080/13669870802086547

Ministry of Ecology and Environment, Ministry of Industry and Information Technology, Ministry of Agriculture and Rural Affairs, Ministry of Commerce, General Administration of Customs, and State Administration for Market Regulation. (2022). Key List of New Pollutants for Priority Control. (2023 Edition). Government of China. Accessed February 15 2022, https://www.gov.cn/zhengce/2022-12/30/content_5734728.htm

Rotterdam Convention. (2025a). Rotterdam Convention - Home Page. Accessed February 18 2025. https://www.pic.int/

Rotterdam Convention. (2025b). Rotterdam Convention. Accessed January 1 2025. https://www.pic.int/TheConvention/Overview/tabid/1044/language/en-US/Default.aspx

Rotterdam Convention. (2025c). How It Works. Accessed January 2 2025. https://www.pic.int/TheConvention/Overview/Howitworks/tabid/1046/language/en-US/Default.aspx

SAICM. (2006). STRATEGIC APPROACH TO INTERNATIONAL CHEMICALS MANAGEMENT SAICM Texts and Resolutions of the International Conference on Chemicals Management. Accessed February 15 2006, www.saicm.org.

SAICM. (2025a). SAICM. Accessed February 18. https://www.saicm.org/Home/tabid/5410/language/en-US/Default.aspx

SAICM. (2025b). SAICM Implementation towards the Achievement of the 2020 Goal. Accessed February 15, https://www.chemicalsframework.org/implementation/saicm-towards-2020

U.S.EPA. (1996). Soil Screening Guidance : Fact Sheet. Accessed February 14. https://nepis.epa.gov/Exe/ZyPURL.cgi?Dockey=1000239P.txt

U.S.EPA. (2005). Environmental Protection Agency. Water Quality Criteria for Nonylphenol. Accessed February 15. https://cfpub.epa.gov/si/si_public_record_report.cfm?Lab=OST&dirEntryId=74614

U.S.EPA. (2016). Summary of the Toxic Substances Control Act. Accessed September 9 2016. https://www.epa.gov/laws-regulations/summary-toxic-substances-control-act

U.S.EPA. (2021). Regulation of persistent, bioaccumulative, and toxic chemicals under TSCA. *Federal Register* 86(5): 866–892. https://www.federalregister.gov/documents/2021/01/07/2021-00025/reimbursement-for-caskets-and-urns-for-burial-of-unclaimed-remains-in-a-national-cemetery-or-a

U.S.EPA. (2024a). Our Mission and What We Do. Accessed May 1 2024. https://www.epa.gov/aboutepa/our-mission-and-what-we-do

U.S.EPA. (2024b). Summary of the Clean Water Act. Accessed June 12 2024. https://www.epa.gov/laws-regulations/summary-clean-water-act

U.S.EPA. (2024c). NPDES Permit Limits. Accessed September 16 2024. https://www.epa.gov/npdes/npdes-permit-limits

U.S.EPA. (2024d). Summary of the Clean Air Act. Accessed September 17 2024. https://www.epa.gov/laws-regulations/summary-clean-air-act

U.S.EPA. (2024e). ENVIRONMENTAL PROTECTION AGENCY 40 CFR Parts 141. Accessed September 17 2024. https://federalregister.gov/d/2024-12645

U.S.EPA. (2025a). Laws and Executive Orders. Accessed January 1 2025. https://www.epa.gov/laws-regulations/laws-and-executive-orders

U.S.EPA. (2025b). National Pollutant Discharge Elimination System. Accessed January 2 2025. https://www.epa.gov/npdes

U.S.EPA. (2025c). Clean Water Act Section 303(d): Impaired Waters and Total Maximum Daily Loads. Accessed January 3, 2025. https://www.epa.gov/tmdl

UNEP. (n.d.). Overview. Stockholm Convention on Persistent Organic Pollutants. Accessed February 10, 2025. https://www.pops.int/TheConvention/Overview/tabid/3351/Default.aspx

UNEP. (2001). Stockholm Convention on Persistent Organic Pollutants. Accessed February15 2001, https://www.pops.int/TheConvention/ThePOPs/The12InitialPOPs/tabid/296/Default.aspx

UNEP. (2023). Stockholm Convention on Persistent Organic Pollutants (POPs): Text and Annexes Revised in 2023. Accessed February16, 2023. https://www.pops.int/TheConvention/ThePOPs/TheNewPOPs/tabid/2511/Default.aspx.

9 Physical Remediation of EOCs in Soil

Qaiser Hussain, Summaya Sajjad, Cheng Liu, Azeem Khalid, Shahzada Sohail Ijaz, Muhammad Akmal, Shahid Mahmood, and Muhammad Ahtisham Ali

9.1 INTRODUCTION

Soil is a naturally occurring medium on Earth. It is composed of organic and inorganic particles. Soil facilitates the growth of terrestrial plants. However, soil faces contamination from various sources. Emerging organic contaminants (EOCs) are a significant concern. EOCs are diverse chemicals not commonly monitored in the environment. They can potentially cause adverse ecological and public health effects (Petrie et al. 2015; Mukhopadhyay et al. 2022). EOCs include pharmaceuticals, pesticides, and industrial chemicals. Personal care products, endocrine-disrupting compounds, and flame retardants are also EOCs. Human activities have increased EOC release into the environment. Industrialization, urbanization, and agricultural intensification are significant contributors (García et al. 2020). EOCs enter soil through various pathways. Direct deposition and accumulation from water and air are common routes. Soil acts as a significant sink for many organic contaminants. It can retain EOCs for extended periods. This retention increases the risk of long-term environmental impacts (Maddela et al. 2022). EOCs in the soil can undergo various transformation processes. They can leach into groundwater systems. Some EOCs accumulate in the food chain through plant uptake (García et al. 2019). The diversity of EOCs poses challenges in understanding their behavior in soil. Some EOCs exhibit high mobility and persistence. This leads to widespread distribution in the environment (Biel-Mareso et al. 2022). Other EOCs are prone to sorption and degradation. These processes can result in harmful transformation products (Biel-Maeso et al. 2019). Multiple EOCs often co-occur in soil. This co-occurrence complicates environmental impact assessments. Additive or synergistic effects between EOCs are possible. The lack of comprehensive toxicological data for many EOCs is a concern. This gap hinders accurate risk evaluation for human and ecological health.

Conventional soil remediation techniques have limitations for EOCs. Methods developed for traditional contaminants may not be suitable. EOCs require innovative and targeted approaches. Physical remediation techniques show promise for EOC contamination in soil. Physical remediation relies on physical processes to remove or immobilize contaminants (Figure 9.1). The well-established physical techniques are

DOI: 10.1201/9781003509523-9

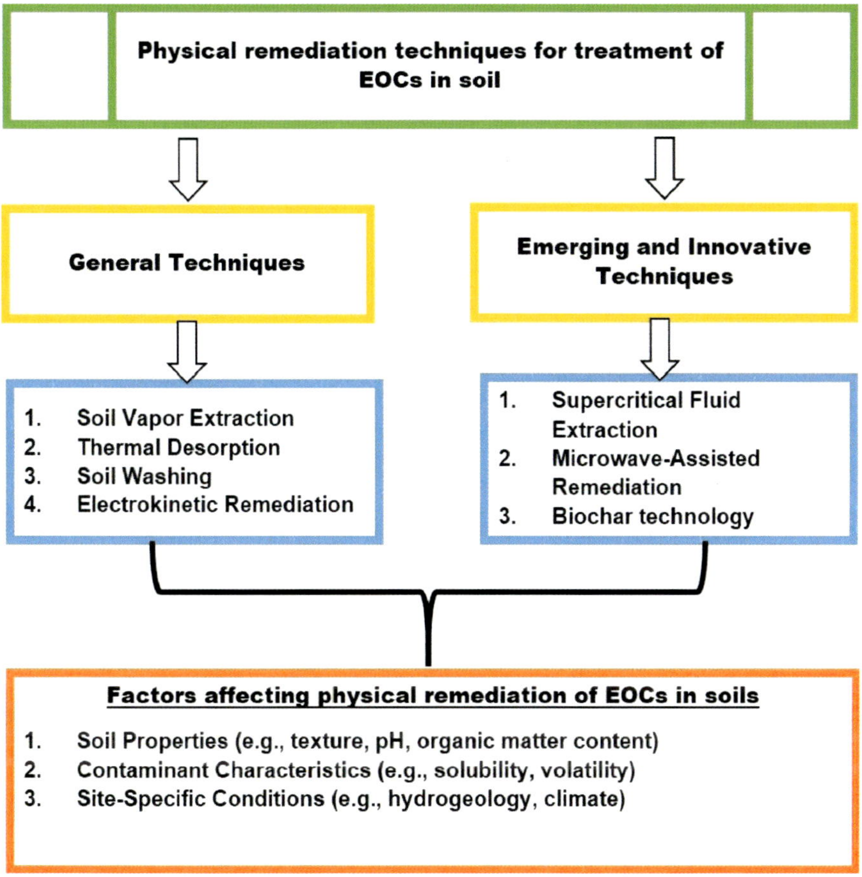

FIGURE 9.1 Flowchart of physical remediation techniques for EOCs in soils and factors affecting the EOCs remediation.

soil vapor extraction, thermal desorption, soil washing, and electrokinetic remediation (EKR) (Kamali et al. 2018; Meng et al. 2021). Emerging and innovative physical remediation techniques are being developed, which include supercritical fluid extraction (SFE), microwave-assisted remediation (MAR), phytoremediation, nanotechnology, and biochar technology (Falciglia et al. 2017; Mustafa et al. 2021; Pourzamani et al. 2021). Each remediation technique has specific applications and limitations. Factors influencing effectiveness vary between methods. Soil type, contaminant properties, and site conditions affect remediation success. Understanding these factors is crucial for selecting appropriate techniques.

Research on EOCs in soil has increased in recent years. However, it remains limited compared with studies on aquatic systems. Most soil-related EOC studies focus on agricultural lands. Biosolids incorporated into the soils are handled in a special way. Soils in urban areas and industrial regions have been found to be the subject of several research works (Li et al. 2018). Some of the recent studies have pointed out the occurrence of different types of EOCs in soil. Among the most popular chemicals that are known to be a menace to human beings, the endocrine-disrupting chemicals are some of the most common. Other emerging pollutants detected in the soil samples include pharmaceuticals, flame retardants, phthalates, perfumes, wetting agents, per- and poly-fluoroalkyl substances (PFAS) (Chiaia-Hernández et al. 2020; Sefiloglu et al. 2021; Maddela et al. 2022).

The EOCs may also be available in the soil and some of these EOCs may be taken up by plants and accumulated within the vegetation system. This plant uptake could lead to bioaccumulation through the food chain. The effects of EOCs on the soil ecosystems are not fully comprehended in the long run. Currently, there are no comprehensive regulations that govern the detection of EOCs in soil matrices. More studies on the toxicological profile of EOC are required with urgency. The European Union, United States, and Australia have begun to address the risks of EOCs but many countries still have no adequate regulation of these chemicals.

In this chapter, information is given regarding how the physical remediation techniques can be useful for managing EOCs in soil. It also looks into the use and nonuse of various methods in soil EOC treatment. Furthermore, factors used to regulate distribution and behavior of EOCs in soils are assessed. The chapter also aims to discuss the current issues and future directions in EOC remediation. The approach of concentrating on the techniques of soil remediation may help in arriving at better EOC mitigation approaches.

9.2 GENERAL TECHNIQUES FOR PHYSICAL REMEDIATION

As the list of EOCs in our soils continues to grow, researchers are giving a lot of thought to the best ways they can physically treat and remediate these problematic compounds. While biological and chemical methods certainly have their place, it is important to also explore the potential of various physical remediation techniques. There are some general and commons physical techniques, such as soil vapor extraction, thermal desorption, soil washing, and EKR, which are used for the remediation of EOCs in various soils. After reviewing some of the latest studies on general physical remediation techniques for EOCs in soils, we have put together a summary in Table 9.1 that highlights the key details.

9.2.1 Soil Vapor Extraction

Soil vapor extraction (SVE) has been established as a realistic in situ method for the treatment of volatile organic compound (VOC)-contaminated unsaturated soils (Chen et al. 2007). SVE involves the use of artificially induced airflow through the contaminated vadose zone to straining and immobilizing VOCs in a vapor phase. Vadose zone wells are used to collect and transport the vaporized contaminants to the surface, where they may be treated by exhaust treatment if needed and then released. SVE has shown the ability to remediate large volumes of VOC contamination from soils within short treatment cycles with minimum risk and maximum effectiveness. This nonintrusive deployment has gained widespread acceptance due to affordability. Subsequent to promising initial laboratory and field-scale feasibility

TABLE 9.1
General Techniques for Physical Remediation of EOCs in Soil

References	Techniques Used	Contaminant Types	Main Results
Ding et al. (2022)	• Thermal oxidation SVE (TOSVE) system • Addition of scoria to oxidant for tail gas treatment	• Halogenated hydrocarbons • 1,2-Dichloroethane (1,2-DCA)	• Thermal extraction improved pollutant extraction from low-permeability soil. • Scoria enhanced tail gas treatment efficiency in the TOSVE system.
Labianca et al. (2020)	• Soil vapor extraction (SVE) • Full-scale remediation system	• TPH, PAHs, solvents, pesticides, VOCs (volatile organic compounds)	• VOC concentration reduced by 73% after 4 years of treatment. • Some soil samples required extended remediation period due to limits.
Jiang and Weng (2022)	• Thermal desorption at different temperatures	• Macromolecular organics, long-chain alkanes, macromolecular acids • Small molecular organics, CO_2, SO_2, NO_x, CO	• Pollutants decompose into small organics at 500°C. • Organics transform into CO_2, SO_2, NO_x, and CO above 500°C.
Rassaei (2023)	• Thermal desorption	• Per- and polyfluoroalkyl substances (PFASs)	• Thermal desorption effectively removes PFASs from calcareous soils. • Optimal removal at 350°C–450°C for 40–60 minutes.
Caetano et al. (2023)	• Ex-situ soil washing with surfactant solutions • Evaluation of washing solution recycling and safe disposal	• Total petroleum hydrocarbons (TPH)	• TPH removal efficiency of $(80.7 \pm 3.2)\%$ with Tween 80 after 5 hours of washing. • TPH removal efficiency of $(90.7 \pm 2.8)\%$ with SDS after 2 hours of washing.
Quinnan et al. (2022)	• Ex situ soil washing • Segregation of fines for alternative treatment	• Per- and polyfluoroalkyl substances (PFAS)	• Successful removal of PFAS from coarse soil fractions • Fines segregated for alternative treatment methods
Zhang et al. (2024)	• Electrokinetic-assisted phytoremediation (EKPR) • Investigation of different electric field intensities	• Dibutyl phthalate (DBP)	• Different electric fields impact DBP removal efficiency by maize. • EKPR shows potential for remediating organic contaminated soils.

(*Continued*)

TABLE 9.1 (*Continued*)
General Techniques for Physical Remediation of EOCs in Soil

References	Techniques Used	Contaminant Types	Main Results
Pasvantu et al. (2023)	• Electrochemical treatment method	• Total petroleum hydrocarbons	• 55% remediation of TPH in soil after 14 days. • Electrochemical method decreased TPH from 5000 to 2228 mg/kg.
Riis et al. (2023)	• EK-BIO technique for in situ bioremediation • Bioaugmentation with dehalorespiring bacteria (KB-1®)	• Tetrachloroethene	• 98.75% reduction in average soil concentrations (PCE eq.). • Met remediation goals within 2 years of system operation.

studies, SVE has been extensively applied. The process involves several sequential steps. Firstly, wells are installed directly amidst or surrounding contaminated soils. Then, the well is isolated at the surface and linked to a vacuum pump. This generates airflow circulation throughout the soil matrix. Volatile fractions transition from the liquid to vapor phase. Vapors are drawn into wells and drawn off. Extracted emissions undergo activated carbon treatment ahead of release. Several factors impact SVE productivity, including operational parameters such as airflow rate and temperature, as well as contaminant qualities such as vapor pressure and solubility. Soil properties such as moisture content and organic matter percentage also exert influence. SVE relies on the movement of contaminants between the phases of solid, liquid, and gases in the soil. It causes movement of aeration through the contaminated soil in the form of pores and pore spaces. It was found that the influential process parameters need to be optimized for making SVE implementation cost-effective. This transfers VOCs to the air stream. Contaminated vapors are collected and treated above ground. The United States Environmental Protection Agency studied contaminants at 1156 superfund soil sites from 1981 to 2020. VOCs were found in 58% of sites, and SVOCs in 70%. In China, a soil pollution report showed that 16.1% of samples exceeded environmental quality standards. Organic pollution accounted for 17.6% of these samples (Zhao et al. 2015).

The SVE process involves three main steps:

1. **Volatilization:** Contaminants move from adsorbed, dissolved, or liquid phases to vapor phase.
2. **Vapor transport:** Vaporized contaminants move through soil pores to extraction wells.
3. **Extraction:** Contaminated vapors are collected and treated above ground.

A typical SVE system has these components: (i) extraction wells or vents, (ii) vacuum blowers or pumps, (iii) piping and manifold system, (iv) off-gas treatment system, and (v) monitoring and control systems. SVE is most effective for chemicals with: boiling points less than 250°C; high vapor pressure (greater than 0.5 mmHg); Henry's Law constant greater than 100 atm.

SVE has been used successfully at various contaminated sites:

- **Kelly Air Force Base, Texas:** Removed chlorinated solvents such as trichloroethylene (TCE) and tetrachloroethylene (PCE) (Robert 1996).
- **Former Dry Cleaner Site, South Carolina:** Addressed PCE and TCE contamination (Simons et al. 2018).
- **Gasoline Station Site, Somerton, Arizona :** Remediated soils with BTEX compounds (Bhuyan and Latin, 2012).

Recent research has explored SVE for EOCs:

- Pharmaceuticals and personal care products (De Melo Henrique et al. 2022)
- PFAS (Hoag et al. 2022)
- Fuel oxygenates and hydraulic fracturing chemicals (Zhang et al. 2023b)

SVE works best for VOCs and semi-volatile organic compound (SVOCs) with high vapor pressures and low soil-water partitioning coefficients. Examples include chlorinated solvents, petroleum hydrocarbons, and some EOCs (Nematollahi et al. 2018; Labianca et al. 2020).

Some of the ways that can enhance the effectiveness of SVE include: drying the wet soil before treatment and raising the temperature of the soil (using microwaves, heated air, or steam injection).

SVE has some limitations:

1. It only works in the unsaturated zone above the water table or groundwater.
2. It is less effective in low-permeability soils such as clays or compacted soil.
3. Nonaqueous phase liquids can reduce efficiency.
4. Inapplicable to samples containing low volatility compounds or those with high K_d (soil-water distribution coefficient) values.
5. Fluctuations in the soil properties can cause uneven distribution of vacuum in the field.

In general, SVE is a popular and efficient technology for treating VOCs and SVOCs in contaminated soil. Some of the benefits include low cost and the ability to install the devices without interfering with the normal functioning of the business. However, its efficiency is influenced by some factors such as the nature of the soil and the contaminants. Current research is being done to determine the effectiveness of the technology in treating emerging contaminants and therefore increasing its uses in environmental cleanup.

9.2.2 Thermal Desorption

Thermal desorption involves heating contaminated soils without oxygen to vaporize the pollutants and separate them from the soil (Zhao et al. 2019). This method is particularly well-suited for treating undegradable organic pollutants. Several review studies have been conducted on thermal desorption to help researchers increase their understanding of the field. For example, Troxler et al. (1993) conducted two reviews to analyze and summarize remediation methods for soils contaminated with petroleum and pesticide. Gao and Jiang (2012) analyzed several heating methods for remediating contaminated soils and discussed their influencing factors. Thermal desorption technologies can be divided into in-situ thermal desorption (ISTD) and ex-situ thermal desorption (ESTD) (Kristanti et al. 2022). ISTD is a process of treating the contaminated soil on-site without excavating it. This is done either by heating the soil directly or indirectly while the contaminants are still in the ground. The gases produced during this process are then collected and purged through treatment. ISTD is beneficial as excavating and transporting contaminated soil is unnecessary, which decreases handling and intrusiveness on the site. However, it may be difficult to heat and treat the subsurface uniformly to ensure that all the required processes are completed. The process of ESTD entails digging the contaminated soil and transporting it to the treatment site. The soil is then subjected to heat in a specific chamber or reactor, making the contaminants gaseous. They are then collected and purified in an off-gas treatment process. ESTD can also be managed more effectively because the soil is enclosed within the treatment system to control the heating and the treatment. However, it requires the extra processes of soil removal, hauling, and management, which can be more invasive and expensive. The decision as to when to use ISTD and when to use ESTD will, therefore, depend on the degree and type of contamination, accessibility of the site, costs, time, and practicality. Each method has its advantages and disadvantages, and the most effective method depends on the characteristics of the contaminated site.

Studies have shown that thermal desorption can significantly reduce concentrations of various contaminants in soils, including polycyclic aromatic hydrocarbons (PAHs), polychlorinated biphenyls (PCBs), dichlorodiphenyltrichloroethane (DDTs), and total petroleum hydrocarbons (TPH). For example, desorption efficiency of PAHs increases with temperature, although residual PAHs may become more strongly immobilized and their bioaccessibility reduced (Cheng et al. 2021). Similarly, thermal desorption has effectively reduced PCB concentrations in soils, although oxygen presence can lead to more toxic byproducts like polychlorinated dibenzofurans (PCDFs) (Zhao et al. 2016). One challenge is off-gas treatment, which can contain volatilized organics. Effective off-gas treatment is necessary to prevent secondary pollution. Various methods have been explored, including pulsed corona discharge plasma showing high DDT degradation efficiency in off-gas (Ma et al. 2019). Additionally, combined processes for wastewater treatment from thermal desorption, such as demulsification, flocculation, ammonia stripping, Fenton oxidation, and reverse osmosis, have achieved high removals of various contaminants (Xiao et al. 2022).

Thermal desorption not only removes contaminants but also impacts soil physicochemical properties. For example, thermal treatment can alter organic carbon speciation and behavior of residual pollutants. Key factors influencing residual PAH desorption include decomposition of aliphatic/carboxylic carbons and aromatic carbon recalcitrance during thermal treatment (Cheng et al. 2021). Additionally, low-temperature indirect thermal desorption has been shown to rapidly recolonize microbial communities in treated soils, suggesting this method can be effective and environmentally sustainable (Sang et al. 2021). Ongoing research seeks to better understand pollutant migration/transformation during thermal desorption and develop more effective off-gas methods. Areas of active investigation also include integrating thermal desorption with other remediation technologies and developing cost-effective and energy-efficient processes (Wang et al. 2021).

Low-temperature thermal desorption (heating temperature <350°C) is suitable for treating VOCs, such as gasoline and benzene. High-temperature thermal desorption (heating temperature between 350°C and 600°C) is suitable for treating SVOCs, such as PAHs and PCBs, or inorganic compounds, such as Hg. Microwave heating technology, with advantages of rapid, selective, and simultaneous heating, is a feasible and attractive thermal method to remediate organic contaminated soil, avoiding overheating the surface of soil substrate.

9.2.3 Soil Washing

Soil washing emerges as the most preferred technique in the treatment of contaminated soils that are loaded with organic materials. This process dissolves contaminants in soil into a water-based solution with the help of different washing agents that make it a versatile method to remove contaminants from soil (Srivastava et al. 2022; Faigle et al. 2023). The success of soil washing depends on the washing agents used, which may be inorganic or organic chelating agents, surfactants, or natural products. Agents extract contaminants to various extents under different circumstances (Liu et al. 2021). For example, synthetic surfactants, such as sodium dodecyl sulfate and Triton X-100, can increase contaminant removal efficiency by up to 90% in soils. However, it is more advisable to use natural surfactants like humic acids because they are advantageous to the environment and increase microbial action (Conte et al. 2005). Cyclodextrins have also been observed to form complexes with pollutants, hence helping in their removal from soils (Mousset et al. 2014).

Combining soil washing with other remedial techniques can increase the overall efficiency and address the limitations of applying soil washing individually. The integration of soil washing with photocatalysis has been proved to degrade aromatic compound present in the polluted soils (Fabbri et al. 2008). Similarly, combining soil washing with electrochemical processes can be beneficial for the removal of chlorinated hydrocarbons (Muñoz-Morales et al. 2021). The majority of the research on soil washing has been done in a controlled environment within a laboratory, and there are limited field tests carried out at the pilot scale. However, the effectiveness of this method in real-life situations has to be established, and that is why, it has to be practiced. Soil characteristics such as the type of soil, soil texture, and washing conditions significantly influence the effectiveness of the soil washing process in the field (Liu et al. 2021). For example, a combination of sulfuric acid and isopropanol was efficient in removing the mixed pollutants from sandy loam soil (Semer & Reddy 1996).

Creating eco-friendly and cost-effective washing agents is a main focus of research. Natural soaps and agents made from waste have shown promise in cutting down on environmental harm and making the soil healthier after cleanup. Sewage sludge-based products have an impact on heavy metal removal and soil quality improvement (Klik et al. 2021). Also, scientists have looked into methods like advanced oxidation and selective adsorption to get back extracting agents from soil washing mixtures (Trellu et al. 2021). Soil washing works well, but it costs a lot. This means we need to fine-tune things such as how much surfactant to use, the liquid-to-solid ratio, how long to wash, how to mix, and how many times to wash. When we pair soil washing with other methods, such as advanced oxidation, we can create a better way to remove specific pollutants. In the future, researchers should try to clean up complex pollution cases and stop new pollution problems. At the same time, they should come up with cheap, eco-friendly technologies that work in real-world settings. Systems that blend soil washing with other techniques show potential to make soil cleanup more effective and sustainable around the globe.

9.2.4 Electrokinetic Remediation

EKR has emerged as a promising technique for removing organic pollutants from contaminated environments. Various studies have highlighted the effectiveness of EKR in treating a wide range of pollutants such as emerging pollutants, heavy metals, chlorophenols, and petroleum hydrocarbons. The EKR process includes basic techniques such as electrolysis, electroosmosis, electrophoresis, diffusion, and electromigration and depends on factors such as pH, electrodes, voltage, and electrolytes (Wen et al. 2021). The efficiency of EKR for the removal of organic pollutants, however, has been enhanced by several augmentation methods. Surfactants, co-solvents, nanoparticles, and oxidant chemicals have been identified to enhance the extraction or destruction of pollutants in situ (Cameselle & Gouveia 2018). For instance, one study was done on the surfactants and the complexing agents to enhance the desorption and solubility of the PAHs. This study revealed that, at 1 M EDTA, greater than 90% of the pollutants were desorbed from kaolinite clay, while 70% were desorbed from sandy soil within 30 days (Alcántara et al. 2012). Similarly, in another study, the authors emphasized that polarity reversal and voltage gradient should be applied periodically in order to sustain the conditions that are conducive to microbial activity and biodegradation of organics; the percentage of diesel removal in this case was 36% (Mena et al. 2016).

The researches of EKR in the past few years have used the computational techniques like density functional theory, molecular dynamics simulations, and quantum mechanics/molecular mechanics (QM/MM) simulations for the enhanced understanding and treatment of EKR. Moreover, research has considered other methods such as separated electrode chambers for increasing the effectiveness of the removal of the organic pollutant. A study on the electrokinetic-bioremediation process revealed that the suggested technique effectively reduced heavy metals by 60% and POPs by 90% of sites polluted by electronic waste recyclers. They used citrate to increase the electrical current in the soil as well as the electroosmotic flow, which in turn, enhanced the removal of the various contaminants from the soil (Chen et al. 2021). However, several issues are still notable when it comes to the use of EKR for the removal of organic pollutants. The diverse nature of the soil matrices, the properties of the contaminants, and the necessity for the development of the program to be suited to the site in question are the main challenges. Furthermore, the energy consumption and the side effects of EKR including changes of the pH and temperature of the soil must be considered. Future research should address the further improvements of techniques like the advanced oxidation technologies and nanotechnology because these technologies have high removal rates of EOCs from the soils according to the laboratory findings. Furthermore, the extension of laboratory results to field applications needs improved understanding of the controlling mechanisms in heterogeneous context of the environment.

9.3 EMERGING AND INNOVATIVE TECHNIQUES

Conventional soil remediation methods, such as excavation and landfill disposal, often come with significant drawbacks, including high costs and the potential for further environmental contamination. In recent years, researchers have been exploring more innovative, sustainable, and effective techniques such as SFE, MAR, and biochar technology for the treatment of EOC-contaminated soils (Table 9.2).

9.3.1 Supercritical Fluid Extraction

SFE is one of the most efficient techniques for the decontamination of organic pollutants in different matrices (Yabalak et al. 2024). The rationale of this method is based on the use of supercritical fluids, which possess characteristics of both liquids, as solvents, gases, and as diffusion media. The most preferred supercritical fluid in environmental applications is carbon dioxide (CO_2) because it is nontoxic, nonflammable, and its critical temperature and pressure are low (Sadatshojaei & Wood 2021). SFE is one of the most beneficial methods due to the fact that it is possible to select the temperature and pressure to achieve the desired solvating power and extract various contaminants. This flexibility is advantageous when dealing with multiple compounds of pollution at the same time (Sunarso & Ismadji 2009). Supercritical fluids such as carbon dioxide and water are used for the extraction of interferents from solid and liquid samples as green solvents with better economy than conventional solvents. SFE has demonstrated potential in treating soil contamination by selectively removing different contaminants without changing the soil's characteristics with the existing prototypes having remediation efficiencies of about 80% (Chen et al. 2022). Furthermore, the application of SFE goes further to the total elimination of highly resistant organic waste pollutants in industrial effluents through mineralization and thus a sustainable and cost-effective means of dealing with hazardous organic waste (Pal & Giri 2021). The current studies call for a better understanding of the SFE parameters such as pressure, temperature, and flow design to improve the efficiency of remediation; more large-scale studies are recommended. Moreover, SFE has no use for organic solvents, thus minimizing the environmental effect and the expenses on solvent disposal. The efficiency of SFE depends on the solubility of the analytes in the supercritical fluid, the interaction between the analytes and the matrix as well as the extraction conditions such as the temperature and pressure (Librando et al. 2004). The use of organic modifiers such as methanol, *n*-hexane, and toluene can also improve the extraction yield since the solubility of the target compounds in the supercritical fluid will be improved (Librando et al. 2005).

TABLE 9.2
Emerging and Innovative Techniques for Physical Remediation of EOCs in Soil

References	Techniques Used	Contaminant Types	Main Results
Jones et al. (2023)	• Subcritical water degradation with hydrogen peroxide • Static wet oxidation of chlorophenols	• Organochlorine	• Complete oxidation and extraction achieved at 100°C. • 100% removal and destruction of DDT at 250°C.
Kariyawasam et al. (2023)	• Supercritical fluid extraction (SFE) with ethanol as the modifier • Microwave-assisted extraction (MAE) • Eucalyptus oil-assisted extraction (EuAE)	• Polycyclic aromatic hydrocarbons	• Comparable recoveries of PAHs using all three methods. • SFE was the most efficient for naturally incurred soils.
Wu et al. (2024)	• Bluecoke-based modifiers ($KHCO_3$@BC and $KHCO_3/MnO_2$@BC) • Microwave remediation assisted by KBC or KMnBC.	• Polycyclic aromatic hydrocarbons/ crude oil	• PHs removal rates: 76.16% and 88.31%. • Soil quality improved, industrial feasibility confirmed for MW remediation.
Chen et al. (2023)	• Microwave heating remediation	• Polycyclic aromatic hydrocarbons/ crude oil	• 98.95% TPH removal in light crude oil soil in 15 minutes. • Heavy crude oil soil removal was less efficient and consumes more energy.
Tian et al. (2021)	• Bluecoke powder as microwave absorber	• Nitrobenzene	• Enhanced removal rate of nitrobenzene achieved with bluecoke powder. • Heating temperature characteristics improved during microwave remediation process.
Qu et al. (2024)	• Biochar supported nZVI (nZVI@BC800) fabricated by carbothermal reduction method	• Benzo[*a*]pyrene	• nZVI (nZVI@BC800) resulted in 71.80% degradation efficiency benzo[*a*]pyrene within 05 minutes.
Lin et al. (2024)	• Rice straw-biochar dosage (0%–2.0%)	• Di-2-ethylhexyl phthalate (DEHP)	• DEHP removal efficiency, maximally 77.53% in treatments with 1.0% of biochar.

Nonetheless, the commercialization of SFE for environmental remediation has the following challenges even though it was effective in laboratory and pilot-scale studies. The efficiency of the process is the major concern here because the process has to run at supercritical conditions and the recovery of the solvent is also very expensive (Anitescu & Tavlarides 2006). However, there is potential for SFE to recover industrial adsorbents such as granular activated carbon (GAC) used in wastewater treatment, which is a possible area of application that may be used to offset some of these costs (Tomasko et al. 1995).

The following are recommendations for future research on SFE for environmental remediation: There is a need to improve the economic dimension of SFE. This means that there may be a need to optimize the methods of solvent recovery as well as possibly looking at the use of other supercritical fluids that may be cheaper. Additionally, there is a need to undertake more research to foster the understanding of the supercritical fluids and the matrix of the environment for the improvement of extraction techniques.

9.3.2 Microwave-Assisted Remediation

MAR is one of the thermal technologies deemed to have the potential for the removal of organic pollutants in soil and water. This technology combines microwave heating and chemical activation to improve the efficiency of separating and removing different types of impurities. Microwave heating ensures that solvents and samples are warmed at a faster rate and evenly since it can penetrate through them. This selective and simultaneous heating ensures that certain areas do not become too hot as is the case with other heating methods (Bose & Kumar 2022).

Microwave-assisted extraction (MAE) and microwave-assisted peroxide-sulfate (MW/PS) are two of the most common techniques used in MAR. MAE irradiates solvents and samples with microwave energy in a short time, thus enhancing the extraction of organic pollutants from solid samples. This method is preferred in that the extraction time is shorter, use of solvent is less, and several samples can be extracted at the same time, which increases sample throughput (Sanchez-Prado et al. 2010). Unlike, MW/PS oxidation that uses microwave energy converts persulfate to sulfate radicals (SO_4^{-1}) that remove organic pollutants. This method has been found to be more effective than the single PS or MW treatments in degrading high concentrations of organophosphorus pesticides such as parathion. The rate of degradation varies with the level of PS used, microwave temperature as well as moisture content in the soil (Bose & Kumar 2022).

In soil remediation, microwave radiation heats water content in the soil to produce steam that helps in steam distillation of both polar and nonpolar compounds at temperatures not exceeding 100°C (Sanchez-Prado et al. 2010). Some of the authors have established that this thermal method is efficient in the removal of organic contaminants such as nitrobenzene from the soil with high removal rates possible within short treatment times. Microwave technology can also initiate chemical oxidants to break down pollutants and convert them into harmless substances. Microwave-activated persulfate oxidation, capable of generating sulfate radicals, has been found to be more effective in degrading complex organophosphorus pesticides compared with a single oxidation treatment (Hu et al. 2020). Optimization of factors including persulfate dosage and soil properties influence the degradation efficiency. Another advantage is the recovery of valuable petroleum hydrocarbons from contaminated soils. Experiments reported that high oil recovery percentages could be obtained after brief microwave heating at mild temperatures. For instance, the nitrobenzene removal rate in soil was observed to be around 80% in 600 seconds of microwave treatment and the process can be fine-tuned to achieve 100% removal (Kan et al. 2021). This rapid heating also helps in the regeneration of hydrocarbons from the petroleum-contaminated soils, with the researches showing up to 91% degradation. The percentage of oil recovery was 6% after the treatment period of 20 minutes at 250°C–300°C (Yang et al. 2009). Pilot applications have evidenced removals at low energy costs, indicating viability for scaling operations.

In situ treatment reduces need for excavation and transportation. The process has been applied to remove nitrobenzene, organophosphorus compounds, and petroleum hydrocarbons from various soil types and contamination levels (Llompart et al. 2019). Amendments like bluecoke powder can improve heating and removal for certain contaminants (Kan et al. 2021). However, soil properties such as texture and moisture content impact efficiency, necessitating site-specific parameter optimization (Chen et al. 2016). Initial equipment costs are high but reduced treatment periods and energy usage offset long-term expenses compared with other remediation methods. Research reveals that decontamination costs are comparable, and microwave-based treatment as a fast, efficient, and economical method of soil washing. Subsequent developments are expected to solve issues with the heterogeneity of the ground and the expansion of technology's marketability.

9.3.3 Biochar Technology

Biochar has recently been studied as an effective adsorbent for the enhancement of organic pollutant removal due to its inherent properties. Biochar is a carbonaceous material produced by thermal decomposition of biomass in a restricted supply of oxygen. It has a large surface area, porosity, and cation exchange capacity (CEC) through which it can effectively remove organic pollutants from the soil and water system (Murtaza et al. 2023). Some of the other researches have also revealed that the biochar effectively reduces the TPH in the oil-contaminated soil. The result also showed that the degradation efficiency of TPH was higher and the content of carbon and nitrogen in the soil were improved when biochar was added in combination with plants such as *Tagetes erecta* and microorganisms (Fang et al. 2024). Furthermore, the biochar derived from the agricultural waste having nanotextured surfaces has been described as a cheap sorbent for the environmental cleaning and waste management due to its pollutant uptake ability (Ghosh et al. 2023). In addition to the organic pollutants, the literature review on biochar has been considered in reducing microplastic pollution in the soils and sediments, and its efficacy has been studied in various environmental matrices (Li et al. 2024).

The mechanisms behind the adsorption or stabilization of organic pollutants in soil may include pore filling, electrostatic attraction, hydrophobic interactions, and other specific interactions such as hydrogen bonding and surface precipitation. The nature of the feedstock and the conditions under which pyrolysis is carried out determine the surface properties of biochar, which define the adsorption capacity of biochar in the context of the organic pollutants (Islam et al. 2021). It has been established by researches that biochar possess the capacity to adsorb a wide range of organic chemicals. For example, the biochar prepared from the wheat straw could effectively remediate the soil with various concentrations of phenanthrene through promoting the microbial immobilization and degradation of PAHs (Zhang et al. 2023b). In another study, biochar enhanced the degradation of pyrene through enhanced aging effects in the sediment and creating a stable habitat for PAH-degrading microorganisms (Chen et al. 2019). Biochar has also been utilised in agrosoils contaminated with agrochemicals and synthetic dyes, demonstrating its high sorption ability (Gautam et al. 2021; Cara et al. 2022).

The enhancement of soil remediation efficiency has also been studied with the use of biochar in combination with phytoremediation and microbial remediation. It also has the ability to hold plant growth-promoting bacteria, which enhance microbial activities in phytoremediation (Zhang et al. 2019). Biochar has also been used in bioelectrochemical systems to enhance the breakdown of organic contaminants by creating conductivity pathways for electron transfer and pollutants removal. This approach further revealed that biochar could boost the efficiency of the innovative remediation strategies by a significant extent (Haider et al. 2022). Yet, the properties of biochar are still dependent on the feedstock and conditions under which pyrolysis was conducted. However, the effectiveness of the technology in the long run and the impact it can have on the environment has to be looked into. The future studies should include large-scale experiments under different environmental conditions, standardized biochar preparation, and application techniques to explore the full potential of biochar in soil remediation. Additional steps need to be taken in order to address the related problems.

9.4 CHALLENGES AND FUTURE PERSPECTIVES

While physical remediation techniques have proven effective for removing EOCs from contaminated soils, there remain various challenges requiring attention to further enhance their application and sustainability. A major challenge lies in the diversity and complexity of EOC mixtures present at contaminated sites in real field conditions. As revealed by previous studies, many EOCs do not exist singly in the environment but rather co-occur in complex mixtures. The interactions between different EOCs within these mixtures, such as additive, antagonistic, or synergistic effects, are still poorly understood. This complexity complicates the selection and optimization of appropriate remediation techniques. Soil properties also affect the treatment efficiencies and they vary from one contaminated site to the other. Variation in the characteristics of the soil such as particle size distribution, mineral content, organic matter content, and moisture content affect the distribution and binding of EOCs in the matrix. This influences their accessibility to be extracted, evaporated, or reacted during remediation processes. One of the issues that have not been solved adequately is the selection of the appropriate physical remediation technology and enhancement strategies in relation to the characteristics of each site.

Monitoring the removal or destruction of all EOC constituents, including potentially toxic transformation products, poses difficulties given the diversity of compounds involved. Many EOCs also lack comprehensive toxicological profiling, hindering risk-based evaluation of remediation completion and residual impacts. Further research is still needed to develop robust analytical methods and toxicity data for nonregulated EOCs. Sustainability considerations must also be addressed. Current physical remediation protocols require optimization to minimize secondary pollution emissions, energy demands, and waste generation. Strategies such as ISTD, a mature technology suitable for remediating various organic compounds with relatively high efficiency, could help alleviate these issues. Combined remediation approaches leveraging physical, chemical, and biological mechanisms in an integrated system may offer synergistic outcomes with improved efficiency. Emerging microwave heating technology, with advantages of rapid, selective, and simultaneous heating avoiding overheating soil surfaces, may provide innovative and sustainable solutions.

Implementation of physical remediation under full-scale, real-world conditions requires further validation. Development of standardized assessment methods would uniformly evaluate risks and remediation completeness for complex EOC mixtures. An integrated, systems-oriented research approach encompassing all relevant technical and sustainability criteria can help propel progress toward optimized solutions globally. Continuous progress in these areas is important for advancing physical remediation as a viable strategy for EOC contamination.

9.5 CONCLUSION

This chapter has presented a discussion on the different physical remediation methods that have been employed to remediate emerging organic pollutants in soils. Thermal desorption and soil washing are well-proven technologies for the removal of VOCs and SVOCs, depending on the properties of the soil and the characteristics of the contaminants. SFE, microwave heating, and biochar application are some of the newer technologies that have been identified as innovative but need further research and development. However, some challenges are still evident when it comes to the remediation of EOC contamination through physical remediation. This is because the treatment success depends on the heterogeneity of the soil, clay, and silt content in particular. Challenges in monitoring transformation byproducts and assessing risks include data limitations for many unregulated EOCs. Sustainability is still a factor of concern because the methods should reduce energy consumption, exhaust emission, and hazardous waste through the treatment of solid wastes. It is possible that the integrated systems that use physical, chemical, and biological methods can be more effective when used together. Rising ISTD and microwave heating may help to reduce sustainability concerns by providing fast and selective treatment without the need to excavate and destroy the soil functionality. Full-scale physical remediation under field conditions is still awaiting further confirmation of its effectiveness. There is a need to develop standard assessment tools that can be used to assess risks and completeness of EOC mixtures uniformly. Progressing remediation for difficult courses such as PFAS is also important. In conclusion, it is crucial to employ a comprehensive and interdisciplinary research strategy to continue advancing optimized, universally-applicable methods for the physical removal of EOC-contaminated soils. Further advancement in the management of present challenges will enhance the physical remediation as a well-established technology solution for EOC contamination globally.

REFERENCES

Alcántara, M., J. Gómez, M. Pazos, and M. Sanromán, 2012. Electrokinetic remediation of lead and phenanthrene polluted soils. *Geoderma*, 173–174, 128–133. https://doi.org/10.1016/j.geoderma.2011.12.009

Anitescu, G., and L. Tavlarides, 2006. Supercritical extraction of contaminants from soils and sediments. *The Journal of Supercritical Fluids*, 38(2), 167–180. https://doi.org/10.1016/j.supflu.2006.03.024

Biel-Mareso, M., C. Corada-Fernández, and P.A. Lara-Martín, 2022. Occurrence and behaviour of contaminants of emerging concern in soil: A review. *Science of Total Environment*, 806, 150300. https://doi.org/10.1016/j.scitotenv.2021.150300

Biel-Maeso, M., C. González-González, P. A. Lara-Martín, and C. Corada-Fernández, 2019. Sorption and degradation of contaminants of emerging concern in soils under aerobic and anaerobic conditions. *Science of Total Environment*, 666, 662–671. https://doi.org/10.1016/j.scitotenv.2019.02.279

Bose, S., and M. Kumar, 2022. Microwave-assisted persulfate/peroxymonosulfate process for environmental remediation. *Current Opinion in Chemical Engineering*, 36, 100826. https://doi.org/10.1016/j.coche.2022.100826

Bhuyan, S, J, and M. Latin, 2012. BTEX Remediation under challenging site conditions using in-situ ozone injection and soil vapor extraction technologies: A case study. *Soil and Sediment Contamination: An International Journal*, 21(4), 545–556. https://doi.org/10.1080/15320383.2012.664187

Caetano, G., R. De Matos Machado, M.J.N. Correia, and I.M. Marrucho, 2023. Remediation of soils contaminated with total petroleum hydrocarbons through soil washing with surfactant solutions. *Environmental Technology*, 1–14. https://doi.org/10.1080/09593330.2023.2198733

Cameselle, C., and S. Gouveia, 2018. Electrokinetic remediation for the removal of organic contaminants in soils. *Current Opinion in Electrochemistry*, 11, 41–47. https://doi.org/10.1016/j.coelec.2018.07.005

Cara, I.G., D. Țopa, I. Puiu, and G. Jităreanu, 2022. Biochar a promising strategy for pesticide-contaminated soils. *Agriculture*, 12(10), 1579. https://doi.org/10.3390/agriculture12101579

Chen, F., Q. Zhang, J. Ma, Q. Zhu, Y. Wang, and H. Liang, 2021. Effective remediation of organic-metal co-contaminated soil by enhanced electrokinetic-bioremediation process. *Frontiers of Environmental Science & Engineering*, 15(6). https://doi.org/10.1007/s11783-021-1401-y

Chen, J., C. Liang, C. Chen, and C. Liu, 2007. Composite analytical solutions for a soil vapour extraction system. *Hydrological Processes*, 21, 1506–1516. https://doi.org/10.1002/hyp.6341

Chen, J., S. Xue, Y. Song, M. Shen, Z. Zhang, T. Yuan, F. Tian, and D.D. Dionysiou, 2016. Microwave-induced carbon nanotubes catalytic degradation of organic pollutants in aqueous solution. *Journal of Hazardous Materials*, 310, 226–234. https://doi.org/10.1016/j.jhazmat.2016.02.049

Chen, L., J. Hasanov, J. Chen, Y. Feng, Y. Kanda, and A. Komiya, 2022. Supercritical fluid remediation for soil contaminants: Mechanisms, parameter optimization and pilot systems. *Journal of Supercritical Fluids*, 189, 105718. https://doi.org/10.1016/j.supflu.2022.105718

Chen, X., L. Li, Y. Zhang, and H. Gu, 2023. Microwave heating remediation of light and heavy crude oil-contaminated soil. *Energy & Fuels*, 37(7), 5323–5330. https://doi.org/10.1021/acs.energyfuels.3c00078

Chen, Z., J. Chen, X. Yang, C. Chen, S. Huang, and H. Luo, 2019. Biochar as an effective material on sediment remediation for polycyclic aromatic hydrocarbons contamination. In *IOP Conference Series: Earth and Environmental Science*, vol. 281, no. 1, p. 012016. https://doi.org/10.1088/1755-1315/281/1/012016

Cheng, Y., H. Sun, E. Yang, J. Lv, B. Wen, F. Sun, L. Luo, and Z. Liu, 2021. Distribution and bioaccessibility of polycyclic aromatic hydrocarbons in industrially contaminated site soils as affected by thermal treatment. *Journal of Hazardous Materials*, 411, 125129. https://doi.org/10.1016/j.jhazmat.2021.125129

Chiaia-Hernández, A., M. Scheringer, A. Müller, G. Stieger, D. Wächter, A. Keller, M. Pintado-Herrera, P. Lara-Martín, T. Bucheli, and J. Hollender, 2020. Target and suspect screening analysis reveals persistent emerging organic contaminants in soils and sediments. *Science of the Total Environment*, 740, 140181. https://doi.org/10.1016/j.scitotenv.2020.140181

Conte, P., A. Agretto, R. Spaccini, and A. Piccolo, 2005. Soil remediation: Humic acids as natural surfactants in the washings of highly contaminated soils. *Environmental Pollution*, 135(3), 515–522. https://doi.org/10.1016/j.envpol.2004.10.006

De Melo Henrique, J.M., J. Isidro, C. Sáez, R. López-Vizcaíno, A. Yustres, V. Navarro, E.V.D. Santos, and M.A. Rodrigo, 2022. Enhancing soil vapor extraction with EKSF for the removal of HCHs. *Chemosphere*, 296, 134052. https://doi.org/10.1016/j.chemosphere.2022.134052

Ding, Y., Y. Zhang, Z. Deng, H. Song, J. Wang, and H. Guo, 2022. An innovative method for soil vapor extraction to improve extraction and tail gas treatment efficiency. *Scientific Reports*, 12(1). https://doi.org/10.1038/s41598-022-08734-8

Fabbri, D., A.B. Prevot, V. Zelano, M. Ginepro, and E. Pramauro, 2008. Removal and degradation of aromatic compounds from a highly polluted site by coupling soil washing with photocatalysis. *Chemosphere*, 71(1), 59–65. https://doi.org/10.1016/j.chemosphere.2007.10.028

Faigle, B., H.G. Edel, and B. Volz, 2023. Successful treatment of PFAS-contaminated soils on large scale: Practical experience with soil washing. In *EGU General Assembly 2023*, Vienna, Austria, 24–28 Apr 2023, EGU23–1373. https://doi.org/10.5194/egusphere-egu23-1373

Falciglia, P.P., F.G.A. Vagliasindi, and A. Catalfo, 2017. Thermal desorption of soils polluted by organic insulating liquids: A review. *Energies*, 10, 1344. https://doi.org/10.3390/en10091344

Fang, X., M. Zhang, P. Zheng, H. Wang, K. Wang, J. Lv, and F. Shi, 2024. Biochar-bacteria-plant combined potential for remediation of oil-contaminated soil. *Frontiers in Microbiology*, 15. https://doi.org/10.3389/fmicb.2024.1343366

Gao, G.L., and J.G. Jiang, 2012. Thermal desorption remediation of soil highly polluted by1#, 2-dichloroethane in China. *Advanced Materials Research*, 356–360, 1131–1134.

García, J., M. García-Galán, J. Day, R. Boopathy, J. White, S. Wallace, and R. Hunter, 2020. A review of emerging organic contaminants (EOCs), antibiotic resistant bacteria (ARB), and antibiotic resistance genes (ARGs) in the environment: Increasing removal with wetlands and reducing environmental impacts. *Bioresource Technology*, 123228. https://doi.org/10.1016/j.biortech.2020.123228

García, M., C. Fernández-López, F. Polesel, and S. Trapp, 2019. Predicting the uptake of emerging organic contaminants in vegetables irrigated with treated wastewater: Implications for food safety assessment. *Environmental Research*, 172, 175–181. https://doi.org/10.1016/j.envres.2019.02.011

Gautam, R.K., M. Goswami, R.K. Mishra, P. Chaturvedi, M.K. Awashthi, R.S. Singh, B.S. Giri, and A. Pandey, 2021. Biochar for remediation of agrochemicals and synthetic organic dyes from environmental samples: A review. *Chemosphere*, 272, 129917. https://doi.org/10.1016/j.chemosphere.2021.129917

Ghosh, S., M. Nandasana, T.J. Webster, and S. Thongmee, 2023. Agrowaste-generated biochar for the sustainable remediation of refractory pollutants. *Frontiers in Chemistry*, 11. https://doi.org/10.3389/fchem.2023.1266556

Haider, F.U., X. Wang, U. Zulfiqar, M. Farooq, S. Hussain, T. Mehmood, M. Naveed, Y. Li, C. Liqun, Q. Saeed, I. Ahmad, and A. Mustafa, 2022. Biochar application for remediation of organic toxic pollutants in contaminated soils: An update. *Ecotoxicology and Environmental Safety*, 248, 114322. https://doi.org/10.1016/j.ecoenv.2022.114322

Hoag, G.E., M.C. Marley, B.L. Cliff, and P. Nangeroni, 2022. Soil vapor extraction research developments. In *Hydrocarbon Contaminated Soils and Groundwater*, eds. Kostecki, P.T., and E.J. Calabrese. Routledge, New York, pp. 187–202. https://doi.org/10.1201/9780203751572-18

Hu, L., P. Wang, T. Shen, Q. Wang, X. Wang, P. Xu, Q. Zheng, and G. Zhang, 2020. The application of microwaves in sulfate radical-based advanced oxidation processes for environmental remediation: A review. *Science of the Total Environment*, 722, 137831. https://doi.org/10.1016/j.scitotenv.2020.137831

Islam, T., Y. Li, and H. Cheng, 2021. Biochars and engineered biochars for water and soil remediation: A review. Sustainability, 13(17), 9932. https://doi.org/10.3390/su13179932

Jiang, S., and S. Weng, 2022. The emission characteristics of pollutants from thermal desorption of soil contaminated by transformer oil. *Atmosphere*, 13(4), 515. https://doi.org/10.3390/atmos13040515

Jones, A.D., A.T. Morehead, and Y. Yang, 2023. Degradation and extraction of organochlorine pollutants from environmental solids under subcritical water conditions. *Molecules/Molecules Online/Molecules Annual*, 28(14), 5445. https://doi.org/10.3390/molecules28145445

Kamali, M., T. Gameiro, M.E.V. Costa, and I. Capela, 2018. Emerging pollutants: Removal approaches and fate in the environment. In *Water Treatment*. IntechOpen, pp. 127–167. https://doi.org/10.5772/intechopen.76582

Kan, H., T. Wang, J. Yu, G. Qu, P. Zhang, H. Jia, and H. Sun, 2021. Remediation of organophosphorus pesticide polluted soil using persulfate oxidation activated by microwave. *Journal of Hazardous Materials*, 401, 123361. https://doi.org/10.1016/j.jhazmat.2020.123361

Kariyawasam, T., G.S. Doran, J.A. Howitt, and P.D. Prenzler, 2023. Optimization and comparison of microwave-assisted extraction, supercritical fluid extraction, and eucalyptus oil–assisted extraction of polycyclic aromatic hydrocarbons from soil and sediment. *Environmental Toxicology and Chemistry*, 42(5), 982–994. https://doi.org/10.1002/etc.5593

Klik, B., Z.M. Gusiatin, and D. Kulikowska, 2021. A holistic approach to remediation of soil contaminated with Cu, Pb and Zn with sewage sludge-derived washing agents and synthetic chelator. *Journal of Cleaner Production*, 311, 127664. https://doi.org/10.1016/j.jclepro.2021.127664

Kristanti, R.A., W. Khanitchaidecha, G. Taludar, P. Karácsony, L.T.T. Cao, T.W. Chen, N.M. Darwish, and B.M. AlMunqedhi, 2022. A review on thermal desorption treatment for soil contamination. *Tropical Aquatic and Soil Pollution*, 2(1), 45–58. https://doi.org/10.53623/tasp.v2i1.68

Labianca, C., S. De Gisi, F. Picardi, F. Todaro, and M. Notarnicola, 2020. Remediation of a petroleum hydrocarbon-contaminated site by soil vapor extraction: A full-scale case study. *Applied Sciences*, 10(12), 4261. https://doi.org/10.3390/app10124261

Li, G., G. Sun, Y. Ren, X. Luo, and Y. Zhu, 2018. Urban soil and human health: A review. *European Journal of Soil Science*, 69. https://doi.org/10.1111/ejss.12518

Li, W, Y. Xing, Y. Guo, D. Zhang, Y. Tang, J. Chen, H. Zhang, and B. Jiang, 2024. The removal and mitigation effects of biochar on microplastics in water and soils: Application and mechanism analysis. *Sustainability*, 16(22), 9749. https://doi.org/10.3390/su16229749

Librando, V., O. Hutzinger, G. Tringali, and M. Aresta, 2004. Supercritical fluid extraction of polycyclic aromatic hydrocarbons from marine sediments and soil samples. *Chemosphere*, 54(8), 1189–1197. https://doi.org/10.1016/j.chemosphere.2003.07.008

Librando, V., G. Tomaselli, and G. Tringali, 2005. Optimization of supercritical fluid extraction by carbon dioxide with organic modifiers of polycyclic aromatic hydrocarbons from urban particulate matter. *Annali Di Chimica*, 95(3–4), 211–216. https://doi.org/10.1002/adic.200590023

Lin, Z., W. Wu, C. Yang, G. Yang, W. Wu, T. Wei, F. Huang, H. Li, L. Ren, Y. Liang, D. Zhang, Z. Li, and Z. Zhen, 2024. Mechanisms of biochar assisted di-2-ethylhexyl phthalate (DEHP) biodegradation in tomato rhizosphere by metabolic and metagenomic analysis. *Chemosphere*, 353, 141520. https://doi.org/10.1016/j.chemosphere.2024.141520

Liu, J., L. Zhao, Q. Liu, J. Li, Z. Qiao, P. Sun, and Y. Yang, 2021. A critical review on soil washing during soil remediation for heavy metals and organic pollutants. *International Journal of Environmental Science and Technology*, 19(1), 601–624. https://doi.org/10.1007/s13762-021-03144-1

Llompart, M., M. Celeiro, and T. Dagnac, 2019. Microwave-assisted extraction of pharmaceuticals, personal care products and industrial contaminants in the environment. *TrAC. Trends in Analytical Chemistry*, 116, 136–150. https://doi.org/10.1016/j.trac.2019.04.029

Ma, F., Y. Zhu, B. Wu, Q. Zhang, D. Xu, J. Xu, B. Wang, Q. Gu, and F. Li, 2019. Degradation of DDTs in thermal desorption off-gas by pulsed corona discharge plasma. *Chemosphere*, 233, 913–919. https://doi.org/10.1016/j.chemosphere.2019.05.292

Maddela, N., B. Ramakrishnan, D. Kakarla, K. Venkateswarlu, and M. Megharaj, 2022. Major contaminants of emerging concern in soils: A perspective on potential health risks. *RSC Advances*, 12, 12396–12415. https://doi.org/10.1039/d1ra09072k

Mena, E., J. Villaseñor, M.A. Rodrigo, and P. Cañizares, 2016. Electrokinetic remediation of soil polluted with insoluble organics using biological permeable reactive barriers: Effect of periodic polarity reversal and voltage gradient. *Chemical Engineering Journal*, 299, 30–36. https://doi.org/10.1016/j.cej.2016.04.049

Meng, P., P. Babakhani, P.M. Clayton, R. Pyon, M.N. Azeez, and P. Loganathan, 2021. Potentials of phytoremediation for emerging organic contaminants: A review. *Science of the Total Environment*, 789, 147958. https://doi.org/10.1016/j.scitotenv.2021.147958

Mousset, E., M.A. Oturan, E.D. Van Hullebusch, G. Guibaud, and G. Esposito, 2014. Soil washing/flushing treatments of organic pollutants enhanced by cyclodextrins and integrated treatments: State of the art. *Critical Reviews in Environmental Science and Technology*, 44(7), 705–795. https://doi.org/10.1080/10643389.2012.741307

Mukhopadhyay, A., S. Duttagupta, and A. Mukherjee, 2022. Emerging organic contaminants in global community drinking water sources and supply: A review of occurrence, processes and remediation. *Journal of Environmental Chemical Engineering*, 10(3), 107560. https://doi.org/10.1016/j.jece.2022.107560

Muñoz-Morales, M., C. Sáez, P. Cañizares, and M.A. Rodrigo, 2021. Electrochemically assisted soil washing for the remediation of non-polar and volatile pollutants. *Current Pollution Reports*, 7(2), 180–193. https://doi.org/10.1007/s40726-021-00179-3

Murtaza, G., Z. Ahmed, S.M. Eldin, I. Ali, M. Usman, R. Iqbal, M. Rizwan, U.K. Abdel-Hameed, A.A. Haider, and A. Tariq, 2023. Biochar as a green sorbent for remediation of polluted soils and associated toxicity risks: A critical review. *Separations*, 10(3), 197. https://doi.org/10.3390/separations10030197

Mustafa, D., U. Höltkemeier, U. Sabeel, A. Yar, M. Herbrich, S.O. Sojinu, T. Poehler, D. Wu, B. Thiele, and P. Kuschk, 2021. Supercritical fluid extraction of soil contaminants: A review. *Molecules*, 26, 5389. https://doi.org/10.3390/molecules26175389

Nematollahi, H., N. Moradi, M. RiyaziNejad, and H. Vahidi, 2018. Removal of aliphatic hydrocarbons from gas oil contaminated clay soil via soil vapor extraction. *Civil Engineering Journal*, 4(8), 1858. https://doi.org/10.28991/cej-03091120

Pal, D.B., and D.D. Giri, 2021. Remediation of industrial organic waste pollutants. In *Sustainable Environmental Clean-up*, eds. Mishra, V.K., and A. Kumar. Elsevier, pp. 295–314. https://doi.org/10.1016/b978-0-12-823828-8.00014-1

Pasvantu, I.C., A.M. Lăcureanu, C. Streche, and D.M. Cocârță, 2023. Application of the electrochemical treatment method for the remediation of petroleum hydrocarbons from contaminated soils. *Annals. Series on Engineering Sciences*, 15(2), 59–68. https://doi.org/10.56082/annalsarscieng.2023.2.59

Petrie, B., R. Barden, and B. Kasprzyk-Hordern, 2015. A review on emerging contaminants in wastewaters and the environment: Current knowledge, understudied areas and recommendations for future monitoring. *Water Research*, 72, 3–27. https://doi.org/10.1016/j.watres.2014.08.053

Pourzamani, H., Y. Hajizadeh, S.S. Koloor, and L. Heidari, 2021. Application of nanomaterials for removal of emerging organic contaminants from soil: Opportunities, challenges and prospects. *Chemosphere*, 270, 128602. https://doi.org/10.1016/j.chemosphere.2020.128602

Qu, J., J. Xue, M. Sun, K. Li, J. Wang, G. Zhang, L. Wang, Z. Jiang, and Y. Zhang, 2024. Superefficient non-radical degradation of benzo[a]pyrene in soil by Fe-biochar composites activating persulfate. *Chemical Engineering Journal*, 481, 148585. https://doi.org/10.1016/j.cej.2024.148585

Quinnan, J., C. Morrell, N. Nagle, and K.G. Maynard, 2022. Ex situ soil washing to remove PFAS adsorbed to soils from source zones. *Remediation*, 32(3), 151–166. https://doi.org/10.1002/rem.21727

Rassaei, F. 2023. Sustainable per- and polyfluoroalkyl substances contaminated soil remediation: Evaluating the potential of thermal desorption. *Environmental Progress & Sustainable Energy*, 43(1). https://doi.org/10.1002/ep.14251

Riis, C., E.E. Cox, J. Wang, D. Gent, et al. 2023. Remediating a PCE source area in clay using electrokinetically enhanced in situ bioremediation. *Ground Water Monitoring & Remediation*, 43(3), 70–78. https://doi.org/10.1111/gwmr.12592

Robert, D. F, 1996. Physical/chemical treatment of organically contaminated soils and sediments. *Journal of the Air & Waste Management Association*, 46(5), 391–413, DOI: 10.1080/10473289.1996.10467473

Sadatshojaei, E., and D.A. Wood, 2021. Applications of supercritical fluids in environmental remediation. In Green Sustainable Process for Chemical and Environmental Engineering and Science, eds. Inamuddin, A.M.A., and K. Suvardhan. Elsevier, pp. 257–265. https://doi.org/10.1016/b978-0-12-821884-6.00003-6

Sanchez-Prado, L., C. Garcia-Jares, and M. Llompart, 2010. Microwave-assisted extraction: Application to the determination of emerging pollutants in solid samples. *Journal of Chromatography*, 1217(16), 2390–2414. https://doi.org/10.1016/j.chroma.2009.11.080

Sang, Y., W. Yu, L. He, et al. 2021. Sustainable remediation of lube oil-contaminated soil by low temperature indirect thermal desorption: Removal behaviors of contaminants, physicochemical properties change and microbial community recolonization in soils. *Environmental Pollution*, 287, 117599. https://doi.org/10.1016/j.envpol.2021.117599

Sefiloglu, F., U. Tezel, and I. Balcioglu, 2021. Validation of an analytical workflow for the analysis of pesticide and emerging organic contaminant residues in paddy soil and rice. *Journal of Agricultural and Food Chemistry*, 69(11), 3298–3306. https://doi.org/10.1021/acs.jafc.0c06111

Semer, R., and K.R. Reddy, 1996. Evaluation of soil washing process to remove mixed contaminants from a sandy loam. *Journal of Hazardous Materials*, 45(1), 45–57. https://doi.org/10.1016/0304-3894(96)82887-1

Simons, R.A., L.V. Dishaw, R.K. Ellison, and G.R. Czaplicki, 2018. Lessons from a record of closure: Remediation of a former dry cleaner site. *Remediation Journal*, 28, 59–74. https://doi.org/10.1002/rem.21546

Srivastava, V., M. Puri, T. Srivastava, P. Nidheesh, and M.S. Kumar, 2022. Integrated soil washing and bioreactor systems for the treatment of hexachlorocyclohexane contaminated soil: A review on enhanced degradation mechanisms, and factors affecting soil washing and bioreactor performances. *Environmental Research*, 208, 112752. https://doi.org/10.1016/j.envres.2022.112752

Sunarso, J., and S. Ismadji, 2009. Decontamination of hazardous substances from solid matrices and liquids using supercritical fluids extraction: A review. *Journal of Hazardous Materials*, 161(1), 1–20. https://doi.org/10.1016/j.jhazmat.2008.03.069

Tian, W., H. Wu, L. Wu, C. Liu, C. Yue, and J. Zhou, 2021. Enhanced microwave remediation of organic contaminated soil: A new way of utilization of bluecoke powder. *International Journal of Environmental Science and Technology*, 19(6), 5451–5460. https://doi.org/10.1007/s13762-021-03439-3

Tomasko, D.L., S.J. Macnaughton, N.R. Foster, and C.A. Eckert, 1995. Removal of pollutants from solid matrices using supercritical fluids. *Separation Science and Technology*, 30(7–9), 1901–1915. https://doi.org/10.1080/01496399508010383

Trellu, C., Y. Pechaud, N. Oturan, et al. 2021. Remediation of soils contaminated by hydrophobic organic compounds: How to recover extracting agents from soil washing solutions? *Journal of Hazardous Materials*, 404, 124137. https://doi.org/10.1016/j.jhazmat.2020.124137

Troxler, W.L., J.J. Cudahy, R.P. Zink, J.J. Yezzi, and S.I. Rosenthal, 1993. Treatment of non-hazardous petroleum-contaminated soils by thermal desorption technologies. *Air & Waste*, 43(11), 1512–1525. https://doi.org/10.1080/1073161x.1993.10467224

Wang, B., A. Wu, X. Li, et al. 2021. Progress in fundamental research on thermal desorption remediation of organic compound-contaminated soil. *Waste Disposal & Sustainable Energy*, 3(2), 83–95. https://doi.org/10.1007/s42768-021-00071-2

Wen, D., R. Fu, and Q. Li, 2021. Removal of inorganic contaminants in soil by electrokinetic remediation technologies: A review. *Journal of Hazardous Materials*, 401, 123345. https://doi.org/10.1016/j.jhazmat.2020.123345

Wu, L., Z. Hu, Y. Gao, et al. 2024. Feasibility of microwave remediation of simulative crude oil-contaminated soil assisted by bluecoke-based modifiers. *Chemosphere*, 362, 142600. https://doi.org/10.1016/j.chemosphere.2024.142600

Xiao, F., J. Yin, D. Shen, T. Chen, and L. Lv, 2022. Treatment of wastewater from thermal desorption for remediation of oil-contaminated soil by the combination of multiple processes. *Journal of Chemistry*, 2022, 1–13. https://doi.org/10.1155/2022/3616050

Yabalak, E., M.T. Aminzai, A.M. Gizir, and Y. Yang, 2024. A review: Subcritical water extraction of organic pollutants from environmental matrices. *Molecules Annual*, 29(1), 258. https://doi.org/10.3390/molecules29010258

Yang, S., P. Wang, X. Yang, G. Wei, W. Zhang, and L. Shan, 2009. A novel advanced oxidation process to degrade organic pollutants in wastewater: Microwave-activated persulfate oxidation. *Journal of Environmental Sciences*, 21(9), 1175–1180. https://doi.org/10.1016/s1001-0742(08)62399-2

Zhang, M., Y. Luo, Y. Zhu, et al. 2023a. Insights into the mechanisms underlying the biodegradation of phenanthrene in biochar-amended soil: From bioavailability to soil microbial communities. *Biochar*, 5(1). https://doi.org/10.1007/s42773-023-00213-1

Zhang, M., J. Wang, S.H. Bai, Y. Zhang, Y. Teng, and Z. Xu, 2019. Assisted phytoremediation of a co-contaminated soil with biochar amendment: Contaminant removals and bacterial community properties. *Geoderma*, 348, 115–123. https://doi.org/10.1016/j.geoderma.2019.04.031

Zhang, S., Z. Liu, R. Sun, W. Liu, and Y. Chen, 2023b. Orthogonal experimental study on remediation of ethylbenzene contaminated soil by SVE. *Sustainability*, 15(2), 1168. https://doi.org/10.3390/su15021168

Zhang, X., J. Yang, L. Qi, et al. 2024. Evaluation of electrokinetic-assisted phytoremediation efficiency of dibutyl phthalate contaminated soil by maize (*Zea mays* L.) under different electric field intensities. *Science of the Total Environment*, 944, 173838. https://doi.org/10.1016/j.scitotenv.2024.173838

Zhao, C., Y. Dong, Y. Feng, Y. Li, and Y. Dong, 2019. Thermal desorption for remediation of contaminated soil: A review. *Chemosphere*, 221, 841–855. https://doi.org/10.1016/j.chemosphere.2019.01.079

Zhao, F., Y. Ma, Y. Zhu, Z. Tang, and S. McGrath, 2015. Soil contamination in China: Current status and mitigation strategies. *Environmental Science & Technology*, 49(2), 750–759. https://doi.org/10.1021/ES5047099

Zhao, Z., M. Ni, X. Li, and A. Buekens, 2016. Thermal desorption for remediating PCB-contaminated soil. *International Journal of Environment and Pollution*, 60(1/2/3/4), 171. https://doi.org/10.1504/ijep.2016.082113

10 Chemical Remediation of EOCs in Soil

Hina Imtiaz, Lena Madden, and Mahtab Ahmad

10.1 INTRODUCTION

Advanced methods for getting rid of pharmaceuticals, insecticides, perfluorinated chemicals, polychlorinated biphenyls (PCBs), and personal care items rely on the chemical remediation of emerging organic contaminants (EOCs). The three main processes for cleaning up EOC-contaminated soils are oxidation, reduction, and adsorption. Adsorption techniques, including the use of biochar and activated carbon (AC), limit the mobility of contaminants by trapping them on their surfaces. Complex pollutants are broken down into less dangerous molecules by oxidative processes, which employ substances such as hydrogen peroxide and ozone. Reductive methods, including zero-valent iron (ZVI) nanoparticles, detoxify contaminants using electron transfer processes. For the treatment of permeable soils with a modest level of heterogeneity, chemical remediation is widely used and highly effective (Aparicio et al. 2022). It aids in the treatment of plumes, contaminated source zones, and focused treatment zones. Successful chemical remediation necessitates a deep comprehension of the hydrogeology, geochemistry, and soil properties. Chemical remediation often works best on moderately saturated soils with low hydraulic conductivities, low levels of natural organic matter, and low levels of reduced metals (Bolan et al. 2023).

10.2 IN-SITU CHEMICAL OXIDATION

Chemical oxidation is an in-situ remediation process that lowers EOCs' oxidation state for treating EOCs (Ugrina & Jurić 2023). By oxidizing the EOCs with an electron, chemical oxidants added to the subsurface soil at a site transform them into less harmful compounds. Therefore, the oxidation status of the EOC is decreased by chemical oxidation. The toxicity of the target species can then be eliminated by its oxidized state. In the end, this will create a soil environment devoid of contamination (McGachy & Sedlak 2023).

10.2.1 Catalyzed H_2O_2 Propagations

One of the unique methods of in-situ chemical oxidation technologies is catalyzed H_2O_2 propagations (CHPs) in which a variety of nucleophiles, reductants, superoxide radicals, and potent oxidant hydroxyl radicals are produced. Because of this, it is the most reactive in-situ chemical oxidation technology. The method relies on soluble

DOI: 10.1201/9781003509523-10

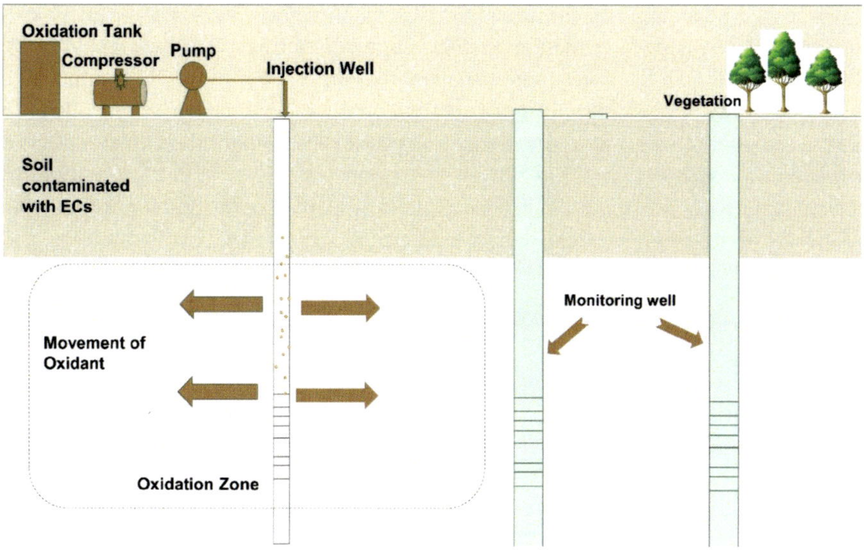

FIGURE 10.1 Injecting oxidants into the soil contaminated with EOCs.

iron, iron chelates, or iron minerals catalyzing the breakdown of hydrogen peroxide (McGachy & Sedlak 2023).

The mechanisms illustrated in Figure 10.1 involves Fenton like reactions generating the reactive oxygen species, leading to the production of potent oxidants such as hydroxyl radicals ($OH^{\bullet}$). Equations 10.1–10.5 show the increment in propagation, and the amplification of reactive oxygen species enhances the effectiveness of the treatment (Amina et al. 2022).

$$H_2O_2 + Fe^{2+} \rightarrow Fe^{3+} + OH^- + OH^{\bullet} \tag{10.1}$$

$$H_2O_2 + OH^{\bullet} \rightarrow H_2O + HO_2^{\bullet} \tag{10.2}$$

$$HO_2^{\bullet} \leftrightarrow H^+ + O_2^- \tag{10.3}$$

$$H_2O_2 \rightarrow OH^{\bullet} + ROH \tag{10.4}$$

$$Fe^{2+} + HO_2^{\bullet} \rightarrow Fe^{3+} + HO_2^- \tag{10.5}$$

The hydroxyl radical, which is the significant and powerful oxidant in CHP, uses additional species like hydroperoxide anion (HO_2^-), superoxide anion ($O_2^{\bullet -}$), and per hydroxyl radical ($HO_2^{\bullet}$) produced by the subsequent chemical interactions of the hydroxyl radical with hydrogen peroxide. These species are generated by CHP reactions and are capable of degrading almost any emerging pollutant. The pH, hydrogen peroxide concentration, and catalyst type all have a big influence on the CHP process. Usually, soluble iron, added as iron chelates, and naturally occurring iron oxide minerals serve as catalysts, but the effectiveness of iron chelates is associated with pH since these materials function best in acidic pH settings (Watts & Teel 2006). Molamahmood and his colleagues reported that the CHP process accelerated the degradation of phenols in the soil as superoxide radicals and hydroxyl radicals both have a significant role in the breakdown of phenol (Molamahmood et al. 2021). It has been reported that CHP has proven to be very successful in the degradation of antibiotics in soils. For example, CHPs remove antibiotics in the soil up to 10 mg/kg efficiently (Amina et al. 2022).

10.2.2 Advanced Fenton Oxidation

Advanced Fenton oxidation extensively uses advanced oxidation processes for the remediation of contaminated soil. It is the most prevalent technique due to its efficiency, rapid processing, and broad range of applications. Modifications of Fenton's reagent enhance the degradation of pharmaceuticals and organic pollutants. Fenton (1894) first used a solution of H_2O_2 and Fe^{3+} salt (later known as Fenton's reagent) to successfully oxidize tartaric acid. Equation 10.6 represents the complete reaction, and as a result, the $HO^{\bullet}$ radical is produced, which is the most reactive species that can oxidize emerging pollutants immediately in a nonselective way.

This catalytic mechanism mostly produces $HO^{\bullet}$ at acidic pH levels.

$$Fe^{2+} + H_2O_2 \rightarrow Fe^{+} + HO^{\bullet} + HO^{-} \tag{10.6}$$

Khan et al. (2023) proved that advanced Fenton oxidation is a novel and effective approach for the treatment of organic pollutants and emerging contaminants, including pesticides, pharmaceuticals, dyes, and surfactants. Mu et al. (2023) also confirmed that Fenton oxidation remains successful in treating polycyclic aromatic hydrocarbons (PAHs) in soils.

10.2.3 Potassium Permanganate Oxidation

Potassium permanganate ($KMnO_4$) is a green oxidant with strong oxidizing properties and greater efficiency than ozonation and chlorination. With $KMnO_4$, oxidation of pollutants occurs based on the formation or transfer of electrons of free radicals including $HO^{\bullet}$, $SO_4^{\bullet-}$, and $O_2^{\bullet-}$. These free radicals attack the pollutant and degrade it into less toxic compounds. Moreover, the formation of reactive manganese species (i.e., Mn(III), Mn(V), and Mn(VI)) has recently gained attention for treating the EOCs. The conceptual design of the in-situ application of $KMnO_4$ in the soil for the removal of EOCs is shown in Figure 10.2.

This method involves the injection of permanganates into the subsurface through specialized wells, where they react with EOCs. The organic pollutants are degraded through oxidative reactions, resulting in the formation of mineral by-products and stable complexes (equation 10.7).

$$MnO_4^{-} + 3e^{-} + 4H^{+} \rightarrow 2H_2O + MnO_2 \tag{10.7}$$

Permanganate is highly reactive with alkanes. It is also efficient in the remedy of dichloroethylene isomers, PCB, vinyl chloride, phenols, and some PAHs. However,

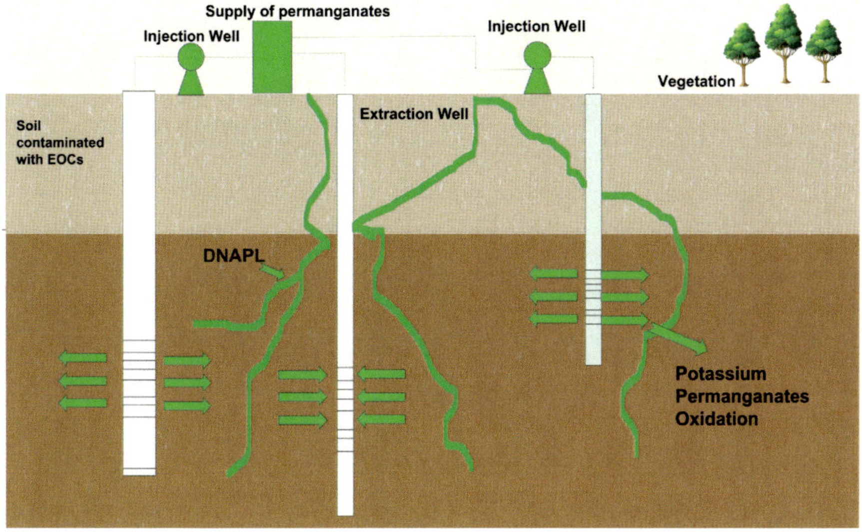

FIGURE 10.2 Conceptual design of potassium permanganate oxidation in soil contaminated with EOCs.

it is least effective for benzene and other aromatic compounds due to low degrees of ring activation. It has been reported that $KMnO_4$ detoxifies polyvinyl chloride and microplastics in contaminated soils (Wang et al. 2023). The benzo[*a*]pyrene was also removed up to 94.3% using a moderate dose of $KMnO_4$ (25 mmol/L) (Chen et al. 2023).

10.2.4 Persulfate Advanced Oxidation Processes

Another method for the remediation of EOCs is persulfate advanced oxidation process. It does not require activation, and the injection of persulfate anion degrades the contaminant directly due to its high oxidation potential. Moreover, the minerals in the soil also promote the breakdown of persulfate into more active substances, which can further eradicate the EOCs. In this process, when persulfate is added, the initial soil pH starts decreasing significantly, which leads to a decrease in the dose of the oxidant. Peroxydisulfate (PDS, $S_2O_8^{2-}$) and peroxymonosulfate (PMS, HSO_5^-) can also be added in small concentrations to persulfate for the remediation of EOCs in soil. Qutob and his colleagues used persulfate advanced oxidation for the elimination of benzo[a]pyrene, phenanthrene, and pyrene in tropical soils. The oxidation process aided with persulfate resulted in the removal of benzo[a]pyrene, phenanthrene, and pyrene by 86%, 81%, and 73%, respectively (Qutob et al. 2024).

10.2.5 Oxidation with Ozone

Ozone (O_3) is an oxidant used for the remediation of EOCs but the application process is quite different because it is used as a gas and is composed of dry air. In the presence of high voltage, O_2 molecules split and react to form O_3. This O_3 generation occurs on-site due to its instability. The chemical reaction of O_3 involves direct or indirect oxidation as shown in equations 10.8–10.17 (Ji et al. 2024).

Direct reaction:

$$RC = CR + O_3 \rightarrow O_2 + RCOCR \quad (10.8)$$

Chain-initiating:

$$O_3 + OH^- \rightarrow O_2 + OH^{\bullet} \quad (10.9)$$

$$H_2O + O_3 \rightarrow 2OH^{\bullet} + O_2 \left(\text{in the presence of ultraviolet light}\right) \quad (10.10)$$

Chain-propagating:

$$2H_2O + OH^{\bullet} \rightarrow OH^- + 3H^+ + HO_2^{\bullet} \quad (10.11)$$

$$HO_2^{\bullet} \rightarrow H^+ + O_2^{-\bullet} \quad (10.12)$$

$$RH + OH^{\bullet} \rightarrow OH^- + R^{\bullet} \quad (10.13)$$

$$H_2O + O_3 + R^{\bullet} \rightarrow O_2 + OH^{\bullet} + ROH \quad (10.14)$$

Chain-terminating:

$$Fe^{2+} + HO_2^{\bullet} \rightarrow O_2 + Fe^{3+} + H^+ \quad (10.15)$$

$$HO_2^{\bullet} + Fe^{2+} \rightarrow Fe^{3+} + HO_2^- \quad (10.16)$$

$$Fc^{3+} + O_2^{-\bullet} \rightarrow \Gamma e^{2+} + O_2 \quad (10.17)$$

In direct reaction oxidation, the targeted pollutant is oxidized by O_3. In contrast, indirect oxidation does not achieve the target by relying on the hydroxyl radical ($OH^{\bullet}$). Indirect oxidation involves chain initiation, chain propagation, and chain termination. The $OH^{\bullet}$ attacks the organic pollutants and splits their organic carbon-to-carbon bonds.

The application of O_3 has two forms: (i) vadose zone injection and (ii) O_3 sparging. In situ O_3 oxidation mixes up the air and O_3 gas directly injected into the soil. The existence of O_3 depends on the concentration, reactivity of reactants, temperature, and pH. The O_3 injection is done by sparging, which does not transport and distribute the air uniformly due to a smaller number of air channels. Therefore, air channels should be closed during sparging, and zones can overlap each other for better transport. The direct method of the gaseous O_3 was used through perforated pipes, and the removal of methomyl by 7.8% was achieved (da Silva et al. 2024).

10.3 IN-SITU CHEMICAL REDUCTION (ISCR)

Emerging pollutants that are persistent in the environment and resistant to natural degradation, such as medicines, personal care items, and per- and poly-fluoroalkyl substances (PFAS), are especially well-suited for this approach. In-situ chemical reduction (ISCR) can break down complicated pollutants and make them less harmful or easier to remove by promoting electron transfer processes (Mouvet et al. 2020).

There are several important steps in the ISCR process. To determine the types and concentrations of contaminants as well as the properties of the soil, including its composition, pH, and permeability, a comprehensive site assessment is first carried out. This aids in choosing the best reducing chemical and application technique (Liu et al. 2024). Hydrogen gas, sodium dithionite, and ZVI are common reducing agents used in ISCR for soil remediation. Because ZVI can dissolve the strong carbon-fluorine bonds that give PFAS their persistence, it is very effective against these pollutants. Following the selection of the reducing agent, it is applied to the contaminated soil using soil mixing, permeable reactive barriers, or injection wells (Kumpiene et al., 2019).

ISCR can greatly enhance the degradation process for new contaminants such as PFAS, medications, pharmaceuticals, and personal care products (PPCPs). For instance, the strong carbon-fluorine linkages in PFAS make them resistant to natural degradation. These compounds can be partially defluorinated or changed into less hazardous forms by employing ZVI. Comparably, when reducing agents are present, medicines and PPCPs may experience chemical changes that weaken their organic structures and reduce their bioactivity or toxicity. Although the ISCR method works well for a variety of organic pollutants, its effectiveness varies depending on the particulars of the contaminants and the soil (Hussain et al. 2022).

The capacity of ISCR to target contaminants directly in the soil, without the need for excavation or removal, is one of its primary benefits in soil remediation (Kumpiene et al. 2019). Compared with conventional remediation techniques that entail moving contaminated soil to treatment facilities, this technique reduces environmental disturbance and provides a more economical option. Furthermore, ISCR works especially well for removing persistent pollutants that are challenging to eliminate using traditional techniques, such as PFAS and PPCPs. Nevertheless, process optimization still has issues, especially when handling complicated contamination combinations. Long treatment periods may be necessary, and reaction rates can occasionally be sluggish. ISCR can occasionally only cause partial degradation, producing hazardous intermediate chemicals or by-products (Mouvet et al. 2020).

ZVI is used for the in-situ remediation of persistent organic contaminants. It was first used to remove organic compounds from soil in 1997. This is a very flexible method as ZVI can be synthesized via several methods by simply adding the chemicals, or ZVI can be modified according to the type of contaminants by doping with metals and using surfactants. This synergistic effect enhances the removal of contaminants without altering the soil properties. Wan et al. (2024) used ZVI for the treatment of nitrochlorobenzene in the soil.

ISCR for soil remediation has limits despite its potential. The selection of appropriate reducing agents is complicated by the variety of developing pollutants. For instance, different PFAS compounds have different structures, and not all of them respond well to the same reducing agent. The efficiency of the treatment may be impacted by soil properties including heterogeneity, which can also affect how well the reducing agent is dispersed. Furthermore, additional research is required to guarantee the long-term viability and safety of ISCR due to the possibility of partial degradation of specific pollutants (Kumpiene et al. 2019; Mouvet et al. 2020).

10.4 PRECIPITATION

It is a simple method to remediate the soil contaminated with EOCs. Chemical reactions that change pollutants from a dissolved phase into a solid phase are the foundation of the precipitation principle used in EOC remediation. The application of suitable chemicals can result in the development of solid precipitates when the soil is contaminated with EOCs, heavy metals, inorganic compounds, or some organic contaminants (Ahmad et al. 2024). These precipitates are easier to separate and remove from the soil since they are typically less mobile and harmful than the initial pollutants. Precipitation frequently aids in lowering the bioavailability of contaminants, decreasing the likelihood that they will be absorbed by microbes, plants, or animals (Liu et al. 2024).

Precipitation in soil remediation is a multi-step procedure. The polluted soil is first treated with a precipitation agent, such as lime, sodium hydroxide, or other salts. The particular pollutants present to determine whether an agent is best. Solid precipitates, frequently in the form of insoluble compounds that are less mobile in the environment, are created when the additional chemical reacts with the pollutants (Ahmad et al. 2024). Washing or letting them settle in the soil profile are two ways to remove these solid precipitates from the soil matrix. To stop the pollutants from returning to the environment, extra stabilizing techniques are sometimes employed, such as encasing the precipitated pollutants. This takes place in alkaline conditions with pH 9–11 and contaminants form complexes or aggregates and remain immobilized in the soil. Sometimes phosphates are added to convert the contaminants into insoluble complex forms (Khan et al. 2022).

Precipitation is less frequently used to treat emerging pollutants such as pesticides, herbicides, and medications. But occasionally, organic pollutants can combine with metal ions to create insoluble complexes, which makes precipitation a good choice for these pollutants when metals are present (Liu et al. 2024). However, precipitation by itself is ineffective for many organic EOCs; instead, it needs to be used in conjunction with other techniques such as chemical oxidation or bioremediation.

In soil remediation, precipitation offers several benefits. It is a straightforward, economical procedure that uses less sophisticated technology, making it appropriate for heavy metal-contaminated areas (Ahmad et al. 2024). When it comes to inorganic contaminants, it works especially well since it reduces their mobility by turning them into insoluble forms that stop them from spreading farther across the environment. Precipitation is also more cost-effective because it often uses less energy than more sophisticated techniques such as thermal treatments. However, there are some limitations to this process such as the slow settlement of aggregates and slow breakdown of pollutants (Sarkar et al. 2021).

10.5 ELECTROCHEMICAL DEGRADATION

This technique uses oxidation and reduction reactions to break down or change contaminants into less dangerous forms by delivering an electric current. Complex organic pollutants can be broken down into mineralized or biodegradable forms like carbon dioxide and water via electrochemical advanced oxidation processes such as electro-Fenton and anodic oxidation, which produce reactive species like hydroxyl

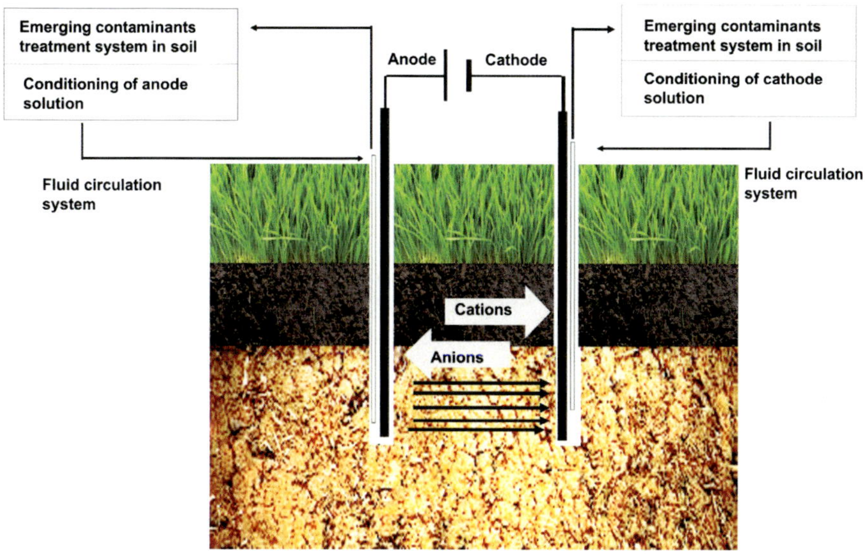

FIGURE 10.3 Electrochemical degradation of EOCs in soil.

radicals. A low-voltage current is delivered to soil pores to cleanse the EOCs using an environmentally beneficial process called electrochemical degradation (Bessaim et al. 2020). The pollutants are carried, pumped out, processed, and decreased by immobilization at electrodes (Ferreira et al. 2020).

To aid in the elimination of contaminants, the anode and cathode, two electrodes, are added to the soil solution in a chamber that is filled with an appropriate solution. The anode and cathode produce acidic and alkaline solutions, respectively (Bessaim et al. 2020). Due to pH shifts, the H^+ ions migrate from the anode and OH^- from the cathode (Figure 10.3). The H^+ ions dominate the solution, and two low- and high-pH zones are developed. This results in the dissolution, precipitation, absorption, or breakdown of the EOC. Sometimes, chemicals are also added to the electrolytic solution to enhance the interaction of soil with pollutants. This led to the elimination of the contaminants by maintaining the same soil structure and characteristics after treatment (Ferreira et al. 2020).

Anode (oxidation):

$$2H_2O \rightarrow 4e^- + O_2 + 4H^+ \tag{10.18}$$

Cathode (reduction):

$$4H_2O + 4e^- \rightarrow 4OH^- + 2H_2 \tag{10.19}$$

10.6 SOLIDIFICATION/STABILIZATION

Solidification and stabilization are remediation methods in which the release of EOCs in the soil is restricted (Haghsheno & Arabani 2023, 2024). It is the potential method for immobilizing emerging contaminants in soils, reducing bioavailability, and lowering environmental risks. Using stabilizing agents such as fly ash, cement, or clay, which react with EOCs to create stable, insoluble compounds, is a part of this process. This method is improved by the electrochemical component, which lowers contaminant mobility by promoting ion migration, electroprecipitation, and adsorption (Zhang et al. 2024).

Sometimes, the mobility of the contaminants is reduced using greener cement and magnesia (MgO) based cement by binding and restricting its release in the soil matrix. This method is cost-effective and environmentally friendly, but it can be revitalized by adopting more sustainable practices. This technique can effectively immobilize emerging pollutants like pharmaceutical residues (antibiotics ciprofloxacin and diclofenac), and PFAS, which are particularly difficult to remediate (Chen et al. 2024).

One of the main benefits of this technique is that it can drastically lower the leachability of pollutants, limiting their bioavailability and reducing their discharge into groundwater. By changing the pH of the soil, encouraging redox processes, and strengthening the binding of pollutants to stabilizing chemicals, electrochemical treatment further increases the stabilization's effectiveness (Tyagi and Annachhatre 2023). Recent developments in the use of nanomaterials and geopolymer-based binders, such as graphene oxide and biochar, have demonstrated encouraging outcomes in enhancing contaminant immobilization, especially for persistent organic pollutants such as bisphenol A (BPA) and PAHs. Haghsheno and Arabani (2023) treated soil polluted with oil near the industrial zone with stabilizers such as fly ash, slag, rice and wheat husk ash, cement dust, limestone, and nanoparticles. They found that cement dust, fly ash, slag, and rice husks continue to be effective at treating soil by 20%–60% (Haghsheno & Arabani 2023).

10.7 SOIL WASHING

Another ex-situ technique for decontaminating soil is soil washing, which involves the use of reagents or extractants. It is a good substitute for traditional soil treatment methods. For a specific amount of time, soil and appropriate extractant solutions are combined (Gusiatin et al. 2020). Precipitation, ion exchange processes, and chelation are involved in the elimination of the pollutants. Once the treated soil satisfies the legal requirements, it is returned to its original site (Liu et al. 2021a). This approach is incredibly successful, produces results quickly, and has no long-term consequences.

In addition, an exceptional type of soil washing is electrochemical soil washing, which is a sophisticated remediation process that employs chemical extractants and an applied electric field to improve the removal of emerging pollutants from soils. To desorb and mobilize pollutants, this method involves washing contaminated soil with reagents such as acids (e.g., hydrochloric acid, citric acid), chelating agents (e.g., ethylenediaminetetraacetic acid, ethylenediamine-*N,N′*-disuccinic acid), or surfactants (e.g., sodium dodecyl sulfate, Tween 80) (Kumar et al. 2022). The use of electrochemical treatment improves pollutant mobilization and degradation by producing reactive oxidants such as hydroxyl radicals and ozone. Emerging pollutants such as PFAS, pharmaceutical residues, and pesticides can be successfully removed by combining soil washing and electrochemical methods (Muñoz-Morales et al. 2021).

Despite its advantages, selecting appropriate extractants and optimizing electrochemical parameters are critical for obtaining high removal efficiency while preserving soil integrity. The long-term environmental impact of residual washing agents must be considered, necessitating research into biodegradable and environmentally friendly reagents such as plant-based biosurfactants and natural chelators (Kumar et al. 2022). Advances in electrochemical reactor design, the use of reactive electrochemical membranes, and the incorporation of renewable energy sources are all being investigated to improve the sustainability and scalability of electrochemical soil washing for the remediation of complicated contaminated sites. The elimination of polybrominated diphenyl ethers using this technique was reported by Ma et al. (2022), who also verified that soil washing is a highly successful way of treating soil that has been contaminated with EOCs.

10.8 ACTIVATED CARBON ADSORPTION

AC, both granular and powdered, is used to remediate organic contaminants in soils. Because of its high sorption capacity, microporous structure, and environmental friendliness, AC binds contaminants through physical and chemical interactions (Edefell et al. 2022). AC can effectively capture emerging contaminants such as per- and poly-fluoroalkyl substances, pharmaceutical residues (e.g., antibiotics such as ciprofloxacin and sulfamethoxazole), and endocrine disrupting compounds such as BPA and phthalates, preventing their migration into groundwater. Modified ACs, such as biochar-based or functionalized AC with metal oxides, boost the adsorption effectiveness of emerging pollutants.

The combination of electrochemical methods and AC adsorption increases clean-up efficiency by encouraging pollutant decomposition. Electrochemical oxidation, for example, can produce reactive species such as hydroxyl radicals and ozone, which degrade adsorbed organic pollutants and regenerate AC. Because regeneration extends the AC's service life, it is also a cost-effective technique (Hiranmai & Neeraj 2024). This integrated strategy is especially efficient for persistent contaminants such as PAHs and pesticides, which are difficult to remove using traditional adsorption methods alone. Furthermore, electrochemical-assisted adsorption can change the surface charge of AC, increasing selectivity and binding strength for various types of developing pollutants (Thakur & Kumar 2024).

Despite its efficiency, issues such as AC saturation, wasted carbon disposal, and the possibility for adsorbed contaminants to leach must be addressed. To boost sustainability, researchers are looking on re-generable AC materials, electrochemical regeneration technologies, and the use of biochar-based adsorbents made from agricultural waste. Advances in nanomaterial-functionalized AC, such as graphene-oxide-modified AC, are being studied to improve adsorption capacity and selectivity. Future advancements aim to combine AC adsorption and electrochemical remediation in large-scale field applications, ensuring long-term contaminant removal while reducing environmental effect (Thakur & Kumar 2024).

Pesticides and medications are among the organic contaminants that granular AC (GAC) adsorbs. Because of its great effectiveness and versatility, this approach is often used.

The GAC immobilizes organic contaminants without generating any unwanted byproducts because of its huge surface area, high adsorption capacity, and more functional groups (Falciglia et al. 2024). For the treatment of organic contaminants in soil, for example, spherical coal- AC, coconut shell-AC, and humic acid granules are preferred. According to Hiranmai and Neeraj (2024), the primary mechanism is the adsorption of organic contaminants in the AC's micropores and mesopores. However, the availability of an adequate quantity of appropriate GAC is the method's limitation. According to recent research, GAC is highly effective at cleaning up soil that has been contaminated by pesticides, medications, and personal hygiene items (Hiranmai & Neeraj 2024). According to Falciglia et al. (2024), liquid-AC can be used to remove PAHs from polluted soil.

10.9 SURFACTANT-ENHANCED REMEDIATION

Compared with traditional approaches, surfactant-enhanced remediation is another efficient and eco-friendly alternative. The chemical structure of surfactants contains both repulsive and appealing components. This special characteristic allows surfactants to increase the solubility of hydrophobic contaminants in water. Anionic, cationic, zwitterionic, and nonionic surfactants have all been studied for soil remediation. The surfactants facilitate the release of contaminants from the soil matrix, which facilitates their extraction or breakdown by other remediation approaches. This combination enables more comprehensive and effective soil remediation. This approach is less invasive and more environmentally friendly than some other methods. It lessens the necessity for moving or excavating the soil, which lessens site disturbance and related expenses. When utilized properly, the surfactants employed in the process are usually biodegradable and present little damage to ecosystems and human health (Tiwari and Tripathy 2023).

When a surfactant is applied to a heterogeneous soil–water system, its molecules attach to the particles of soil and initiate interactions, as shown in Figure 10.4. The tail of the surfactant micelle forms a bond with the hydrophobic pollutant due to its lipophilic property, whereas its hydrophilic entity develops solid–liquid or liquid–liquid interactions (Alamooti et al. 2024). The polarity in the aqueous phase decreases as the concentration of surfactant increases, as a result, the surface tension is reduced due to the substitution, and after attaining the critical micelle-concentration, the round micelle-type structure is formed (Fu et al. 2022). These micelles aid in desorption of the emerging pollutants from the soil because of their hydrophobic cores and lyophobic surfaces (Figure 10.4). They have the power to disperse nonaqueous-phase liquids and significantly increase their water solubility (Alamooti et al. 2024). Since the pollutants dissolved in water are more mobile, later biotic or abiotic processes such as phytoremediation and microbial decomposition are used to extract the organic pollutants. For instance, nonionic surfactants such as Brij-35 and Triton X-100 are more effective in treating naphthalene-contaminated soil as compared with sodium lauryl benzene sulfonate and sodium lauryl sulphate (Liu et al. 2021b).

Furthermore, hydrocarbon pollutants such as phenanthrene and petroleum components (diesel, kerosene, and petrol) have been treated using Tween surfactant

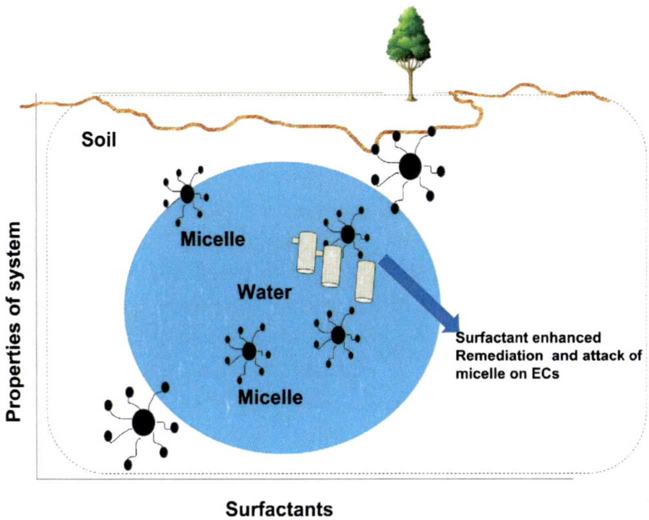

FIGURE 10.4 Mechanism of surfactant-enhanced soil remediation for EOC degradation.

(Saxena et al. 2023). Surfactants must have a high dissolution rate and be able to bind to the soil to effectively clean it. The kind of surfactant and its molecular makeup can have a big influence on the molar solubilization ratio of contaminants. This ratio is determined by the quantity of soluble contaminants relative to the number of soluble surfactant molecules, expressed in moles (Karthick et al. 2019). Tween 80 is the most effective treatment for hydrophobic organic contaminants inside the hydrophobic centers of surfactant micelles. It also lowers the interfacial tension that leads to the mobilization of the contaminants (Karthick et al. 2019).

10.10 NANO-ENHANCED REMEDIATION

Toxicity of EOCs requires the development of innovative methods for the long-term, safe removal of a range of EOCs from soil, including phthalates, microplastics, and persistent organic compounds (Rajput et al. 2024). Nano-enhanced remediation is a novel approach to treating EOCs in soil. Additionally, by improving soil nutrients and immobilizing toxins through absorption and adsorption, nano-enhanced remediation for EOCs offers a dual action (Imtiaz et al. 2024). First, by binding pollutants, the nano-biochar lowers their mobility and bioavailability for microorganisms and plants. Second, the nano-biochar improves soil fertility in addition to remediating it because of its huge surface area, larger adsorption capacity, and nano-sized particles (Rajput et al. 2024). Manufactured nanoparticles have a high capacity for the remediation of EOCs in soils. Their high capacity, stability, and reactivity effectively eliminate contaminants from soil by adsorption mechanisms. Iron (oxides), palladium, titanium dioxide, and amphiphilic polymers are the main types of manufactured nanoparticles (Li et al. 2016). Nano-phytoremediation is a novel method for the clean-up of EOCs from soil. This practice combines nanomaterials with genetically modified plants or naturally occurring hyperaccumulators to remediate soil environments (Biswal 2023). Nanomaterials used in such methods should have characteristics to ensure expected results. Nano-catalysts are also utilized to eliminate PCBs and pesticides in soil by photocatalytic degradation (Zhao et al. 2020).

Ex-situ washing with nonpolar solvents and nano photocatalysts was used to treat EOC-polluted soils (Rajput et al. 2024). However, the in-situ method necessitates the direct application of nanomaterials to the targeted soils in the presence of light, making this technique time-consuming. Nanoparticles can be introduced into the genetic makeup of microorganisms and used to break down EOCs (Li et al. 2016; Rajput et al. 2024). These nanoparticles allow microorganisms to live in extreme settings such as high temperatures, low pH levels, fluctuations in oxygen concentration, salinity, nutritional deficiency, and drought. This microbial nanoremediation approach is sustainable and can help to improve the soil environment by removing EOC contamination (Li et al. 2016).

10.11 DISCHARGE PLASMA TECHNIQUE

One popular method for cleaning up soil is discharge plasma. Langmuir and his associates first proposed it in 1928 when they presented the fourth state of matter (Guo et al. 2022). It is a fluid applied to destroy the EOCs, indicating electrical neutrality. The basic principle is the inelastic collision of particles and the active oxidation and degradation of the organic pollutant (He et al. 2022). These inelastic collisions between carrier gas and electrons transform the electron energy into the energy of the carrier gas, which dissociates the molecules to produce strong oxidizing materials. However, high-energy electrons have a direct strike with contaminants and activate them in their dissociated state. This ultimately leads to the oxidation and degradation of contaminants (Figure 10.5). These active molecules also produce effects such as high temperatures, and ultraviolet radiation, which contribute to the degradation of contaminants (Hatzisymeon et al. 2021). Currently, different methods are used to generate plasma for soil remediation, including corona discharge, dielectric barrier discharge (DBD), and sliding arc discharge.

10.11.1 Corona Discharge

When the intensity of the electric field intersects and exceeds the ionization intensity of the gas, the gas ionizes and produces a well-stable corona discharge. Corona discharge has two patterns for treating contaminants in soil. Needle-plate corona discharge is very famous for ensuring the degradation of PAHs and organic pollutants in soil. The distance between the needles and the plate will affect the breakdown voltage strength (Guo et al. 2022). The needle-plate corona discharge can guarantee good degradation in treating organic pollutants, especially for high concentrations of PAHs in the soil. In acidic soil, the corona plasma was applied and the degradation efficacy of perfluorooctanoic acid was attained by 77% in 2 hours of contact time (Mahyar et al. 2019). The soils contaminated with petroleum and oil were treated by corona discharge, and degradation efficiency was achieved by 76.93% within 1 hour (Li et al. 2017). Similarly, the degradation efficacy of phenanthrene in soil (200 mg/kg)

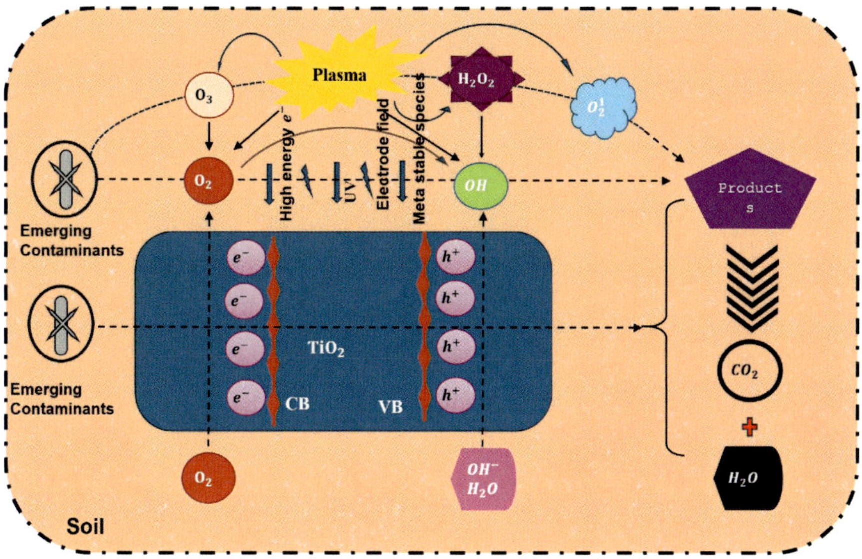

FIGURE 10.5 Mechanism of plasma with TiO_2 for the remediation of EOC-contaminated soils.

was achieved at 91.0% (Guo et al. 2022). The needle plate in the corona plasma degrades the pollutants according to their molecular structure. However, voltage application and power supply also affect the degradation efficacy of EOC.

10.11.2 Dielectric Barrier Discharge

DBD is fabricated by implanting the medium into a discharge tube. The active particles for the reaction are acquired from the atmosphere, as it does not require any vacuum (Guo et al. 2022). The dielectric limits the change in charged particles, which inhibits the excessive discharge current. DBD can be performed in a high-frequency range and at high pressure (Lu et al. 2019). This high current intensity creates a constant discharge mode and produces high electron energy and active atoms. These particles, along with electrons, are responsible for breaking down and oxidizing the contaminants in soils. Recently, DBD was used for treating soil contaminated with PCBs, organochlorine agrochemicals, herbicides, and PAHs. It is also effective in the treatment of herbicide residues between soil pore structures. The removal efficacy is 42.5% when O_3 is 0.5 L/min and the treatment time is 70 minutes (Guo et al. 2022). In addition to the additives mentioned above, DBD can also be used to treat alkanes (Guo et al. 2022). In a study by Li et al. (2016), 87.3% of soil phenanthrene was removed with an airflow of 0.6 L/min. The removal rates of the different treatments ranged from 40.1% to 84.6% and were significantly affected by soil particle size, electrical power, gas flow, and reaction time (Wang et al. 2016). Compared with DBD, the newly designed double DBD can produce more ozone due to two uniform plasma zones (Guo et al. 2022).

10.11.3 Sliding Arc Discharge

Sliding arc discharge is a type of plasma generated by the pressure of the atmosphere. The ionization of gas takes place by applying a high voltage to both sides of the electrode. This ionized gas diffuses and forms a sliding arc plasma (Guo et al. 2022). As a result, the electrons collide and transfer energy, creating a series of reactions. This technique allows the application of high voltage as compared with others and allows the treatment of soils by introducing high-power electrodes (Guo et al. 2022). Currently, sliding arc is used for the treatment of EOCs, and recently, a novel fluidized plasma with a nonthermoelectric bed reactor was suggested for phenol remediation in soils (Du et al. 2017).

10.12 INTEGRATED REMEDIATION APPROACH

The integrated approach requires the combination of plants, biochar, and microbes to remediate soils contaminated with EOCs. It is essential to step up efforts to find sustainable methods for managing and restoring soil that sustainably works around the globe. The synergy between biochar, bacteria, and plants is a highly efficient and sustainable method for removing organic contaminants from soil (Bala et al. 2022). This integration is designed to get maximum benefits, like phytoremediation, which is not very effective for high pollutant concentration; thus the introduction of microbes and biochar amplifies the removal of pollutants. Biochar has functional groups, a porous structure, and a high surface area that improve organic pollutant sorption, lowering their toxicity to soil microorganisms and plants. Biochar-resident plant growth-promoting bacteria (PGPB) may further break down adsorbed organic contaminants, while plants give carbon through root exudates, promoting microbial co-metabolism. PGPB and biochar can promote root development and activity, which help plants absorb contaminants (Figure 10.6) (Xiang et al. 2022). Once, in the plant endospore, pollutants can build up in various plant tissues, be destroyed by plant enzymes and endophytic bacteria, or be discharged into the atmosphere via volatilization. The combined strategy increased plant antioxidant defences while also removing PAHs from the soil. The potential of this technology for environmental pollutant remediation holds promise for the long-term management of substandard soils. By employing various plant species and PAH-degrading bacterial strains, such as *Acinetobacter junii HS29*, *Enterobacter cloacae HS32*, *Enterobacter aerogenes HS39*, *Enterobacter asburiae HS22*, *Brevibacillus reuszeri HS37*, and *Stenotrophomonas* sp. *HS16*, the combined effects of biochar, plants, and bacteria provide a means of eliminating EOCs, such as PCBs and PAHs (Xiang et al. 2022). By using renewable materials, this method provides an environmentally friendly and sustainable means of remediating EOCs (Pathak et al. 2024).

10.13 CONCLUSION

EOCs are recalcitrant and hazardous to the environment, animals, and human health. One possible way to lessen the harm that EOCs pose to the environment and human health is through chemical treatment. To break down EOCs, highly reactive

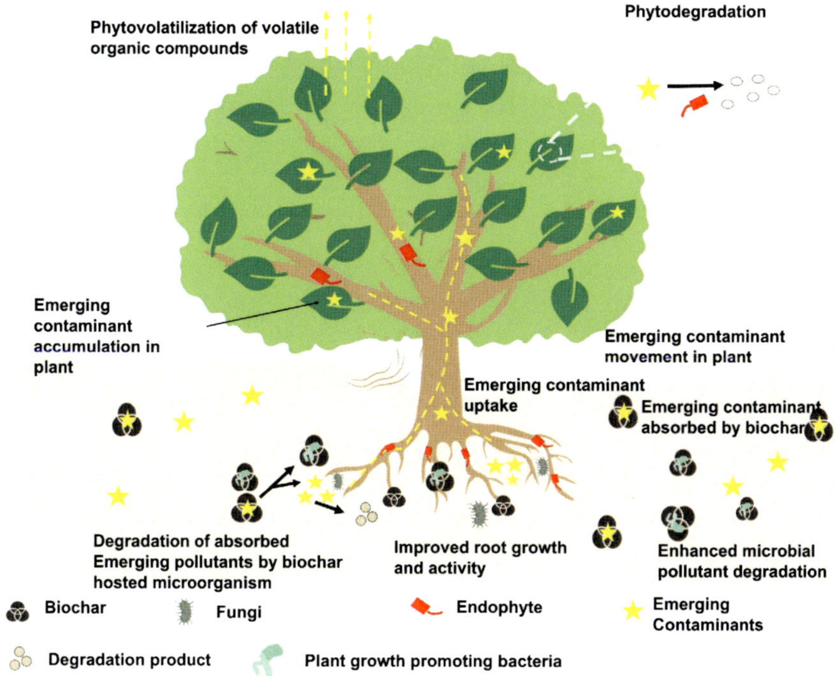

FIGURE 10.6 An integrated approach for the remediation of EOC-contaminated soil.

radicals are used in advanced oxidation procedures, soil washing, electrochemical degradation, AC adsorption, and plasma degradation. Additionally, surfactant-based soil washing effectively eliminates EOCs from soil; however, subsequent wastewater must be managed. EOCs are adsorbed and their bioavailability is inhibited by biochars, nanobiochars, and nanomaterials. EOC type, soil hydrogeochemistry, and soil physiochemical characteristics must all be considered before implementing chemical remediation site conditions. These chemical remediation methods offer integrated approaches to remove EOCs from the soil environment and are sustainable and reasonably priced.

REFERENCES

Ahmad, H.W., H.A. Bibi, M. Chandrasekaran, S. Ahmad, and G.L. Kyriakopoulos, 2024. Sustainable wastewater treatment strategies in effective abatement of emerging pollutants. *Water*, 16(20):2893. https://doi.org/10.3390/w16202893

Alamooti, A., S. Colombano, A. Shoker, et al. 2024. Enhancing remediation of residual DNAPL in multilayer aquifers: Post-injection of alcohol-surfactant-polymer mixtures. *Science of the Total Environment*, 918:170680. https://doi.org/10.1016/j.scitotenv.2024.170680

Amina, Q. Abbas, A. Shakoor, M. Naushad, and B. Yousaf. 2022. In-situ oxidative degradation of sulfamethoxazole by calcium peroxide/persulfate dual oxidant system in water and soil. *Process Safety and Environmental Protection*, 164:696–705. https://doi.org/10.1016/j.psep.2022.06.052

Aparicio, J.D., E.E. Raimondo, J.M. Saez, S.B. Costa-Gutierrez, A. Alvarez, C.S. Benimeli, and M.A. Polti. 2022. The current approach to soil remediation: A review of physicochemical and biological technologies, and the potential of their strategic combination. *Journal of Environmental Chemical Engineering*, 10(2):107141. https://doi.org/10.1016/j.jece.2022.107141

Bala, S., D. Garg, B.V. Thirumalesh, M. Sharma, K. Sridhar, B.S. Inbaraj, and M. Tripathi. 2022. Recent strategies for bioremediation of emerging pollutants: A review for a green and sustainable environment. *Toxics*, 10:484. https://doi.org/10.3390/toxics 10080484

Bessaim, M.M., H. Missoum, K. Bendani, N. Laredj, and M.S. Bekkouche, 2020. Effect of processing time on removal of harmful emerging salt pollutants from saline-sodic soil during electrochemical remediation. *Chemosphere*, 253:126689. https://doi.org/10.1016/j.chemosphere.2020.126688

Biswal, T. 2023. Nano-phytoremediation and its applications. In *Phytoremediation: Management of Environmental Contaminants*, eds. L. Newmann, A.A. Ansari, S.S. Gill, R. Gill, and M. Naeem. 7:335–364. Cham: Springer International Publishing.

Bolan, S., L.P. Padhye, C.N. Mulligan, et al. 2023. Surfactant-enhanced mobilization of persistent organic pollutants: Potential for soil and sediment remediation and unintended consequences. *Journal of Hazardous Materials*, 443:130189. https://doi.org/10.1016/j.jhazmat.2022.130189

Chen, B., J. Xu, H. Lu, and L. Zhu, 2023. Remediation of benzo[a]pyrene contaminated soils by moderate chemical oxidation coupled with microbial degradation. *Science of the Total Environment*, 871:161801. https://doi.org/10.1016/j.scitotenv.2023.161801

Chen, T., B. He, D. Chu, L. Wang, X. Peng, X. Nie, and X. Bai, 2024. Synergy between Fenton reagent and solid waste-based solidifying agents during the solidification/stabilization of lead (II) and arsenic (III) contaminated soils. *Journal of Environmental Management*, 370:122601. https://doi.org/10.1016/j.jenvman.2024.122601

da Silva, L.M., I.F. Mena, C. Sáez, A.J. Motheo, and M.A. Rodrigo, 2024. Remediation of soils contaminated with methomyl using electrochemically produced gaseous oxidants. *Chemosphere*, 362:142653. https://doi.org/10.1016/j.chemosphere.2024.142653

Du, C.M., C. Shang, T. Wang, X.L. Xiao, J.J. Ruan, D.Y. Ma, Y.T. Tang, and R.L. Qiu, 2017. Study of the process and mechanism of the remediation of phenol contaminated soil by plasma vibrated bed. *Plasma Chemistry and Plasma Processing*, 37:1635–1653. https://doi.org/10.1007/s11090-017-9850-6

Edefell, E., O. Svahn, P. Falås, E. Bengtsson, M. Axelsson, R. Ullman, and M. Cimbritz, 2022. Digging deep into a GAC filter–Temporal and spatial profiling of adsorbed organic micropollutants. *Water Research*, 218:118477. https://doi.org/10.1016/j.watres.2022.118477

Falciglia, P.P., G.D. Guidi, F. Vento, G. Finocchiaro, E. Castro, M. Granetto, T. Tosco, and F.G. A. Vagliasindi, 2024. Liquid activated carbon (LAC)–Enhanced microwave remediation of PAH–Contaminated soils. *Separation and Purification Technology*, 331:125628. https://doi.org/10.1016/j.seppur.2023.125628

Fenton, H.J.H., 1894. Oxidation of tartaric acid in presence of iron. *Journal of the Chemical Society, Transactions*, 65:899–910. http://dx.doi.org/10.1039/ct8946500899

Ferreira, A.R., P. Guedes, E.P. Mateus, A.B. Ribeiro, and N. Couto, 2020. Emerging organic contaminants in soil irrigated with effluent: Electrochemical technology as a remediation strategy. *Science of the Total Environment*, 743:140544. https://doi.org/10.1016/j.scitotenv.2020.140544

Fu, Y., C. Qin, S. Gao, C. Lv, C. Zhang, and Y. Yao, 2022. Aquifer flushing using a SDS/1-butanol based in-situ microemulsion: Performance and mechanism for the remediation of nitrobenzene contamination. *Journal of Hazardous Materials*, 424:127409.https://doi.org/10.1016/j.jhazmat.2021.127409

Guo, H., Y. Wang, L. Liao, et al. 2022. Review on remediation of organic-contaminated soil by discharge plasma: Plasma types, impact factors, plasma-assisted catalysis, and indexes for remediation. *Chemical Engineering Journal*, 436:135239. https://doi.org/10.1016/j.cej.2022.135239

Gusiatin, Z.M., D. Kulikowska, and B. Klik, 2020. New-generation washing agents in remediation of metal-polluted soils and methods for washing effluent treatment: A review. *International Journal of Environmental Research and Public Health*, 17:6220. https://doi.org/10.3390/ijerph17176220

Haghsheno, H., and M. Arabani, 2023. The effect of primary stabilizers for stabilization/solidification of oil-polluted soils: A review. *Environmental Technology Reviews*, 12:337–358. https://doi.org/10.1080/21622515.2023.2215460

Haghsheno, H., and M. Arabani, 2024. Stabilization and solidification of oil-polluted soils using secondary stabilizers and industrial wastes. *International Journal of Environmental Science and Technology*, 21:2129–2162. https://doi.org/10.1007/s13762-023-05285-x

Hatzisymeon, M., D. Tataraki, C. Tsakiroglou, G. Rassias, and C.A. Aggelopoulos, 2021. Highly energy-efficient degradation of antibiotics in soil: Extensive cold plasma discharges generation in soil pores driven by high voltage nanopulses. *Science of the Total Environment*, 786:147420. https://doi.org/10.1016/j.scitotenv.2021.147420

He, Y., W. Sang, W. Lu, W. Zhang, C. Zhan, and D. Jia, 2022. Recent advances of emerging organic pollutants degradation in environment by non-thermal plasma technology: A Review. *Water*, 14:1351. https://doi.org/10.3390/w14091351

Hiranmai, R.Y., and A. Neeraj, 2024. Activated carbon-mediated adsorption of emerging contaminants. *Physical Sciences Reviews*, 9:3509–3528. https://doi.org/10.1515/psr-2023-0042

Hussain, A., F. Rehman, H. Rafeeq, M. Waqas, A. Asghar, N. Afsheen, and H.M. Iqbal, 2022. In-situ, ex-situ, and nano-remediation strategies to treat polluted soil, water, and air: A review. *Chemosphere*, 289:133252. https://doi.org/10.1016/j.chemosphere.2021.133252

Imtiaz, H., M. Khan, B.A. Khan, S. Shahid, A.U. Rajapaksha, and M. Ahmad, 2024. Uncovering nano-bonechar for attenuating fluoride in naturally contaminated soil. *Chemosphere*, 353:141490. https://doi.org/10.1016/j.chemosphere.2024.141490

Ji, Z., N. Zhang, C. Huang, X. Duan, D. Ren, and Z. Huo, 2024. The degradation of polycyclic aromatic hydrocarbons (PAHs) by ozone-based advanced oxidation processes: A review. *Ozone: Science & Engineering*, 46:26–42.

Karthick, A., B. Roy, and P. Chattopadhyay, 2019. A review on the application of chemical surfactant and surfactant foam for remediation of petroleum oil contaminated soil. *Journal of Environmental Management*, 243:187–205. https://doi.org/10.1016/j.jenvman.2019.04.092

Khan, S., M. Naushad, M. Govarthanan, J. Iqbal, and S.M. Alfadul, 2022. Emerging contaminants of high concern for the environment: Current trends and future research. *Environmental Research*, 207:112609. https://doi.org/10.1016/j.envres.2021.112609

Khan, Z.H., N.S. Gul, S. Sabahat, et al. 2023. Removal of organic pollutants through hydroxyl radical based advanced oxidation processes. *Ecotoxicology and Environmental Safety*, 267:115564. https://doi.org/10.1016/j.ecoenv.2023.115564

Kumar, M., N. Bolan, T. Jasemizad, et al. 2022. Mobilization of contaminants: Potential for soil remediation and unintended consequences. *Science of the Total Environment*, 839:156373. https://doi.org/10.1016/j.scitotenv.2022.156373

Kumpiene, J., J. Antelo, E. Brännvall, I. Carabante, K. Ek, M. Komárek, and L. Wårell, 2019. In situ chemical stabilization of trace element-contaminated soil: Field demonstrations and barriers to transition from laboratory to the field: A review. *Applied Geochemistry*, 100:335–351. https://doi.org/10.1016/j.apgeochem.2018.12.003

Li, Q., X. Chen, J. Zhuang, and X. Chen, 2016. Decontaminating soil organic pollutants with manufactured nanoparticles. *Environmental Science and Pollution Research*, 23:11533–11548. https://doi.org/10.1007/s11356-016-6255-7

Li, R., Y. Liu, R. Mu, W. Cheng, and S. Ognier, 2017. Evaluation of pulsed corona discharge plasma for the treatment of petroleum-contaminated soil. *Environmental Science and Pollution Research*, 24:1450–1458. https://doi.org/10.1007/s11356-016-7929-x

Li, R., Y. Liu, W. Cheng, W. Zhang, G. Xue, and S. Ognier, 2016. Study on remediation of phenanthrene contaminated soil by pulsed dielectric barrier discharge plasma: the role of active species. *Chemical Engineering Journal*, 296:132–140. https://doi.org/10.1016/j.cej.2016.03.054

Liu, J., L. Zhao, Q. Liu, J. Li, Z. Qiao, P. Sun, and Y. Yang, 2021a. A critical review on soil washing during soil remediation for heavy metals and organic pollutants. *International Journal of Environmental Science and Technology*, 19:1–24. https://doi.org/10.1007/s13762-021-03144-1

Liu, J.W., K.H. Wei, S.W. Xu, et al. 2021b. Surfactant-enhanced remediation of oil-contaminated soil and groundwater: A review. *Science of the Total Environment*, 756:144142. https://doi.org/10.1016/j.scitotenv.2020.144142

Liu, X., K. Sathishkumar, H. Zhang, et al. 2024. Frontiers in environmental cleanup: Recent advances in remediation of emerging pollutants from soil and water. *Journal of Hazardous Materials Advances*, 16:100461. https://doi.org/10.1016/j.hazadv.2024.100461

Lu, W., Y. Abbas, M.F. Mustafa, C. Pan, and H. Wang, 2019. A review on application of dielectric barrier discharge plasma technology on the abatement of volatile organic compounds. *Frontiers of Environmental Science & Engineering*, 13:1–19. https://doi.org/10.1007/s11783-019-1108-5

Ma, Y., H. Li, C. Xie, X. Du, X. Tao, and G. Lu, 2022. Treatment of PBDEs from soil-washing effluent by granular-activated carbon: Adsorption behavior, influencing factors and density functional theory calculation. *Processes*, 10:1815. https://doi.org/10.3390/pr10091815

Mahyar, A., H. Miessner, S. Mueller, et al. 2019. Development and application of different non-thermal plasma reactors for the removal of perfluorosurfactants in water: A comparative study. *Plasma Chemistry and Plasma Processing*, 39:531–544. https://doi.org/10.1007/s11090-019-09977-6

McGachy, L., and D.L. Sedlak, 2023. From theory to practice: Leveraging chemical principles to improve the performance of peroxydisulfate-based in situ chemical oxidation of organic contaminants. *Environmental Science & Technology*, 58:17–32. https://doi.org/10.1021/acs.est.3c07409

Molamahmood, H.V., J. Qin, S. Yu, and M. Long, 2021. Hydrogen peroxide decomposition into oxygen in different soils: Kinetic analysis, mechanism and implication in catalyzed hydrogen peroxide propagations. *Journal of Cleaner Production*, 304:127116. https://doi.org/10.1016/j.jclepro.2021.127116

Mouvet, C., B. Collet, J.M. Gaude, L. Rangon, S. Bristeau, M. Senergues, and T. Woignier, 2020. Physico-chemical and agronomic results of soil remediation by in situ chemical reduction applied to a chlordecone-contaminated nitisol at plot scale in a French Caribbean banana plantation. *Environmental Science and Pollution Research*, 27:41063–41092. https://doi.org/10.1007/s11356-020-07603-z

Mu, S., X. Chen, G. Jiang, and W. Chen, 2023. Fenton oxidation system for treating petroleum-contaminated solid waste: Advances and prospects. *Science of the Total Environment*, 893:164793. https://doi.org/10.1016/j.scitotenv.2023.164793

Muñoz-Morales, M., C. Sáez, P. Cañizares, and M.A. Rodrigo, 2021. Electrochemically assisted soil washing for the remediation of non-polar and volatile pollutants. *Current Pollution Reports*, 7:180–193. https://doi.org/10.1007/s40726-021-00179-3

Pathak, H.K., P.K. Chauhan, C.S. Seth, G. Dubey, and S.K. Upadhyay, 2024. Mechanistic and future prospects in rhizospheric engineering for agricultural contaminants removal, soil health restoration, and management of climate change stress. *Science of the Total Environment*, 927:172116. https://doi.org/10.1016/j.scitotenv.2024.172116

Qutob, M., M. Rafatullah, S.A. Muhammad, M.R. Siddiqui, and M. Alam, 2024. Advanced oxidation of polycyclic aromatic hydrocarbons in tropical soil: Self-catalytic utilization of natural iron contents in an oxygenation reactor supported with persulfate. *Science of the Total Environment*, 926:171843. https://doi.org/10.1016/j.scitotenv.2024.171843

Rajput, P., P. Kumar, A.K. Priya, et al. 2024. Nanomaterials and biochar mediated remediation of emerging contaminants. *Science of the Total Environment*, 916:170064. https://doi.org/10.1016/j.scitotenv.2024.170064

Sarkar, B., R. Mukhopadhyay, S. Ramanayaka, N. Bolan, and Y.S. Ok, 2021. The role of soils in the disposition, sequestration and decontamination of environmental contaminants. *Philosophical Transactions of the Royal Society*, B376:20200177. https://doi.org/10.1098/rstb.2020.0177

Saxena, N., M.M. Islam, S. Baliyan, and D. Sharma, 2023. A comprehensive review on removal of environmental pollutants using a surfactant based remediation process. *RSC Sustainability*, 1:2148–2161. https://doi.org/10.1039/D2SU00069E

Thakur, A., and A. Kumar, 2024. Unraveling the multifaceted mechanisms and untapped potential of activated carbon in remediation of emerging pollutants: A comprehensive review and critical appraisal of advanced techniques. *Chemosphere*, 346:140608. https://doi.org/10.1016/j.chemosphere.2023.140608

Tiwari, M., and D.B. Tripathy, 2023. Soil contaminants and their removal through surfactant-enhanced soil remediation: A comprehensive review. *Sustainability*, 15:13161. https://doi.org/10.3390/su151713161

Tyagi, S., and A.P. Annachhatre, 2023. A review on recent trends in solidification and stabilization techniques for heavy metal immobilization. *Journal of Material Cycles and Waste Management* 25:733–757. https://doi.org/10.1007/s10163-022-01583-7

Ugrina, M., and A. Jurić, 2023. Current trends and future perspectives in the remediation of polluted water, soil and air a review. *Processes*, 11:3270. https://doi.org/10.3390/pr11123270

Wan, J., Y. Guo, Z. Zhang, et al. 2024. Persulfate activation with biochar supported nanoscale zero-valent iron: Engineering application for effective degradation of NCB in soil. *Science of the Total Environment*, 933:173053. https://doi.org/10.1016/j.scitotenv.2024.173053

Wang, K., K. Wang, Y. Chen, et al. 2023. Adsorption–desorption behavior of malachite green by potassium permanganate pre-oxidation polyvinyl chloride microplastics. *Environmental Technology & Innovation*, 30:103138. https://doi.org/10.1016/j.eti.2023.103138

Wang, T., J. Ren, G. Qu, D. Liang, and S. Hu, 2016. Glyphosate contaminated soil remediation by atmospheric pressure dielectric barrier discharge plasma and its residual toxicity evaluation. *Journal of Hazardous Materials*, 320:539–546. https://doi.org/10.1016/j.jhazmat.2016.08.067

Watts, R. J. & A.L. Teel, 2006. Treatment of contaminated soils and groundwater using ISCO. *Practice Periodical of Hazardous, Toxic, and Radioactive Waste Management*, 10:2–9. https://doi.org/10.1061/(ASCE)1090-025X(2006)10:1(2)

Xiang, L., J.D. Harindintwali, F. Wang, et al. 2022. Integrating biochar, bacteria, and plants for sustainable remediation of soils contaminated with organic pollutants. *Environmental Science & Technology*, 56:16546–16566. https://doi.org/10.1021/acs.est.2c02976

Zhang, Y., P. Fu, W. Ni, et al. 2024. A review of solid wastes-based stabilizers for remediating heavy metals co-contaminated soil: Applications and challenges. *Science of the Total Environment*, 920:170667. https://doi.org/10.1016/j.scitotenv.2024.170667

Zhao, Q., X. Zhao, and J. Cao, 2020. Advanced nanomaterials for degrading persistent organic pollutants. In *Advanced Nanomaterials for Pollutant Sensing and Environmental Catalysis*, ed. Q. Zhao, 249–305. Elsevier. https://doi.org/10.1016/B978-0-12-814796-2.00007-1

11 Bioremediation and Related Interfacial Processes of EOCs in Soil

Dong Zhang, Hongting Zhao, Zhengfeng Hu, and Ming Zhang

11.1 INTRODUCTION: BACKGROUND OF BIOLOGICAL INTERFACIAL PROCESSES IN BIOREMEDIATION

Bioremediation has been widely recognized to be a promising technique for remediating organic contaminated soils, in which living organisms, predominantly microorganisms, are employed to degrade, reduce, and/or detoxify pollutants (Arora 2018; Bala et al. 2022; Roy et al. 2024). It can be conducted either in-situ (at the site of contamination) or ex-situ (removing the contaminated material to another location for treatment), by using indigenous microorganisms with enriched nutrients (called *biostimulation*) and/or exogenous engineered microorganisms (called *bioaugmentation*), under either aerobic or anaerobic conditions (Azubuike et al. 2016; Megharaj et al. 2011; Mohanty and Rath 2020; Zhou et al. 2024). Bioremediation is widely recognized as an environmentally friendly and practical approach for soil cleanup, offering key advantages over chemical and physical remediation methods, including lower costs, minimized secondary pollution, and greater social acceptability. Therefore, bioremediation has been explored to eliminate the eco-risks of soils contaminated with various conventional pollutants, including polycyclic aromatic hydrocarbons, petroleum hydrocarbons, and pesticides (Burns et al. 2013; Ebsa et al. 2024; Samanta et al. 2002; Zhao et al. 2024; Zhou et al. 2024). In addition, bioremediation also shows promise in addressing emerging organic contaminants (EOCs) in soils, such as perfluorinated compounds and pharmaceutical and personal care products (PPCPs) (Aguilar-Romero et al. 2022; Bala et al. 2022; Guo et al. 2016; Kucharzyk et al. 2017; Leung et al. 2022; Li et al. 2022). For examples, triclosan (TCS), a typical PPCPs, could be degraded by indigenous microorganisms in a water-sediment system, and the presence of biosurfactants (rhamnolipids) significantly enhanced the bioremediation efficiency (Guo et al. 2016). Kwon et al. (2014) isolated a strain of *Pseudomonas aeruginosa*, which could break down perfluorooctane sulfonic acid (PFOS) effectively (~67% removal).

The fundamental principle of bioremediation involves the use of microorganisms to break down pollutants into less toxic forms (Pande et al. 2020; Sharma et al. 2019). Diverse contaminants can be degraded depending on different microorganisms and

DOI: 10.1201/9781003509523-11

related enzymes. For bioremediation to be successful, certain conditions must be met, irrespective of the microbial species or target contaminants (Alexander 1994; Ganesan et al. 2022; Pande et al. 2020): (i) The microorganism must have the potential to survive in the polluted environment; (ii) the organisms must have an effective enzymatic system, which is responsible for contaminant conversion; (iii) there must be a close intra- or extracellular contact between the microorganism/enzyme and contaminants. In soil systems, especially historical contaminated soil sites, the poor potential for microorganisms to get access to the contaminant limits the efficiency of bioremediation, owing to strong sorption of contaminant on solid surface and/or not soluble in soil solution. If the enzyme's active site involved in bioremediation cannot access the substrate, efficient pollutant biodegradation is unlikely to occur. Therefore, we pay special attention to the accessibility of pollutants and related processes in this chapter.

The direct degradation of solid or crystalline organic pollutants in soil may occur, owing to the contribution of extracellular enzymes (e.g., laccases, manganese peroxidases, lignin peroxidases) secreted by microorganisms (Johnsen et al. 2005). However, compared with intracellular hydrolysis, the likelihood of extracellular degradation is lower. Moreover, the major "purpose" of extracellular enzyme is to dissolve solid pollutant into "soluble phase," facilitating the uptake into the intracellular environment to support the growth of microorganism. We therefore are more focused on the form of intracellular degradation for organic pollutants, which is the more common pathway for bioremediation. In order to achieve the intracellular degradation of EOCs, a series of consecutive processes must occur, referring to EOC desorption from soil particles and transport within soil matrix, uptake/sorption of EOCs on microbial surface from soil solution, transmembrane process, and intracellular enzymatic degradation (as seen in Figure 11.1). In addition, to facilitate the attachment of microbes with EOCs, the transport of microbes within the soil matrix should also be our concern.

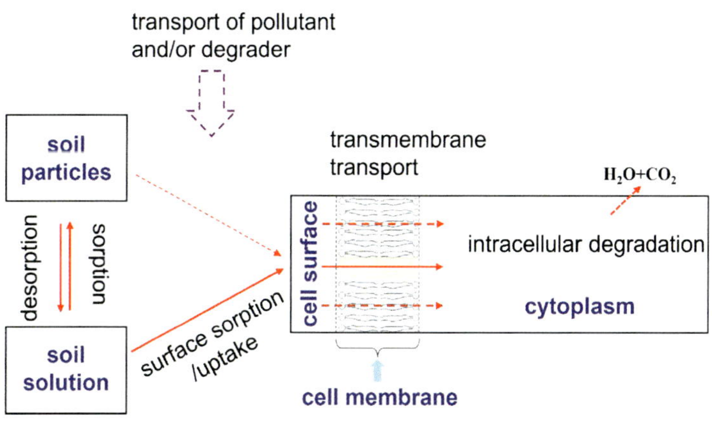

FIGURE 11.1 Interfacial processes during the bioremediation of organic pollutants.

11.2 TRANSPORT OF EOCs AND THEIR DEGRADERS WITHIN SOIL MATRIX

As Harms and Wick (2006) suggested, "assure the well-being of the microorganisms" would be the most important guideline for the bioremediation of EOC-contaminated soil, followed by "bring the contaminant and the microorganisms together." In other words, if there are suitable EOC degraders, the attachment of EOC molecules and their degraders (specifically biological surface/membrane/active intracellular enzyme site) dominantly determines the possibility and efficiency of bioremediation. It is a common sense that contaminants (e.g., PFOS, PFAS, pesticides) patchily distribute in the soil matrix, due to either historical contamination of separate phase contaminants or preferential sorption onto specific constitutions (i.e., organic matter). Considering high bioremediation efficiency, various approaches, including surfactants, organic solvents, eutectic solvents, and electrochemical devices, have been conducted to homogenize contaminants in soil matrix and thus improve the bioavailability and biodegradation of organic compounds (Bolan et al. 2023; Fernández-Cascán et al. 2023; Ishtaweera and Baker 2024; Mahinroosta and Senevirathna 2020). In addition, microbial populations (primarily bacteria serving as EOC degraders) frequently exhibit heterogeneous spatial distribution patterns, attributable to either microbial microcolony formation with solid phases or limited cellular mobility. Even in ideal systems (e.g., quartz sand-packed medium and controlled conditions), low mobilities of microorganisms are very common (Liu et al. 2020; Zhang et al. 2024). This is because the long distances (between microbes and target EOC molecules) and physical steric hindrance result in low migration ability of microorganisms within porous media (i.e., soil).

Theoretically, regardless of indigenous microorganisms or exogenous/engineered microorganisms (in the scenarios of *bioaugmentation*), it is difficult for microbes to transport to their target pollutants in the soil matrix. In terms of the migration of pollutants, it is the same. Therefore, one of the most important roles of enhancement for bioremediation is to mitigate the heterogeneity of the soil matrix and thus facilitate the contact of EOCs and their degraders. As shown in Figure 11.2, treatments that homogenize microorganisms and/or contaminants will facilitate the contact of microorganisms and contaminants, resulting enhanced bioremediation. The environmental behaviors of EOCs in soil, including sorption-desorption and transport have been discussed in details in the above chapters (i.e., Chapters 4 and 5). Therefore, in this chapter, we will talk about the transport of microorganisms (i.e., bacteria) within the soil matrix to facilitate the contact of microbes with EOCs.

1. **The Role of Homogenizing Microorganisms in Bioremediation:** As we mentioned above, the successful bioremediation of organic contaminated soil depends on the robust growth of microorganisms, efficient degradation enzyme system, and the direction contact between pollutants and microbes (Wick et al. 2007). Soil is generally considered to be a good habitat for microbes (e.g., bacteria). If EOC degraders are present in soil system, no matter indigenously or augmented, the contact of EOCs and their degraders seems to be critical and first process for successful bioremediation

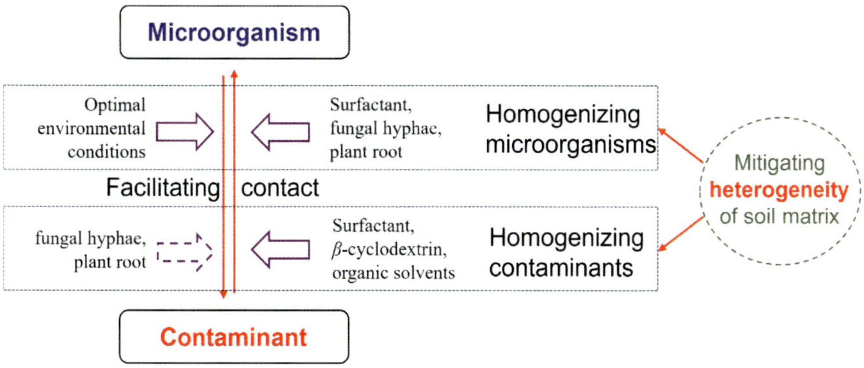

FIGURE 11.2 Optimal bioavailability of contaminants can be achieved by mitigating the heterogeneity of heterogeneously distributed pollutants and microorganisms in the soil matrix.

procedure. In fact, soils are desolate environments for microbes with patchy distribution pattern and so are EOCs. In many cases, it appears that EOC desorption from soil particles (or soil constituents) is not a limiting process for bioavailability and thus bioremediation (Ren et al. 2018), implying that the bacterial transport plays a critical role in the bioremediation. However, a centimetre-range or even several meters distance is required for bacteria to move to the "mission site" for initializing degradation, owing to the heterogeneity of soil (e.g., particles size, constitutes, surface roughness) and distribution of EOCs (Chen et al. 2023; Harms and Wick 2006; Yang et al. 2012). These observations are well-justified, as soil-dwelling bacteria can be immobilized through various mechanisms including straining, blocking, ripening, and soil sorption, consequently exhibiting minimal active motility (Zhong et al. 2017). In addition, the competition of "food" and eco-niche with native bacteria, especially in the bioaugmentation scenario, can inevitably affect the survival, growth, transport of EOC degraders and thus influence the bioremediation efficiency (Gao et al. 2022; Yang et al. 2021). Therefore, approaches are needed to facilitate the migration of microbes in soil.

2. **Surfactant-Assisted Microorganism Homogenizing:** It is essential for a successful bioremediation (including bioaugmentation and improved-indigenous microbes) to mobilize the active transport of bacteria to or throughout the contaminated hotspot (Zhang et al. 2024; Zhong et al. 2017). Great interest has been drawn to find environment-friendly strategies to reduce the sorption/deposition and thus facilitate the transport of bacteria in the soil porous media. As mentioned above, physical and/or chemical conditions (e.g., ionic strength (IS), soil particle size, soil constitute, temperature, pH, etc.) can significantly alter the interactions between bacteria and soil solid surfaces and affect bacterial attachment or migration. These interactions occur through various mechanisms, such as electrostatic attraction,

hydrophobic interactions, and Ca^{2+}-linked bridging (Ghorbanzadeh et al. 2021; Guo et al. 2021; He et al. 2019; Kwak et al. 2023; Sepehrnia et al. 2023; Zhang et al. 2021b). Surfactants are recognized as promising candidates, both theoretically and practically, due to strong ability to modify surface properties of both porous media and bacterial cell surfaces. Tao et al. (2021) reported that the presence of anionic biosurfactant rhamnolipid (1 g/L) promoted the transport of both hydrophilic *Pseudomonas putida* (contact angle θ_c of 36.2°) and hydrophobic *Rhodococcus erythropolis* (θ_c of 94.7°) in saturated sand, by modifying the surface tension. Brown and Jaffé (2001) assumed the sorption of surfactants (Brij 30, and Brij 35) displaced the counterions, expanded the electric double layer of bacterial cell surface, and thus enhanced the transport of bacteria. Yang et al. (2021) found that the biosurfactant changed the wettability of the solid surface and facilitated the generation of corner flows and transport of a hydrophobic bacteria strain (θ_c of ~120°). Liu et al. (2020) observed the presence of both anionic (i.e., SDBS) and nonionic (Triton X-100) surfactants significantly affected bacterial transport depending on bacterial cell surface hydrophobicity (CSH). However, in some cases, inhibited transport of bacteria in the presence of surfactants was reported. For example, Harvey et al. (2010) investigated the influence of imbentin, a nonylphenol-based nonionic surfactant, on the attachment and transport of native bacteria in quartz sand aquifer. They found that the surfactant increased the bacterial attachment, owing to increase in hydrophobic interactions between the bacterial and grain surfaces. Zhong and colleagues (2016) demonstrated that rhamnolipid at low concentrations impeded the transport of both hydrophilic (glucose-grown) and hydrophobic (hexadecane-grown) *Pseudomonas aeruginosa* strains, primarily through alterations in CSH with secondary effects from solution chemistry changes. They found a good linear relation between bacterial deposition rate and surfactant-regulated CSH (Liu et al. 2017, 2020; Zhong et al. 2016). In general, different or even contrasting results have been reported. In some cases, a nonmonotonic pattern has been observed. Several mechanisms for such effects of surfactants are summarized and illustrated in Figure 11.3.

a. Competitive sorption of surfactant with bacterial cells on solid surface. (i) The presorption or competitive replacement of surfactants on porous media can reduce available sorption sites on solid surfaces, weakening the bacterial sorption onto the surface of porous media. Zhang et al. (2024) designed a series of bacterial migration experiments, including the presorption of Tween 80 (a nonionic surfactant) and bacterial postsorption with Tween 80 flushing, to investigate the roles of surfactant in improved bacterial transport in porous media (i.e., soil, and quartz sand). The prior sorption of Tween 80 and subsequent Tween 80 flushing resulted in significant decreases of solid sorbed bacteria and enhancement of transport. These observations and further theoretical calculation indicated that the Tween 80 occupied or displaced the available sorption sites on solid surfaces for bacteria, therefore facilitating bacterial

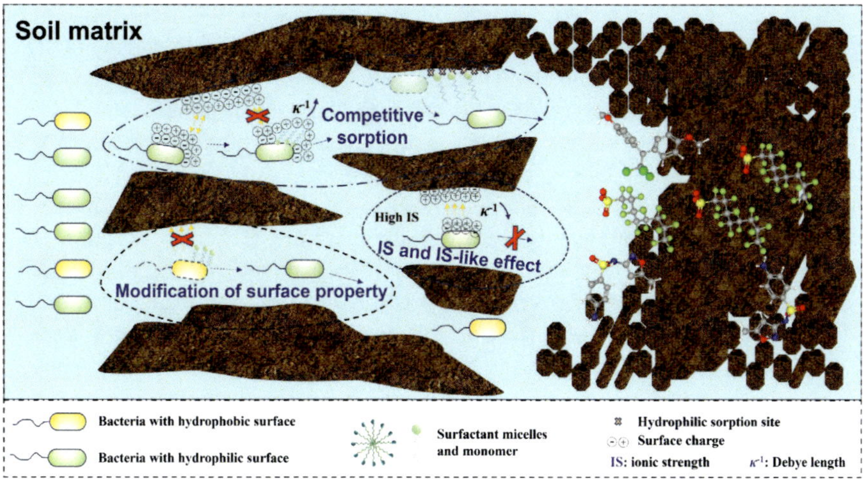

FIGURE 11.3 Schematic diagram of surfactant-assisted bacterial transport in the soil matrix.

migration. Streger et al. (2002) reported similar results that the presence of Tween 80 reduced bacterial adhesion to sand and increased bacterial transport. Anti-adhesion activity for *Staphylococcus aureus*, *Listeria monocytogenes*, and *Micrococcus luteus* of surfactin and rhamnolipid were reported by Zeraik and Nitschke (2010). (ii) Surface hydrophobicity of porous media can also be altered by the sorption of surfactants. The changes further affect the hydrophobic interactions between bacterial cells and solid surfaces, and thus shifts the sorption and transport of bacteria. Yang et al. (2021) reported that biosurfactants (e.g., rhamnolipid) could facilitate the dissemination of bacteria to "new territory" by changing the bacterial solid surface wettability (meaning a decrease of surface contact angle θ_c and hydrophobicity) and corner geometry. Nitschke et al. (2009) found that surfactin reduced the adhesion of two pathogenic bacteria *Listeria monocytogenes* and *Salmonella enteritidis* on stainless steel and polystyrene surfaces. The hydrophobicity of both solid surfaces was tested to be decreased by the presorption of surfactin, where the contact angles (θ_c) reduced from 53.9° to 33.8° for stainless steel and from 102.7° to 98.7° for polystyrene. The authors believed that the alterations of media surface hydrophobicity may be the reason for the results observed. (iii) The sorption of surfactants on the solid surface can prevent the bacterial cells from reaching these surfaces due to the surfactant molecule or micelle-induced steric hindrance. Bai et al. (1997) calculated the length of monorhamnolipid molecule to be 1.2 nm and believed the sorbed surfactant monomer was large enough to sterically hinder the approach of bacteria to the solid surface. The transport, therefore, could be enhanced. For the cases of surfactant micelles, the diameter of micelles in simple terms are thought to be in the range of 2–4 nm, and can reach micron-range in practical cases (Volkring et al. 1998). This phenomenon indicates that the sorbed surfactant on the solid surface may have a steric effect on bacterial sorption.

b. Modification of bacterial surface properties. In fact, the involvement of surfactants in altering the CSH has been widely documented, although the direction of these changes (toward increased hydrophilicity or hydrophobicity) remains inconsistent in the literature. Generally, sorption of the surfactant can either increase or decrease the CSH determined by initial hydrophobicity (Zhang and Zhu 2014). Typically, bacterial strains with "hydrophobic" surface (CSH >40% as measured by bacterial adhesion to hydrocarbon assay) demonstrate reduced hydrophobicity when exposed to surfactants, consequently enhancing their mobility within soil solution. Liu et al. (2020) reported that the sorption of surfactants (SDBS and Triton X-100) on bacterial cells caused more exposure of polar heads to aqueous environment and resulted in a lower energy state. These changes of bacterial surface property facilitated the desorption of bacteria from solid phase and transport through soil matrix. Several observations have showed that surfactants induced CSH reduction in bacteria with high initial CSH (27.0%–30.0%) and very high initial CSH (43.4%–74.0%), and then detached bacteria from solid surfaces (e.g., quartz sand) and facilitated bacterial migration (Gross and Logan 1995; Liu et al. 2017, 2020; Sodagari et al. 2013; Zhang and Miller 1994). The alteration of CSH involves several potential mechanisms, such as surfactant sorption through Van der Waals and hydrophobic interactions between the surfactant hydrophobic moiety and hydrophobic sites of the cell surface (Zhang et al. 2024). These interactions resulted in the exposure of hydrophilic heads or hydrophobic tails of surfactants, causing CSH reduction or increase (Górna et al. 2011; Zhang et al. 2013). In addition, the solubilization of cell components (i.e., lipopolysaccharide, LPS) by surfactant micelles can also happen, which leads to the loss of LPS or uncharged O-antigen of LPS. This phenomenon typically causes CSH increase (LPS as hydrophilic moiety of gram-negative bacteria) and/or extra exposure of charged groups (Li and Zhu 2012).

Bacterial surface contains negative groups such as carboxyl, hydroxyl, and phosphate groups and therefore is typically negatively charged. The sorption of ionic surfactants on bacterial cell surface or loss of LPS moiety by surfactant micelle solubilization may change the bacterial cell surface charge. In the surfactant-related bioremediation strategies, anionic surfactants (e.g., biosurfactant such as rhamnolipid and saponin, SDBS) and nonionic surfactants (e.g., Tween series, Brij series, and sometime Triton X-100) are commonly used, rather than cationic surfactants. In the cases of anionic surfactant sorption, the surface charge can be neutralized or enhanced, whereas the solubilization of LPS may decrease the electronegativity of bacterial surface. Chen et al. (2004) observed that the addition of rhamnolipid decreased the zeta potential of both bacterial strains (i.e., *Lactobacillus casei* and *Streptococcus mitis*) and porous media surface with a correlation with surfactant concentrations. The decrease of the zeta potential indicated possible increase of bacteria–sand repulsive electrostatic interactions, which theoretically was beneficial for the bacterial transport. However, the authors believed that the interaction-free energy (ΔG_{132}) rather than surfactant-induced surface charge modification was the dominant reason for bacterial desorption. The repulsive electrostatic interactions contributed to the migration of desorbed bacteria. Zhang et al. (2024) used Tween 80 as model surfactant and observed a slight increase of zeta potential of bacterial surface from −7.26 to −5.60 mV. Li et al. (2023) found that the presence of 100 mg/L rhamnolipid decreased the zeta potential of bacteria for 17% and claimed rhamnolipid just slightly increased the electrostatic repulsion. In general, the surfactant-induced surface charge modification seems to not be a major contributor to the bacterial desorption and transport, although it may slightly affect the electrostatic interactions between bacteria and solid surfaces.

c. IS and IS-like effects.

Higher IS can lower the energy barrier for bacterial sorption on solid surfaces (e.g., quartz sand, soil) and then facilitate the sorption and deposition of bacteria on porous media, meaning limited transport. This phenomenon has been widely accepted and many studies provided solid evidence (Balthazard-Accou et al. 2014; He et al. 2019; Kwak et al. 2023; Yang, Kim and Tong 2012; Yang, Tong and Kim 2012; Zhang et al. 2024). Li et al. (2018) investigated the effects of IS on the transport of *Bacillus subtilis* cells in saturated porous media. It was found that the Gram-positive bacteria could be readily transported (mass recovery of 80%) under lower IS conditions (i.e., 1 mmol/L), whereas the transport was inhibited (60% mass recovery) under higher IS stress (i.e., 100 mmol/L). The calculated XDLVO interaction energy profiles also confirmed the phenomenon. The increased IS deepened the secondary minimum wells significantly, from less than 0.001 k_BT at 1 mmol/L to −0.0761 k_BT at 100 mmol/L. Yang, Kim and Tong (2012) also reported similar results that the transport of both Gram-negative *Escherichia coli* and Gram-positive *Rhodococcus* sp. QL2 was slightly trapped when the IS (NaCl) increased from 5 to 20 mmol/L. Increased deposition rate parameter k_d and decreased surface zeta potential were also observed, which was consistent with the calculation results of classic DLVO theory. In addition to the abovementioned effects, bivalent cations (e.g., Ca^{2+}, Mg^{2+}) may play extra roles in the deposition and transport of bacteria in porous media. For example, Ghorbanzadeh et al. (2021) reported Ca^{2+}-linked bridging sorption enhanced the straining of bacteria and therefore reduced the bacterial migration. The authors speculated that the precipitation of calcium carbonate can directly retard bacterial transport through electrostatic sorption between positively-charged calcium carbonate and negatively-charged bacterial surface, or indirectly limit the transport of bacteria by physically increasing soil structural stability, narrowing pore structures of soil and reducing soil permeability. Yang, Tong and Kim (2012) observed increased cell deposition under higher IS conditions. The $CaCl_2$ solution with the same IS as NaCl showed superior bacterial retained effect. One possible mechanism for $CaCl_2$-increased attachment of bacteria, as NaCl provided IS, was the codeposition of bacteria with bentonite particles. In addition, the formation of bacteria-bentonite-bacteria cluster was recognized to be another mechanism for bivalent cations (i.e., Ca^{2+}). The bivalent cations may also interact with extracellular polymeric substances (EPS) on the bacterial surface and alter the surface property and conformation, thus affecting bacterial migration in porous media (Chen and Walker 2007; Zhong et al. 2017).

One possible explanation for IS-affected bacteria-solid interaction is the alteration of the thickness of the electric double layer surrounding a charged surface such as bacterial surface, soil particle surface, and quartz sand surface, which can be defined as Debye length (κ^{-1}) (Zhang et al. 2024). Debye length generally relates to the IS inversely, described as $\kappa^{-1} \propto IS^{-0.5}$ (Brown and Jaffé 2001; Wang et al. 2022; Zhang et al. 2024), as described by the following equation (11.1).

$$\kappa^{-1} = \left(\frac{\varepsilon \varepsilon_0 K_B T}{N_A e^2 \sum C_i Z_i^2} \right)^{1/2} \tag{11.1}$$

where ε is the dielectric constant of water, 78.5 (dimensionless); ε_0 is the dielectric permittivity of a vacuum, 8.854×10^{-12} C/(V·m); K_B is Boltzmann constant, 1.38×10^{-23} J/K; T is the absolute temperature taken as 298 K here; N_A is the Avogadro's number, 6.02×10^{23} mol^{-1}; e is the electric unit charge, 1.602×10^{-19} C; C_i is the molar concentration of one species ions (i), mol/L; and Z_i is the valence of the ion (i). IS relates to $\sum C_i Z_i^2$.

Ionic surfactants may act as salts partially and alter the IS of the solution. Zhong et al. (2016) revealed that the low concentration of anionic surfactant rhamnolipid changed the transport of *Pseudomonas aeruginosa* cells in an ideal porous matrix. The authors assumed that the anionic surfactant acted as salts to increase solution IS, which played an additional minor role in affecting bacterial migration. High concentration of (anionic) surfactants may change the solution chemistry such as IS increase, which impedes the bacterial migration through porous media (Liu et al. 2020). Nonionic surfactants can affect bacterial transport with an IS-like effect, where the surfactant in the solution leads to a slight increase in the electrical conductivity similar to the increase of IS (Zhang et al. 2024). However, it seems to contribute minorly in the bacterial transport system.

3. **Fungal Hyphae-Bacteria Cooperation for Microorganism Homogenizing:** Two possible ways of hyphae-assisted bioavailability enhancement of organic pollutants for bacteria have been reported, which refer to "fungal highways" and "fungal pipelines" (Banitz et al. 2013). In the "fungal pipelines", fungi uptake and translocate pollutants to degrading bacteria through mycelial network (Furuno et al. 2012). For the "fungal highways", the active movement of degrading bacteria is facilitated along fungal hyphae and therefore enhance the possibility of remote pollutant uptake. To date, Wick's group has made attempts to provide direct or indirect evidence for enhanced bacterial transport facilitating bioremediation, focusing on fungi–bacteria interactions. For example, Wick et al. (2007) reported that the co-presence of fungal mycelia (*Pythium ultimum*) efficiently improved the translocation of bacterial strain *Pseudomonas putida* PpG7 up to a centimeter range by bridging air-filled pores and providing bacteria a continuous network of water paths. The created water paths enabled strain *P. putida* PpG7 to spread in the soil and access the soil-associated phenanthrene. In such cases, almost all phenanthrene was biodegraded. In contrast, no phenanthrene biodegradation was observed when the fungus was absent, even with the presence of an inoculum of 10^8 catabolically active bacteria. Banitz et al. (2011) developed a spatially explicit bacterial colony model based on controlled laboratory experiments to simulate the correlation between bacterial movement/dispersion and organic compound biodegradation. Simulated results indicated that conditions of limited bacterial movement/dispersion also limited the biodegradation performance of organic compounds. Otto et al. (2016) also obtained similar results by using glass fibers as a model to mimic a fungal mycelial network. In these conditions, with the help of hyphae (acting as "fungal highway"), bacteria can use liquid films around hyphae to overcome the mobility restrictions and to reach remote sites via quick dispersion (Banitz et al. 2013). Worrich et al. (2016) reported that soil-borne bacterial strain *Pseudomonas putida* KT2440 could efficiently degrade benzoate at consistently high percentages (average of 70.7% ± 7.8%) along "fungal highways" under different strength conditions.

Generally, there are several mechanisms for fungal mycelium-related bacterial dispersion through so-called "fungal highways," as illustrated in Figure 11.4. (i) Chemotaxis-driven dispersal of bacteria alongside the "new" water paths. This mechanism has been recognized as the dominant pattern, and the property of hyphae (e.g., hydrophobicity, growth) is identified as a key factor determining bacterial movement (Kohlmeier et al. 2005; Xiong et al. 2022). The presence of hyphae enhances the movement of active bacteria by either driving the dispersal of bacteria toward a contaminated soil layer (Wick et al. 2007) or by facilitating the chemotactic swimming of bacteria in soil (Furuno et al. 2010). (ii) Horizontal gene transfer mechanism. The hyphae-derived mycosphere can be identified as a gene transfer arena (Zhang et al. 2014), where high frequency of horizontal gene transfer between spatially separated bacteria occurs. Therefore, the transfer of

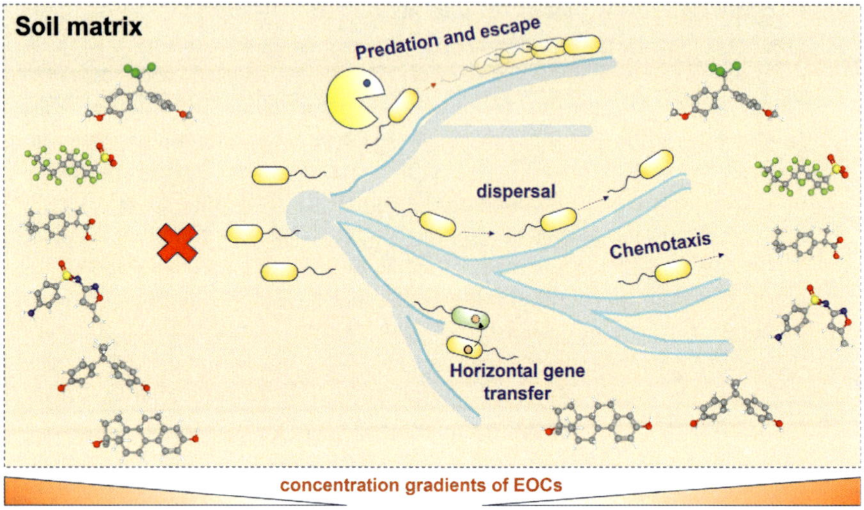

FIGURE 11.4 Mechanisms of bacterial transport along "fungal highway." (Modified from Worrich et al. 2018.)

genetic elements and abilities for EOC degradation can be enhanced. (iii) Escape from predation. The fungal highways can also provide routes for prey bacteria (also acted as EOC degraders) to quickly escape from predators (Otto et al. 2017).

11.3 SORPTION AND ACCUMULATION OF EOCs ON DEGRADERS' SURFACE

Sorption of contaminants by microorganism-based materials has been identified as one of the most promising strategies for removing pollutants from wastewater, and is usually defined as "biosorption." In these cases, biomaterials showed excellent sorption performance for various pollutants because microbial components contain plenty of functional groups (e.g., carboxyl, hydroxyl and phosphate groups, and aromatic structures) and are rich in organic carbon. In addition, they receive wide attention due to low costs and easy modifiability. In the scenarios of soil bioremediation, the sorption of pollutants by live microorganisms plays another crucial role in facilitating biodegradation process. Sorption here can be recognized as a pre-step for subsequent uptake and intracellular metabolism. Theoretically, microbial sorbed pollutants are readily available and a positive correlation between sorption and biodegradation is commonly proposed (Ma et al. 2018; Stylianou et al. 2019; Zhang 2013; Zhang et al. 2013). For example, Muñoz et al. (2020) evaluated the biotransformation of ampicillin, a widely used β-lactam antibiotic, by a nitrifying consortium. After 24 hours of incubation, approximately 72.4% of ampicillin (initial concentration of 50 mg/L) was removed, where the biosorption contributed significantly (~36.7%). Butzen et al. (2020) investigated the sorption behaviors of PFOS and

PFOA onto Gram-negative *Pseudomonas putida* and Gram-positive *Bacillus subtilis* and observed that up to 40% ± 5% and 67% ± 11% of PFOA and PFOS, respectively, were sorbed on bacterial surfaces under most conditions tested. Ye's groups carried out several studies on the biosorption and biodegradation of triphenyltin (TPT) by bacterial species (i.e., *Brevibacillus brevis*, *Stenotrophomonas maltophilia*, *Bacillus thuringiensis*). Their results indicated that the biosorption happened initially and quickly reached their peaks. Then the biosorption rates decreased and the sorbed TPT was further biodegraded (Gao et al. 2014; Huang et al. 2014; Ye et al. 2013). Sethunathan et al. (2004) reported that the inoculation of soil with algae could effectively degrade α-endosulfan, an endocrine disrupting insecticide. The authors also demonstrated that sorption of the endocrine disrupting chemical onto microbial surfaces significantly influenced biodegradation, with α-endosulfan being rapidly adsorbed from soil solution, thereby affecting its subsequent degradation.

The underlying mechanisms for microbial sorption of EOCs vary depending on the property of microbial surfaces as well as pollutant types. Generally, some important mechanisms are listed below. (i) Hydrogen bonding. Hydrogen binding is one of the common mechanisms for microbial sorption of organic pollutants. The surface hydroxyl groups (C-OH) can be involved for hydrogen bonds formation. For example, Lv et al. (2024) believed that the oxygen atoms in the hydroxyl group provided lone electron pairs to form hydrogen bonds for sorption of bisphenol A on a yeast surface. (ii) Electrostatic interactions. Biosurfaces typically contain negative charges originating from functional groups (e.g., oxygen-containing groups, phosphate groups), negatively-charged components of LPS, or teichoic acid. EOCs, especially many PPCPs, have different acid dissociation constants (pK_a) and can be existed as different ionic species. The biosurfaces and EOC molecules may interact with each other through electrostatic forces. Vasiliadou et al. (2013) observed negligible sorption of five PPCPs (i.e., caffeine, sulfamethoxazole, ranitidine, carbamazepine, and ibuprofen) in most conditions due to hydrophilic property (log K_{ow} < 2.5). At pH 7, ranitidine (pK_a of 8.2) carried a net positive charge, which enhanced the sorption onto negatively charged biomass surfaces through electrostatic attraction. Xie et al. (2020) concluded electrical neutralization (electrostatic attraction) as a dominant mechanism for the sorption of positively charged pharmaceutical ciprofloxacin on negatively charged surface of microalgae strain Tai-03. This is because the PPCP has two acid dissociation constants (pK_a of 6.1 and 8.2) and can be stated in zwitterionic form in common environmental conditions (pH ranging from 6.5 to 8.0). (iii) Hydrophobic interactions. The surfaces of microorganisms generally are heterogeneity containing both hydrophobic and hydrophilic sites. These properties allow microbial surfaces to sorb hydrophobic molecules or hydrophobic components, such as pesticides and amphiphilic PFAS. Butzen et al. (2020) revealed a pH-independent sorption of PFOA on *Pseudomonas putida* and indicated hydrophobic bonding as a major mechanism for bacterial sorption of PFOA. The highly hydrophobic fluorinated tail of PFOA makes it to be high hydrophobicity with a log K_{ow} of 6.3. Therefore, hydrophobic interactions occurred between bacterial hydrophobic sites and the fluorinated tail of PFOA, a mechanism that remained largely unaffected by pH variations. EPS on microbial surfaces can also provide sorption sites for hydrophobic EOCs through hydrophobic interactions (Xie et al. 2020). For example, Xu et al. (2013) reported the successful sorption of sulfamethazine on activated sludge surface by the formation of stable EPS-PPCP complexes. (iv) Physical entrapment via surface components such as polysaccharides. Many microorganisms can secrete EPS containing proteins and polysaccharides. The EPS forms a complicated matrix structure and is abundant in carboxyl, amino, hydroxyl groups, enabling it to sorb organic pollutants through physical entrapment (Liao et al. 2021). For example, Xu and Sheng (2020) reported that EPS components could act as a reservoir and effectively intercept sulfamethazine.

11.4 TRANSMEMBRANE PROCESS OF EOCs

Transmembrane transport of dissolved organic pollutants from outside of the microbial membrane into intracellular cytoplasm plays a critical role in the biodegradation of target contaminants. This process is often identified as a rate-limiting step, as these organic molecules are ultimately degraded intracellularly by microbial enzymes (Huang et al. 2014; Ortiz et al. 2005; Song et al. 2010; Zhang et al. 2013; Zhang and Zhu 2014). The cell membrane and cell wall (in some cases, e.g., outer membrane of Gram-negative bacteria) act as "barrier" for specific compounds to enter into intracellular space. The semipermeable property of the cell membrane plays a major role of "barrier" in limiting free transport into intracellular environment and selectively allows nutrient and/or organic pollutant uptake into the cells (Al-Awqati 1995). The cell membrane (inner membrane) contains a phospholipid bilayer and anchored proteins (see Figure 11.5) surrounding the cell and controls the movement of substances in and out. The phospholipid is composed of a hydrophilic head (phosphorus group, water-attracting) and two hydrophobic tails (fatty

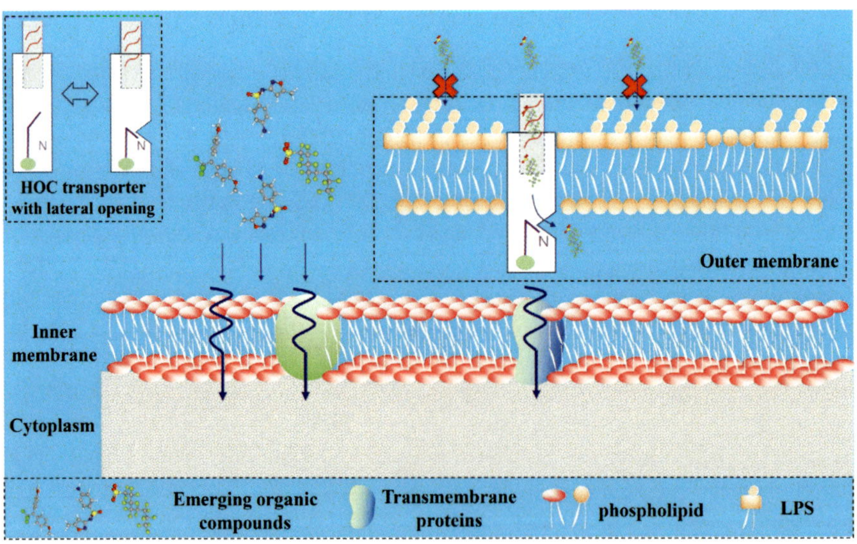

FIGURE 11.5 Possible transmembrane transport routes for emerging organic compounds across bacterial cell membranes and outer membranes in the case of Gram-negative bacteria.

acid chains, water-repelling). Two layers of such amphiphilic molecules create a three-region structure (water attracting head–water repelling tail–water attracting head), impermeable to most water-soluble molecules. Integral proteins, one of the proteins embedded with the lipid bilayer, span the entire membrane and can act as channels, carriers, or receptors, transporting specific substances through the lipid bilayer selectively. In addition, the outer membrane of Gram-negative bacteria can also provide an extra and efficient barrier for hydrophobic organic compounds, dominantly due to the presence of LPS. The transmembrane uptake of organic contaminants may thus suffer a rough task, determining the "real" bioavailability of target pollutants (Bressleer and Gray 2003). The transmembrane route and flux vary depending on the contaminant structure and concentration outside the cell membrane, and protein channels for specific molecules (Vander et al. 2001).

11.4.1 Transport Routes

Two possible transport routes exist for organic contaminants transporting across the cell membrane and cell wall barrier: direct diffusion through the lipid bilayer and protein channel-assisted movement (see Figure 11.5). (i) Diffusion through lipid bilayer. As mentioned above, the phospholipid bilayer creates a unique region with three domains of different polarity. The hydrophobic nature of phospholipid tails makes the central region nonpolar and hydrophobic. This structure limits the free diffusion of relatively hydrophilic organic pollutants, whereas hydrophobic compounds can readily move across the phospholipid bilayer into the intracellular cytoplasm. For example, Song et al. (2010) investigated sorption and transmembrane transport of two types of pharmaceutical chemicals (i.e., hydrophilic kanamycin with log K_{ow} of −6.70 and relatively hydrophobic chloramphenicol with log K_{ow} of 1.14) on the phospholipid bilayer. The sorption of kanamycin occurred through electrostatic attraction and hydrogen bond formation, whereas the chloramphenicol sorption happened via hydrophobic interactions. Furthermore, over 89% of sorbed hydrophilic kanamycin stayed outside the out surface of the phospholipid surface, while more than 80% of sorbed chloramphenicol transmembrane transported into the internal matrix. Bateman et al. (1986) revealed that hydrophobic naphthalene (log K_{ow} of 3.30) could move across the cell membrane of *Pseudomonas putida* PpG1064 into the cell through a simple diffusion route. (ii) Protein channel-assisted transport. There are also at least two types of protein-assisted transport: active transport requiring energy and energy-free diffusion-based transport. (a) In an active transport scenario, pollutants or nutrients can be transported against the concentration gradient, energy-consuming, transport protein-dependent, and high selectivity and specificity (Wang et al. 2020). The transmembrane transport of hydrophobic organic pollutants labelled with isotopic tracers (e.g., C^{14}-octadecane, C^{14}-fluoranthene) across bacterial cell membranes (e.g., *Pseudomonas* sp. DG17, *Rhodococcus* sp. BAP-1) were believed to be active transport process which required energy. The addition of an energy inhibitor (e.g., NaN_3) significantly hindered the transmembrane process, indicating the energy requirement for the transmembrane transport (Hua and Wang 2014; Li et al. 2014). (b) In the case of Gram-negative bacteria, the cell wall contains a layer of membrane-similar structure (called outer membrane), exposing hydrophilic component LPS outside. The LPS is composed of polar oligosaccharides, capable of extending up to 30 Å outside the outer membrane surface (Wiener and Horanyi 2011). The LPS-based domain acts as an efficient barrier for the passage of hydrophobic pollutants. Therefore, to successfully uptake hydrophobic pollutants (e.g., pesticides, PFAS, and other HOCs), a certain protein-based channel is needed. Currently, the well-known protein family involved in the hydrophobic molecule transmembrane transport is the *FadL* family, named after the archetypal long-chain fatty acid (LCFA) transporter FadL of *E. coli* (Hearn et al. 2009). Similar hydrophobic proteins are reported, including *OprG*, *AlkL*, *NapQ*, and *ExFadLO*, which can passage hydrophobic compounds into cell and expel hydrophobic metabolites out of the cell (Eaton 1994; Julsing et al. 2012; Martínez et al. 2013; Touw et al. 2010). The *FadL* family is conserved with 14-stranded β-barrel structure with an inward-facing kink in the third β-strand (S3) as a lateral opening (Hearn et al. 2009; Wiener and Horanyi 2011). More importantly, the *FadL* contains a series of ordered and bond detergent molecules inside the barrel lumen (red lines in the illustration in the upper left corner in Figure 11.5). Foe example, in *PaFadL*, there are three complete tetraethylene glycol monooctyl ether (C_8E_4) molecules. The detergent molecules create a long substrate passageway (~55 Å) for hydrophobic compounds, connecting the extracellular surface and the lateral opening, and thus bypassing the hydrophilic barrier of LPS. The 14-stranded barrel indicates the substrate passageway is wide enough to allow large molecules (e.g., LCFA, PFAS) to transport. The binding of hydrophobic compounds with detergent molecules (especially the third molecule) will trigger a conformational change of the *N* terminus and open the lateral opening. In addition, the transport mechanism seems to be diffusion-based and does not require energy (DiRusso and Black 2004).

Besides the abovementioned transport routes, antibiotics or other chemicals (e.g., biosurfactants) may also disrupt or disorder the cell membrane lipid orientation or structure, resulting in the uptake of these pharmaceutical chemicals. For example, John et al. (2017) reported the presence of kanamycin A, a representative aminoglycoside antibiotic, which led to disordered membrane structure and potential uptake of antibiotics into the bacterial cellular environment. Sotirova et al. (2008) claimed that the biosurfactant could result in transmembrane pores in the cell membrane by forming molecular aggregates. The newly formed pores may then provide extra channels to the periplasm and facilitate chemical transport.

11.4.2 Quantitative of the Transport Flux

To quantitatively evaluate the transmembrane transport rate, the conception of membrane-crossing flux (J) is introduced and can be written as equation 11.2 (Bressleer and Gray 2003):

$$J = k_m\left(C_L^{\text{out}} - C_L^{\text{in}}\right) \tag{11.2}$$

where C_L^{in} is the intracellular concentration of target compound, typically equaling to 0 due to active intracellular enzyme and biodegradation; C_L^{out} is the external concentration of target compound; k_m is the permeability of the membrane, positively correlating to oil–water coefficient of the compound ($\propto 0.1K_{ow}$) and negatively relating to the square root of molecular weight (MW) of the compound ($\propto MW^{-0.5}$).

If the surface sorption and/or accumulation of the target compound is not included, the extracellular concentration of compound (C_L^{out}) can be identified as the water solubility (S_w). In this case, the maximum membrane-crossing flux (J_{max}) can be evaluated as equation 11.3:

$$J_{max} = 0.003K_{ow} \cdot S_w \cdot MW^{-0.5} \tag{11.3}$$

where 0.03 is the empirical constant (Bressleer and Gray 2003).

Bressleer and Gray (2003) established a positive correlation of possible biodegradation rate (k_b) with J_{max}, which can be calculated from the physicochemical properties (i.e., K_{ow}, S_w, and MW) of the target pollutants (16 chemicals of aromatics, chlorine and oxygenated compounds) based on equation 11.3. They proposed a theoretical framework to assess possible biodegradation rate and bioremediation efficiency for specific organic pollutants. However, its universality still needs to be verified and refined. We evaluated the relationship between k_b and J_{max} of 17 PPCPs and a poor correlation was found ($R^2 = 0.12$, $n = 50$) (Zhang 2019). The selected PPCPs in our evaluation contain various negatively and/or positively charged polar groups (e.g., carbamazepine, celestolide, diazepam, diclofenac, erythromycin, estrone, fluoxetine), resulting in various surface sorption and accumulation. This situation may significantly affect the available concentrations of target compounds (C_L^{out}). Therefore, a modified equation considering the sorption and accumulation before transmembrane process is proposed, as described in equation 11.4.

$$J_{max}^{*} = 0.003K_{ow} \cdot K_d \cdot S_w \cdot MW^{-0.5} \tag{11.4}$$

where J_{max}^{*} is the modified membrane-crossing flux; K_d is the sorption coefficient of the target compound on the outside of the membrane. In this situation, the C_L^{out} does not equal to S_w, but equal to $K_d \cdot S_w$, considering the affinity of target compound between biosurfaces and water.

Using equation 11.4, a much higher positive correlation between k_b and J_{max}^{*} was obtained ($R^2 = 0.66$, $n = 50$) than that of k_b-J (Zhang 2019). The results indicate that surface sorption plays a crucial role in the transmembrane transport and thus intracellular degradation. We can theoretically predict the biodegradation rate by calculation based on fundamental physicochemical and sorption parameters.

11.5 INTRACELLULAR DEGRADATION OF EOCs

Microorganisms, including bacteria, fungi, archaea, and microbial consortium, capable of growth on and biodegradation of certain emerging chemicals have been widely reported. Examples of microorganisms with the ability of biotransformation of EOCs (e.g., PFAS, PPCPs, organo-tin, endocrine disrupting chemicals) are summarized in Table 11.1. With the help of EOC-degraders (specifically, enzymes), individually and co-metabolized, the biotransformation processes lead to complete or partial degradation into CO_2 or intermediates. For fungi, three different types of enzymes

TABLE 11.1
Selected Degrading Strains for EOCs and Underlying Key Enzymes

Compound	Strain and Key Enzymes	References
PFOS and PFOA	Bacterial consortium (*Paracoccus, Hyphomicrobium, Micromonosporaceae*)	Sorn et al. (2023)
	Pseudomonas aeruginosa	Kwon et al. (2014)
	Acidimicrobium sp.	Huang and Jaffé (2019)
	Ensifer adhaerens	Chetverikov and Loginov (2019)
	Pseudomonas plecoglossicida	Chetverikov et al. (2017)
Ibuprofen	*Phanerochaete chrysosporium*/MnP[a]	Rodarte-Morales et al. (2012)
	Trametes versicolor/Lac[b], *Ganoderma lucidum*/MnP	Vasiliadou et al. (2016)
	Firmicutes, Aeromonas, Nitrospira, Betaproteobacteria	Zhao et al. (2015)
	Sphingomonas sp. Ibu-2/*ipf*ABDEF	Murdoch and Hay (2005, 2013)
	Patulibacter sp. I11	Almeida et al. (2013)
	Aquabacterium, Gammaproteobacteria	Sanchez-Huerta et al. (2023)
Sulfamethoxazole	*Trametes versicolor*/Lac, *Ganoderma lucidum*/MnP	Vasiliadou et al. (2016)
	Rhodococcus rhodochrous	Bouju et al. (2012) and Gauthier et al. (2010)
	Pseudomonas aeruginosa	Bouju et al. (2012)
	Acinetobacter sp.W1	Wang and Wang (2018)
	Dyella sp. WW1	Wang et al. (2018)
	Firmicutes, Aeromonas, Nitrospira, Betaproteobacteria	Zhao et al. (2015)
	Microbacterium sp. BR1	Fenu et al. (2015)
	Bacillus subtilis	Larcher and Yargeau (2011)
	Hydrogenotrophic archaea (*Methanobacterium subterraneum, Methanobacterium formicicum, Methanolinea mesofila*)	Carneiro et al. (2020)
	Activated sludge	Alvarino et al. (2016)
Carbamazepine	*Pseudomonas* sp. CBZ-4	Li et al. (2013)
	Acinetobacter sp. HY-7	Cui et al. (2009)
	Trametes versicolor/Lac, *Ganoderma lucidum*/MnP	Vasiliadou et al. (2016)
	Starkeya sp. C11, *Rizobium* sp. C12	Bessa et al. (2017)
	Dyella sp. WW1	Wang et al. (2018)
	Unidentified basidiomycete BNI/Lac, LiP[c], MnP	Santosa et al. (2012)
	Phodococcus rhodochrous	Gauthier et al. (2010)
Triclosan	*Pseudomonas putida* TriRY, *Alcaligenes xylosoxidans* TR1	Meade et al. (2011)
	Dyella sp. WW1	Wang et al. (2018)
	Rhodotorula mucilaginosa, Penicillium sp.	Taştan et al. (2016)
	Aspergillus versicolor	Taştan and Dönmez (2015)

(Continued)

TABLE 11.1 (*Continued*)
Selected Degrading Strains for EOCs and Underlying Key Enzymes

Compound	Strain and Key Enzymes	References
	Nannochloris sp.	Bai and Acharya (2016)
	Sphingopyxis sp. KCY1	Lee et al. (2012)
	Nitromonas europaea	Roh et al. (2009)
	Sphingomonas sp. YL-JM2C	Mulla et al. (2015)
	Burkholderia xenovorans LB400, *Sphingomonas*	Kim et al. (2011)
	Georgenia, Soehngenia, Comamonas	Zhang et al. (2021a)
triphenyltin	*Bacillus thuringiensis*/ Na^+/K^+-ATPase, Ca^{2+}/Mg^{2+}-ATPase	Yi et al. (2017)
bisphenol A	*Aquabacterium, Gammaproteobacteria*	Sanchez-Huerta et al. (2023)

[a] Lac, laccase.
[b] MnP, manganese-dependent peroxidases.
[c] LiP, lignin-peroxidases.

(i.e., laccase (Lac), lignin-peroxidases (LiP), and manganese-dependent peroxidases (MnP)) are assumed to response to the biotransformation of EOCs (Asif et al. 2017). For instance, the white-rot fungi *Trametes versicolor* secreted Lac, LiP, and MnP, which could efficiently degrade various EOCs such as ibuprofen, 17β-estradio, 17-α-ethynylestradio, ketoprofen, acetaminophen, and carbamazepine, under different conditions (Blánquez and Guieysse 2008; Cruz-Morató et al. 2013; Jelic et al. 2012). The LiP and MnP secreted by *Phanerochaete chrysosporium*, another commonly used white-rot fungi, were believed to be the major contributors for high bioremoval efficiency of diclofenac, ibuprofen, and naproxen (Rodarte-Morales et al. 2012). Bacterial enzymes, by contrast, are barely mentioned. Yi et al. (2017) analyzed the activities of Na^+/K^+- and Ca^{2+}/Mg^{2+}-ATPase during the dearylation of TPT and believed that these ATPases were correlated to the biodegradation of TPT. Murdoch and Hay (2005, 2013) identified a cluster of genes (*ipfABDEF*) encoding enzymes in *Sphingomonas* for aromatic dioxygenation and acyl removal of ibuprofen and phenylacetic acid.

Considering the existence of various potential "target" for EOC-degraders, different strategies may be adopted. Here are four commonly used strategies and initialization for the biodegradation of different EOCs with different functional groups.

1. **Cleavage of Nitrogen-Related Bonds:** Many PPCPs contain nitrogen-based functional groups, including amino, piperazine ring, N-O bond, and N-S bond. The strategies for their degraders take may be initiated via the cleavage of nitrogen-based bonds. For example, in an MABR fed, the acetaminophen underwent initial oxidative attack at the amino group, resulting in the formation of 4-aminophenol as primary transformation product (Sanchez-Huerta et al. 2023). The piperazine ring can also

be the nitrogen-related target where the bacteria initially decompose. Jia et al. (2018) identified the piperazine ring as the most probable initial cleavage site through intermediate analysis. The molecule of sulfamethoxazole (SMX) contains three nitrogen atoms in the forms of amino (NH_2-), N-O in a heterogenic ring, and another amino (S-NH-C). The three nitrogen-based bonds can all be the target by the SMX-degraders, as reported by Carneiro et al. (2020), Jia et al. (2017), and summarized by Zhang (2019). Ibuprofen is a pharmaceutical chemical containing a benzene ring and its biodegradation pathway usually involves the hydroxylation and ring cleavage (Murdoch and Hay 2005).

2. **Ether Bond Cleavage:** The ether bond (C-O-C) in some pharmaceutical chemicals can also be the target for specifical enzymes. For example, following partial hydroxylation and aromatic ring cleavage, the ether bond of TCS became accessible to chlorohydroquinone dehydrogenase. This enzymatic action converted the resulting intermediate, chlorohydroquinone, into 2,4-dichlorophenol via oxidative degradation (Lee et al. 2012; Mulla et al. 2015).
3. **Dearylation:** For EOCs like TPT, the whole aromatic rings can be removed from the core molecule without ring cleavage. TPT consists of three benzene rings, which are separately linked to the tin molecule, and can be biotransformed through the cleavage of these tin-carbon bonds sequentially or synchronously (Yi et al. 2017). Ye's group (Huang et al. 2014; Yi et al. 2017) determined the possible intermediates of TPT biodegradation by the *Bacillus* sp. and inferred that fast and successive dearylation occurred.
4. **Defluorination and Oxidation of C-C Bonds:** For PFOS, oxidation of the C-S bond initially seems to be a commonly used strategy for PFOS degraders (Chetverikov and Loginov 2019; Chetverikov et al. 2017; Huang and Jaffé 2019). Then the C-C bond is oxidized along with defluorination and shorter-chain fluorinated compounds are formed (same in the case of PFAS without S-C bond). For example, Huang and Jaffé (2019) detected the production of SO_4^{2-} initially and measured the formation of shorter-chain fluorinated compounds such as PFBS and HFBA when they incubated PFOS with *Acidimicrobium* sp. strain A6. Sorn et al. (2023) detected PFOS metabolites such as 3,3,3-trifluoropropionic acid and PFBA, indicating the presence of defluorination. However, in some cases, defluorination does not always happen, and an alternative strategy is the partial cleavage of C-C bonds in the PFOS molecule. For example, Kwon et al. (2014) tried several attempts at fluoride ion measurement during the PFOS biodegradation, and no change in fluoride level was observed. The ambiguous results compel us to conduct further research. This is acceptable because many EOCs are pollutants with complex structure containing several groups such as aromatic rings, heterogenic rings, and varied nitrogen-related bonds, and the biodegradation is conducted under different conditions. For example, Sanchez-Huerta et al. (2023) investigated the bioremoval of ten organic micropollutants (such as nonionic hydrophilic acetaminophen, nonionic hydrophobic bisphenol A, TCS, ionic hydrophobic ibuprofen) using a membrane-aerated biofilm

reactor. The authors found that heterotrophic bacteria was the driving force for hydrophilic compound biodegradation, whereas for those compounds with medium hydrophobicity, a combination consortium of heterotrophic and enriched nitrifying bacteria were needed for bioremoval.

11.6 SUMMARY OF ROLES OF BIO-INTERFACIAL PROCESSES IN THE BIOREMEDIATION OF EOCs

The surface sorption, transmembrane transport, and intracellular enzyme-based degradation of EOCs can be identified to be sequential processes during the biodegradation of EOCs by microorganisms. The enhanced surface sorption (sorption affinity coefficient, K_d) and accelerated transmembrane transport (apparent transmembrane flux, J^*) will provide more available EOC molecules to intracellular active enzymes and facilitate the biodegradation rate (k_b). There seems to be positive correlations between k_b and J^*, as well as K_d. However, due to the variety of EOCs, more work needs to be conducted. In another point of view, we should draw much more attention to the transport of EOC degraders (mainly bacteria) as well as EOCs themselves in soil, of which the latter has been extensively studied. The hindered transport of bacteria limits the availability of EOCs and thus the bioremediation strategies (including both the biostimulation and bioaugmentation). Several approaches (i.e., addition of surfactant, coexisting of hyphae or plant roots) can facilitate the bacterial transport in soil and are beneficial for the bioremediation process.

REFERENCES

Aguilar-Romero I., van Dillewijn P., Nesme J., Sørensen S.J., Nogales R., Delgado-Moreno L., Romero E., 2022, A novel and affordable bioaugmentation strategy with microbial extracts to accelerate the biodegradation of emerging contaminants in different media, *Science of the Total Environment* 834, 155234.

Al-Awqati Q., 1995, Membrane permeability, *Current Opinion in Cell Biology* 7, 463–464.

Alexander M., 1994, *Biodegradation and Bioremediation*, Academic Press, New York.

Almeida B., Kjeldal H., Lolas I., Knudsen A.D., Carvalho G., Nielsen K.L., Crespo M.T.B., Stensballe A., Nielsen J.L., 2013, Quantitative proteomic analysis of ibuprofen-degrading *Patulibacter* sp. strain I11, Biodegradation 24, 615–630.

Alvarino T., Nastold P., Suarez S., Omil F., Corvini P.F.X., Bouju H., 2016, Role of biotransformation, sorption and mineralization of ^{14}C-labelled sulfamethoxazole under different redox conditions, *Science of the Total Environment* 542, 706–715.

Arora N.K., 2018, Bioremediation: a green approach for restoration of polluted ecosystems, *Environmental Sustainability* 1, 305–307.

Asif M.B., Hai F.I., Singh L., Price W.E., Nghiem L.D., 2017, Degradation of pharmaceuticals and personal care products by white-rot fungi: a critical review, *Current Pollution Reports* 3, 88–103.

Azubuike C.C., Chikere C.B., Okpokwasili G.C., 2016, Bioremediation techniques-classification based on site of application: principles, advantages, limitations and prospects, *World Journal of Microbiology and Biotechnology* 32, 180.

Bai G.Y., Brusseau M.L., Miller R.M., 1997, Influence of a rhamnolipid biosurfactant on the transport of bacteria through a sandy soil, *Applied and Environmental Microbiology* 63, 1866–1873.

Bai X.L., Acharya K., 2016, Removal of trimethoprim, sulfamethoxazole, and triclosan by the green alga *Nannochloris* sp., *Journal of Hazardous Materials* 315, 70–75.

Bala S., Garg D., Thirumalesh B.V., Sharma M., Sridhar K., Inbaraj B.S., Tripathi M., 2022, Recent strategies for bioremediation of emerging pollutants: a review for a green and sustainable environment, *Toxics* 10, 484.

Balthazard-Accou K., Fifi U., Agnamey P., Casimir J.A., Brasseur P., Emmanuel E., 2014, Influence of ionic strength and soil characteristics on the behavior of *Cryptosporidium* oocysts in saturated porous media, *Chemosphere* 103, 114–120.

Banitz T., Fetzer I., Johst K., Wick L.Y., Harms H., Frank K., 2011, Assessing biodegradation benefits from dispersal networks, *Ecological Modelling* 222, 2552–2560.

Banitz T., Johst K., Wick L.Y., Schamfuß S., Harms H., Frank K., 2013, Highways versus pipelines: contributions of two fungal transport mechanisms to efficient bioremediation, *Environmental Microbiology Reports* 5, 211–218.

Bateman J.N., Speer B., Feduik L., Hartline R.A., 1986, Naphthalene association and uptake in *Pseudomonas putida*, *Journal of Bacteriology* 166, 155–161.

Bessa V.S., Moreira I.S., Tiritan M.E., Castro P.M.L., 2017, Enrichment of bacterial strains for the biodegradation of diclofenac and carbamazepine from activated sludge, *International Biodeterioration & Biodegradation* 120, 135–142.

Blánquez P., Guieysse B., 2008, Continuous biodegradation of 17β-estradiol and 17α-ethynylestradiol by *Trametes versicolor*, *Journal of Hazardous Materials* 150, 459–462.

Bolan S., Padhye L.P., Mulligan C.N., Alonso E.R., Saint-Fort R., Jasemizad T., Wang C.S., Zhang T., Rinklebe J., Wang H.L., Siddique K.H.M., Kirkham M.B., Bolan N., 2023, Surfactant-enhanced mobilization of persistent organic pollutants: potential for soil and sediment remediation and unintended consequences, *Journal of Hazardous Materials* 443, 130189.

Bouju H., Richen B., Beffa T., Corvini P.F.X., Kolvenbach B.A., 2012, Isolation of bacterial strains capable of sulfamethoxazole mineralization from an acclimated membrane bioreactor, *Applied and Environmental Microbiology* 78, 277–279.

Bressleer D.C., Gray M.R., 2003, Transport and reaction processes in bioremediation of organic contaminants. 1. Review of bacterial degradation and transport, *International Journal of Chemical Reactor Engineering* 1(R3), 1–16.

Brown D.G., Jaffé P.R., 2001, Effects of nonionic surfactants on bacterial transport through porous media, *Environmental Science & Technology* 35, 3877–3883.

Burns R.G., DeForest J.L., Marxsen J., Sinsabaugh R.L., Stromberger M.E., Wallenstein M.D., Weintraub M.N., Zoppini A., 2013, Soil enzyme in a changing environment: current knowledge and future directions, *Soil Biology & Biochemistry* 58, 216–234.

Butzen M.L., Wilkinson J.T., McGuinness S.R., Amezquita S., Peaslee G.F., Fein J.B., 2020, Sorption and desorption behavior of PFOS and PFOA onto a gram-positive and a gram-negative bacterial species measured using particle-induced gamma-ray emission (PIGE) spectroscopy, *Chemical Geology* 552, 119778.

Carneiro R.B., Mukaeda C.M., Sabatini C.A., Santos-Neto Á.J., Zaiat M., 2020, Influence of organic loading rate on ciprofloxacin and sulfamethoxazole biodegradation in anaerobic fixed bed biofilm reactors, *Journal of Environmental Management* 273, 111170.

Chen F.X., Zhou B., Yang L.Q., Che X.J., Zhuang J., 2023, Predicting bacterial transport through saturated porous media using an automated machine learning model, *Frontiers in Microbiology* 14, 1152059.

Chen G., Qiao M.Q., Zhang H.Y., Zhu H.L., 2004, Bacterial desorption in water-saturated porous media in the presence of rhamnolipid biosurfactant, *Research in Microbiology* 155, 655–661.

Chen G., Walker S.L., 2007, Role of solution chemistry and ion valence on the adhesion kinetics of groundwater and marine bacteria, *Langmuir* 23, 7162–7169.

Chetverikov S.P., Loginov O.N., 2019, A new *Ensifer adhaerens* strain M1 is capable of transformation of perfluorocarboxylic acids, *Microbiology* 88, 115–117.

Chetverikov S.P., Sharipov D.A., Korshunova T.Y., Loginov O.N., 2017, Degradation of perfluorooctanyl sulfonate by strain *Pseudomonas plecoglossicida* 2.4-D, *Applied Biochemistry and Microbiology* 53, 533–538.

Cruz-Morató C., Ferrando-Climent L., Rodriguez-Mozaz S., Barceló D., Marco-Urrea E., Vicent T., Sarrà M., 2013, Degradation of pharmaceuticals in non-sterile urban wastewater by *Trametes versicolor* in a fluidized bed bioreactor, *Water Research* 47, 5200–5210.

Cui C.Z., Hu H.Y., Yu Y.Q., Yu Y., 2009, Isolation of a bacterial strain capable of carbamazepine-degrading and biodegradation characteristics, *Microbiology* 36, 557–562. (in Chinese).

DiRusso C.C., Black P.N., 2004, Bacterial long-chain fatty acid transport: gateway to a fatty acid-responsive signaling system, *Journal of Biological Chemistry* 279, 49563–49566.

Eaton R.W., 1994, Organization and evolution of naphthalene catabolic pathways: sequence of the DNA encoding 2-hydroxychromene-2-carboxylate isomerase and trans-o-hydroxybenzylidenepyruvate hdyratase-aldolase from the NAH7 plasmid. *Journal of Bateriology* 176, 7757–7762

Ebsa G., Gizaw B., Admassie M., Degu T., Alemu T., 2024, The role and mechanisms of microbes in dichlorodiphenyltrichloroethane (DDT) and its residues bioremediation, *Biotechnology Reports* 42, e00835.

Fenu A., Donckels B.M.R., Beffa T., Bemfohr C., Weemaes M., 2015, Evaluating the application of *Microbacterium* sp. strain BR1 for the removal of sulfamethoxazole in full-scale membrane bioreactors, *Water Science & Technology* 72, 1754–1761.

Fernández-Cascán J., Isidro J., Guadaño J., Sáez C., Rodrigo M.A., 2023, Electrochemically-assisted remediation of silt polluted with aged HCHs, *Electrochimica Acta* 464, 142934.

Furuno S., Foss S., Wild E., Jones K.C., Semple K.T., Harms H., Wick L.Y., 2012, Mycelia promote active transport and spatial dispersion of polycyclic aromatic hydrocarbons, *Environmental Science & Technology* 46, 5463–5470.

Furuno S., Pazolt K., Rabe C., Neu T.R., Harms H., Wick L.Y., 2010, Fungal mycelia allow chemotactic dispersal of polycyclic aromatic hydrocarbon-degrading bacteria in water-unsaturated systems, *Environmental Microbiology* 12, 1391–1398.

Ganesan M., Mani R., Sai S., Kasivelu G., Awasthi M.K., Rajagopal R., Azelee N.I.W., Selvi P.K., Chang S.W., Ravindran B., 2022, Bioremediation by oil degrading marine bacteria: an overview of supplements and pathways in key processes, *Chemosphere* 303, 134956.

Gao D.W., Zhao H., Wang L.T., Li Y., Tang T., Bai Y.H., Liang H., 2022, Current and emerging trends in bioaugmentation of organic contaminated soils: a review, *Journal of Environmental Management* 320, 115799.

Gao J., Ye J.S., Ma J.W., Tang L.T., Huang J., 2014, Biosorption and biodegradation of triphenyltin by *Stenotrophomonas maltophilia* and their influence on cellular metabolism, *Journal of Hazardous Materials* 276, 112–119.

Gauthier H., Yargeau V., Cooper D.G., 2010, Biodegradation of pharmaceuticals by *Rhodococcus rhodochrous* and *Aspergillus niger* by co-metabolism, *Science of the Total Environment* 408, 1701–1706.

Ghorbanzadeh N., Shokati R., Farhangi M.B., Shabanpour M., Unc A., 2021, Effect of the biogenic precipitation of calcium carbonate on bacterial transport in sand columns, *Ecohydrology & Hydrobiology* 21, 280–291.

Górna H., Ławniczak Ł., Zgoła-Grześkowiak A., Kaczorek E., 2011, Differences and dynamic changes in the cell surface properties of three *Pseudomonas aeruginosa* strains isolated from petroleum-polluted soil as a response to various carbon sources and the external addition of rhamnolipids, *Bioresource Technol*ogy 102, 3028–3033.

Gross M.J., Logan B.E., 1995, Influence of different chemical treatments on transport of *Alcaligenes paradoxus* in porous media, *Applied and Environmental Microbiology* 61, 1750–1756.

Guo Q., Yan J., Wen J.J., Hu Y.Y., Chen Y.B., Wu W.J., 2016, Rhamnolipid-enhanced aerobic biodegradation of triclosan (TCS) by indigenous microorganisms in water-sediment systems, *Science of the Total Environment* 571, 1304–1311.

Guo S., Liu X.M., Zhao H., Wang L., Tang J.C., 2021, High pyrolysis temperature biochar reduced the transport of petroleum degradation bacteria *Corynebacterium variabile* HRJ4 in porous media, *Journal of Environmental Sciences* 100, 228–239.

Harms H., Wick L.Y., 2006, Dispersing pollutant-degrading bacteria in contaminated soil without touching it, *Engineering in Life Sciences* 6, 252–260.

Harvey R.W., Metge D.W., Barber L.B., Aiken G.R., 2010, Effects of altered groundwater chemistry upon the pH-dependency and magnitude of bacterial attachment during transport within an organically contaminated sandy aquifer, *Water Research* 44, 1062–1071.

He L., Wu D., Tong M.P., 2019, The influence of different charged poly(amido amine) dendrimer on the transport and deposition of bacteria in porous media, *Water Research* 161, 364–371.

Hearn E.M., Patel D.R., Lepore B.W., Indic M., van den Berg B., 2009, Transmembrane passage of hydrophobic compounds through a protein channel wall, *Nature* 458, 367–371.

Hua F., Wang H.Q., 2014, Uptake and trans-membrane transport of petroleum hydrocarbons by microorganisms, *Biotechnology, Biotechnological Equipment* 28, 165–175.

Huang J., Ye J.S., Ma J.W., Gao J., Chen S.Q., Wu X.L., 2014, Triphenyltin biosorption, dephenylation pathway and cellular responses during triphenyltin biodegradation by *Bacillus thuringiensis* and tea saponin, *Chemical Engineering Journal* 249, 167–173.

Huang S., Jaffé P.R., 2019, Defluorination of perfluorooctanoic acid (PFOA) and perfluorooctane sulfonate (PFOS) by *Acidimicrobium* sp. strain A6, *Environmental Science & Technology* 53, 11410–11419.

Ishtaweera P., Baker G.A., 2024, Progress in the application of ionic liquids and deep eutectic solvents for the separation and quantification of per- and polyfluoroalkyl substances, *Journal of Hazardous Materials* 465, 132959.

Jelic A., Cruz-Morató C., Macro-Urrea E., Sarrà M., Perez S., Vicent T., Petrovic M., Barcelo D., 2012, Degradation of carbamazepine by *Trametes versicolor* in an air pulsed fluidized bed bioreactor and identification of intermediates, *Water Research* 46, 955–964.

Jia Y.Y., Khanal S.K., Shu H.Y., Zhang H.Q., Chen G.H., Lu H., 2018, Ciprofloxacin degradation in anaerobic sulfate-reducing bacteria (SBR) sludge system: mechanism and pathways, *Water Research* 136, 64–74.

Jia Y.Y., Khanal S.K., Zhang H.Q., Chen G.H., Lu H., 2017, Sulfamethoxazole degradation in anaerobic sulfate-reducing bacteria sludge system, *Water Research* 119, 12–20.

John T., Thomas T., Abel B., Wood B.R., Chalmers D.K., Martin L.L., 2017, How kanamycin A interacts with bacterial and mammalian mimetic membranes, *Biochimica et Biophysica Acta – Biomembranes* 1859, 2242–2252.

Johnsen A.R., Wick L.Y., Harms H., 2005, Principles of microbial PAH-degradation in soil, *Environmental Pollution* 133, 71–84.

Julsing M.K., Schrewe M., Cornelissen S., Hermann I., Schmid A., Bhüler B., 2012, Outer membrane protein AlkL boosts biocatalytic oxyfunctionalization of hydrophobic substrates in *Escherichia coli*, *Applied and Environmental Microbiology* 78, 5724–5733.

Kim Y.M., Murugesan K., Schmidt S., Bokare V., Jeon J.R., Kim E.J., Chang Y.S., 2011, Triclosan susceptibility and co-metabolism: a comparison for three aerobic pollutant-degrading bacteria, *Bioresource Technology* 102, 2206–2212.

Kohlmeier S., Smits T.H.M., Ford R.M., Keel C., Harms H., Wick L.Y., 2005, Taking the fungal highway: mobilization of pollutant-degrading bacteria by fungi, *Environmental Science & Technology* 39, 4640–4646.

Kucharzyk K.H., Darlington R., Benotti M., Deeb R., Hawley E., 2017, Novel treatment technologies for PFAS compounds: a critical review, *Journal of Environmental Management* 204, 757–764.

Kwak E., Kim J.H., Choi J.W., Lee S., 2023, Injection strategy for effective bacterial delivery in bioaugmentation scheme by controlling ionic strength and pore-water velocity, *Journal of Environmental Management* 328, 116971.

Kwon B.G., Lim H.J., Na S.H., Choi B.I., Shin D.S., Chung S.Y., 2014, Biodegradation of perfluorooctane sulfonate (PFOS) as an emerging contaminant, *Chemosphere* 109, 221–225.

Larcher S., Yargeau V., 2011, Biodegradation of sulfamethoxazole by individual and mixed bacteria, *Applied Microbiology and Biotechnology* 91, 211–218.

Lee D.G., Zhao F., Rezenom Y.H., Russell D.H., Chu K.H., 2012, Biodegradation of triclosan by a wastewater microorganism, *Water Research* 46, 4226–4234.

Leung S.C.E., Shukla P., Chen D.C., Eftekhari E., An H.J., Zare F., Ghasemi N., Zhang D.K., Nguyen N.T., Li Q., 2022, Emerging technologies for PFOS/PFOA degradation and removal: a review, *Science of the Total Environment* 827, 153669.

Li A., Cai R., Cui D., Qiu T., Pang C.L., Yang J.X., Ma F., Ren N.Q., 2013, Characterization and biodegradation kinetics of a new cold-adapted carbamazepine-degrading bacterium, *Pseudomonas* sp. CBZ-4, *Journal of Environmental Sciences* 25, 2281–2290.

Li F., Zhu L.Z., 2012, Effect of surfactant-induced cell surface modifications on electron transport system and catechol 1,2-dioxygenase activities and phenanthrene biodegradation by *Citrobacter* sp. SA01, *Bioresource Technology* 123, 42–48.

Li J.H., Li X.H., Da Y.B., Yu J.L., Long B., Zhang P., Bakker C., McCarl B.A., Yuan J.S., Dai S.Y., 2022, Sustainable environmental remediation via biomimetic multifunctional lignocellulosic nano-framework, *Nature Communication* 13, 4368.

Li Q., Yang J.K., Fan W., Zhou D.D., Wang X.Y., Zhang L.L., Huo M.X., Crittenden J.C., 2018, Different transport behaviors of *Bacillus subtilis* cells and spores in saturated porous media: implications of contamination risks associated with bacterial sporulation in aquifer, *Colloids and Surfaces B: Biointerfaces* 162, 35–42.

Li X., Yan N., Sun J., Zhao M.M., Zheng X.L., Zhang W.D., Zhang Z.Y., 2023, Rhamnolipid-induced alleviation of bioclogging in Managed Aquifer Recharge (MAR): interactions with bacteria and porous media, *Journal of Environmental Management* 345, 118635.

Li Y., Wang H., Hua F., Su M., Zhao Y., 2014, Trans-membrane transport of fluoranthene by *Rhodococcus* sp. BAP-1 and optimization of uptake process, *Bioresource Technology* 155, 213–219.

Liao Q., Rong H.W., Zhao M.H., Luo H.Y., Chu Z.R., Wang R.D., 2021, Interaction between tetracycline and microorganisms during wastewater treatment: a review, *Science of the Total Environment* 757, 143981.

Liu G.S., Fang Z.D., Zhong H., Shi L.S., Yang X., Liu Z.F., 2020, Transport of *Pseudomonas aeruginosa* in porous media mediated by low-concentration surfactants: the critical role of surfactant to change cell surface hydrophobicity, *Water Resources Research* 56, e2019WR026103.

Liu G.S., Zhong H., Jiang Y.B., Brusseau M.L., Huang J.S., Shi L.S., Liu Z.F., Liu Y., Zeng G.M., 2017, Effect of low-concentration rhamnolipid biosurfactant on *Pseudomonas aeruginosa* transport in natural porous media, *Water Resources Research* 53, 361–375.

Lv Z.K., Wang Z.Y., Wang H.G., Li J.B., Li K., 2024, Adsorption of cationic/anionic dyes and endocrine disruptors by yeast/cyclodextrin polymer composites, *RSC Advances* 14, 6627.

Ma Z., Liu J., Dick R.P., Li H., Shen D., Gao Y.Z., Waigi M.G., Ling W.T., 2018, Rhamnolipid influences biosorption and biodegradation of phenanthrene by phenanthrene-degrading strain *Pseudomonas* sp. Ph6, *Environmental Pollution* 240, 359–367.

Mahinroosta R., Senevirathna L., 2020, A review of the emerging treatment technologies for PFAS contaminated soils, *Journal of Environmental Management* 255, 109896.

Martínez E., Estupiñán M., Pastor F.I.J., Busquets M., Díaz P., Manresa A., 2013, Functional characterization of ExFadLO, an outer membrane protein required for exporting oxygenated long-chain fatty acids in *Pseudomonas aeruginosa*, *Biochimie* 95, 290–298

Meade M.J., Waddell R.L., Callahan T.M., 2011, Soil bacteria *Pseudomonas putida* and *Alcaligenes xylosoxidans* subsp. denitrificans inactivate triclosan in liquid and solid substrates, *FEMS Microbioly Letters* 204, 45–48.

Megharaj M., Ramakrishnan B., Venkateswarlu K., Sethunathan N., Naidu R., 2011, Bioremediation approaches for organic pollutants: a critical perspective, *Environmental International* 37, 1362–1375.

Mohanty I., Rath A., 2020, Bioremediation: current research trends and applications, In: Kashyap B. K., Solanki M. K., Kamboj D. V., Pandey A. K. (eds) *Waste to Energy: Prospects and Applications*, Springer, Singapore.

Mulla S.I., Wang H., Sun Q., Hu A., Yu C.P., 2015, Characterization of triclosan metabolism in *Sphingomonas* sp. YL-JM2C, *Scientific Reports* 6, 21965–21966.

Muñoz A.J.R., de María Cuervo López F., Texier A.C., 2020, Ampicillin biotransformation by a nitrifying consortium, *World Journal of Microbiology and Biotechnology* 36, 21.

Murdoch R.W., Hay A.G., 2013, Genetic and chemical characterization of ibuprofen degradation by *Sphingomonas* Ibu-2, *Microbiology* 159, 621–632.

Murdoch R.W., Hay A.G., 2005, Formation of catechols via removal acid side chains of ibuprofen and related aromatic acids, *Applied and Environmental Microbiology* 71, 6121–6125.

Nitschke M., Araújo L.V., Costa S.G.V.A.O., Pires R.C., Zeraik A.E., Fernandes A.C.L.B., Freire D.M.G., Contiero J., 2009, Surfactin reduces the adhesion of food-borne pathogenic bacteria to solid surfaces, *Letters in Applied Microbiology* 49, 241–247.

Ortiz A., Teruel J.A., Aranda F.J., 2005, Effect of triorganotin compounds on membrane permeability, *Biochimica et Biophysica Acta* 1720, 137–142.

Otto S., Banitz T., Harms H., Wick L.Y., 2017, Catch me if you can: dispersal and foraging of *Bdellovibrio bacteriovorus* 109J along mycelia, *ISME Journal* 11, 386–393.

Otto S., Banitz T., Thullner M., Harms H., Wick L.Y., 2016, Effects of facilitated bacterial dispersal on the degradation and emission of a desorbing contaminant, *Environmental Science & Technology* 50, 6320–6326.

Pande V., Pandey S.C., Sati D., Pande V., Samant M., 2020, Bioremediation: an emerging effective approach towards environment restoration, *Environmental Sustainability* 3, 91–103.

Ren X.Y., Zeng G.M., Tang L., Wang J.J., Wan J., Liu Y.N., Yu J.F., Yi H., Ye S.J., Deng R., 2018, Sorption, transport and biodegradation: an insight into bioavailability of persistent organic pollutants in soil, *Science of the Total Environment* 610–611, 1154–1163.

Rodarte-Morales A., Feijoo G., Moreira M., Lema J., 2012, Biotransformation of three pharmaceutical active compounds by the fungus *Phanerochaete chrysosporium* in a fed batch stirred reactor under air and oxygen supply, *Biodegradation* 23, 145–156.

Roh H., Subramanya N., Zhao F., Yu C.P., Sandt J., Chu K.H., 2009, Biodegradation potential of wastewater micropollutants by ammonia-oxidizing bacteria, *Chemosphere* 77, 1084–1089.

Roy A., Dubey P., Srivastava A., Kaur I., Shrivastava A., Vajpayee P., Srivastava S., Srivastava P.K., 2024, Exploring the potential of *Meyerozyma caribbica* and its combined application with bacteria for lindane bioremediation, *Chemosphere* 361, 142413.

Samanta S.K., Singh O.V., Jain R.K., 2002, Polycyclic aromatic hydrocarbons: environmental pollution and bioremediation, *Trends in Biotechnology* 20, 243–248.

Sanchez-Huerta C., Medina J.S., Wang C.Z., Fortunato L., Hong P.Y., 2023, Understanding the role of sorption and biodegradation in the removal of organic micropollutants by membrane aerated biofilm reactor (MABR) with different biofilm thickness, *Water Research* 236, 119935.

Santosa I.J., Grossmana M.J., Sartorattob A., Ponezib A.N., Durranta L.R., 2012, Degradation of the recalcitrant pharmaceuticals carbamazepine and 17α-ethinylestradiol by ligninolytic fungi, *Chemical Engineering Transactions* 27, 169–174.

Sepehrnia N., Gorakifard M., Hallett P.D., Hajabbasi M.A., Shokri N., Coyne M., 2023, Contrasting transport and fate of hydrophilic and hydrophobic bacteria in wettable and water-repellent porous media: straining or attachment? *Colloids and Surfaces B: Biointerfaces* 228, 113433.

Sethunathan N., Megharaj M., Chen Z.L., Williams B.D., Lewis G., Naidu R., 2004, Algal degradation of a known endocrine disrupting insecticide, α-endosulfan, and its metabolite, endosulfan sulfate, in liquid medium and soil, *Journal of Agricultural and Food Chemistry* 52, 3030–3035.

Sharma A., Kalyani P., Trivedi V.D., Kapley A., Phale P.S., 2019, Nitrogen-dependent induction of atrazine degradation pathway in *Pseudomonas* sp. strain AKN5, *FEMS Microbiology Letters* 366, fny277.

Sodagari M., Wang H., Newby B.Z., Ju L.K., 2013, Effect of rhamnolipids on initial attachment of bacteria on glass and octadecyltrichlorosilane-modified glass, *Colloids and Surfaces B: Biointerfaces* 103, 121–128.

Song C., Gao N.Y., Gao H.W., 2010, Transmembrane distribution of kanamycin and chloramphenicol: insights into the cytotoxicity of antibacterial drugs, *Molecular BioSystems* 6, 1901–1910.

Sorn S., Hara-Yamamura H., Vet S., Xiao M.H., Hoek E.M.V., Honda R., 2023, Biological treatment of perfluorooctanesulfonic acid (PFOS) using microbial capsules of a polysulfone membrane, *Chemosphere* 329, 138585.

Sotirova A.V., Spasova D.I., Galabova D.N., Karpenko E., Shulga A., 2008, Rhamnolipid-biosurfactant permeabilizing effects on gram-positive and gram-negative bacterial strains, *Current Microbiology* 56, 639–644.

Streger S.H., Vainberg S., Dong H.L., Hatzinger P.B., 2002, Enhancing transport of *Hydrogenophaga flava* ENV735 for bioaugmentation of aquifers contaminated with methyl tert-butyl ether, *Applied and Environmental Microbiology* 68, 5571–5579.

Stylianou M., Björnsdotter M.K., Olsson P.E., Jogsten I.E., Jass J., 2019, Distinct transcriptional response of *Caenorhabditis elegans* to different exposure routes of perfluorooctane sulfonic acid, *Environmental Research* 168, 406–413.

Tao W., Song Y.T., Singhal N., McGoverin C., Vanholsbeeck F., Swift S., 2021, A novel optical biosensor for in situ and small-scale monitoring of bacterial transport in saturated columns, *Journal of Environmental Management* 289, 112452.

Taştan B.E., Dönmez G., 2015, Biodegradation of pesticide triclosan by *A. versicolor* in simulated wastewater and semi-synthetic media, *Pesticide Biochemistry and Physiology* 118, 33–37.

Taştan B.E., Ozdemir C., Tekinay T., 2016, Effects of different culture media on biodegradation of triclosan by *Rhodotorula mucilaginosa* and *Penicillium* sp., *Water Science & Technology* 74, 473–481.

Touw D.S., Patel D.R., Berg B., 2010, The crystal structure of OprG from *Pseudomonas aeruginosa*, a potential channel for transport of hydrophobic molecules across the outer membrane, *PLoS One* 5, e15016.

Vander J., Sherman J., Luciano D., 2001, Movement of molecules across cell membranes, In: Vander, A. J., Sherman, J. H., Luciano, D. S. (eds) *Human Physiology: The Mechanisms of Body Function*, eighth ed., McGraw-Hill Higher Education, New York.

Vasiliadou I.A., Molina R., Martínez F., Melero J.A., 2013, Biological removal of pharmaceutical and personal care products by a mixed microbial culture: sorption, desorption and biodegradation, *Biochemical Engineering Journal* 181, 108–119.

Vasiliadou I.A., Sánchez-Vázquez R., Molina R., Martínez F., Melero J.A., Bautista L.F., Iglesias J., Morales G., 2016, Biological removal of pharmaceutical compounds using white-rot fungi with concomitant FAME production of the residual biomass, *Journal of Environmental Management* 180, 228–237.

Volkring F., Breure A.M., Rulkens W.H., 1998, Microbiological aspects of surfactant use for biological soil remediation, *Biodegradation* 8, 401–417.

Wang H.Q., Jiang R.H., Kong D.K., Liu Z.L., Wu X.X., Xu J., Li Y., 2020, Transmembrane transport of polycyclic aromatic hydrocarbons by bacteria and functional regulation of membrane proteins, *Frontiers of Environmental Sciences and Engineering* 14, 9.

Wang J.L., Wang S.Z., 2018, Microbial degradation of sulfamethoxazole in the environment, *Applied Microbiology and Biotechnology* 102, 3573–3582.

Wang S.Z., Yin Y.N., Wang J.L., 2018, Microbial degradation of triclosan by a novel strain of *Dyella* sp., *Applied Microbiology and Biotechnology* 102, 1997–2006.

Wang Y.H., Xu L.H., Chen H.E., Zhang M., 2022, Retention and transport behavior of microplastic particles in water-saturated porous media, *Science of the Total Environment* 808, 152154.

Wick L.Y., Remer R., Würz B., Reichenbach J., Braun S., Schäfer F., Harms H., 2007, Effect of fungal hyphae on the access of bacteria to phenanthrene in soil, *Environmental Science & Technology* 41, 500–505.

Wiener M.C., Horanyi P.S., 2011, How hydrophobic molecules traverse the outer membranes of gram-negative bacteria, *Proceedings of the National Academy of Sciences of the United States of America* 108, 10929–10930

Worrich A., Köbug S., Banitz T., Centler F., Frank K., Thullner M., Harms H., Miltner A., Wick L.Y., Kästner M., 2016, Bacterial dispersal promotes biodegradation in heterogeneous systems exposed to osmotic stress, *Frontiers in Microbiology* 7, 1214.

Worrich A., Wick L.Y., Banitz T., 2018, Ecology of contaminant biotransformation in the mycosphere: role of transport processes, *Advances in Applied Microbiology* 104, 93–133.

Xie P., Chen C., Zhang C.F., Su G.Y., Ren N.Q., Ho S.H., 2020, Revealing the role of adsorption in ciprofloxacin and sulfadiazine elimination routes in microalgae, *Water Research* 172, 115475.

Xiong B.J., Kleinsteuber S., Straiiber H., Dusny C., Harms H., Wick L.Y., 2022, Impact of fungal hyphae on growth and dispersal of obligate anaerobic bacteria in aerated habitats, *mBio* 13(3), e0076922.

Xu J., Sheng G.P., 2020, Microbial extracellular polymeric substances (EPS) acted as a potential reservoir in responding to high concentrations of sulfonamides shocks during biological wastewater treatment, *Bioresource Technology* 313, 123654.

Xu J., Sheng G.P., Ma Y., Wang L.F., Yu H.Q., 2013, Roles of extracellular polymeric substances (EPS) in the migration and removal of sulfamethazine in activated sludge system, *Water Research* 47, 5298–5306.

Yang H.Y., Kim H.J., Tong M.P., 2012, Influence of humic acid on the transport behavior of bacteria in quartz sand, *Colloids and Surfaces B: Biointerfaces* 91, 122–129.

Yang H.Y., Tong M.P., Kim H.J., 2012, Influence of bentonite particles on representative gram negative and gram positive bacterial deposition in porous media, *Environmental Science & Technology* 46, 11627–11634.

Yang J.Q., Sanfilippo J.E., Abbasi N., Gitai Z., Bassler B.L., Stone H.A., 2021, Evidence for biosurfactant-induced flow in corners and bacterial spreading in unsaturated porous media, *Proceedings of the National Academy of Sciences of the United States of America* 118, e2111060118.

Ye J.S., Yin H., Peng H., Bai J.Q., Xie D.P., Wang L.L., 2013, Biosorption and biodegradation of triphenyltin by *Brevibacillus brevis*, *Bioresource Technology* 129, 236–241.

Yi W.Y., Li C.S., Ye J.S., Long Y., Qin H.M., 2017, Correlation between triphenyltin degradation and cellular metabolic responses of *Bacillus thuringiensis*, *International Biodeterioration & Biodegradation* 122, 61–68.

Zeraik A.E., Nitschke M., 2010, Biosurfactants as agents to reduce adhesion of pathogenic bacteria to polystyrene surfaces: effect of temperature and hydrophobicity, *Current Microbiology* 61, 554–559.

Zhang D., 2013, Surfactant Controlled Bacterial Interfacial Behaviors of PAHs and Its Mechanisms (Ph.D. dissertation), Zhejiang University (in Chinese).

Zhang D., 2019, The role of microorganisms in the removal of pharmaceutical and personal care products, In: Prasad M. N. V., Vithanage M., Kapley A. (eds) *Pharmaceuticals and Personal Care Products: Waste Management and Treatment Technology*, Butterworth-Heinemann, Oxford.

Zhang D., Jiang J.C., Shi H.D., Lu L., Zhang M., Lin J., Lü T., Huang J.G., Zhong Z.S., Zhao H.T., 2024, Nonionic surfactant Tween 80-facilitated bacterial transport in porous media: a nonmonotonic concentration-dependent performance, mechanism, and machine learning prediction, *Environmental Research* 251, 118670.

Zhang D., Zhu L.Z., 2014, Controlling microbiological interfacial behaviors of hydrophobic organic compounds by surfactants in biodegradation process, *Frontiers of Environmental Science and Engineering* 8, 305–315.

Zhang D., Zhu L.Z., Li F., 2013, Influences and mechanisms of surfactants on pyrene biodegradation based on interactions of surfactant with a *Klebsiella oxytoca* strain, *Bioresource Technology* 142, 454–461.

Zhang L., Wu D., Liang J.L., Wang L., Zhou Y., 2021a, Triclosan transformation and impact on an elemental sulfur-driven sulfidogenic process, *Chemical Engineering Journal* 421, 129634.

Zhang M.Y., He L., Jin X., Bai F., Tong M.P., Ni J.R., 2021b, Flagella and their properties affect the transport and deposition behaviors of *Escherichia coli* in quartz sand, *Environmental Science & Technology* 55, 4964–4973.

Zhang M.Z., Pereira e Silva M.C., de Mares Maryam C., van Elsas J.D., 2014, The mycosphere constitutes an arena for horizontal gene transfer with strong evolutionary implications for bacterial-fungal interactions, *FEMS Microbiology Ecology* 89, 516–526.

Zhang Y.M., Miller R.M., 1994, Effect of a *Pseudomonas* rhamnolipid biosurfactant on cell hydrophobicity and biodegradation of octadecane, *Applied and Environmental Microbiology* 60, 2101–2106.

Zhao S., Li L.L., Wang Y.J., Liu Z.W., Yang S., Gao X., Zhang C.Y., Yu A.F., 2024, Remediation of petroleum-contaminated site soil by bioaugmentation with immobilized bacterial pellets stimulated by a controlled-release oxygen composite, *Environmental Pollution* 356, 124253.

Zhao X., Wang X.C., Chen Z.L., Xu H., Zhang Q.F., 2015, Microbial community structure and pharmaceuticals and personal care products removal in a membrane bioreactor seeded with aerobic granular sludge, *Applied Microbiology and Biotechnology* 99, 425–433.

Zhong H., Liu G.S., Jiang Y.B., Brusseau M.L., Liu Z.F., Liu Y., Zeng G.M., 2016, Effect of low-concentration rhamnolipid on transport of *Pseudomonas aeruginosa* ATCC 9027 in an ideal porous medium with hydrophilic or hydrophobic surfaces, *Colloids and Surfaces B: Biointerfaces* 139, 244–248.

Zhong H., Liu G.S., Jiang Y.B., Yang J.Z., Liu Y., Yang X., Liu Z.F., Zeng G.M., 2017, Transport of bacteria in porous media and its enhancement by surfactants for bioaugmentation: a review, *Biotechnology Advances* 35, 490–504.

Zhou X., Luo X.S., Liu K.Z., Zheng T., Ling P., Huang J., Chen W.J., Huang Q.Y., 2024, Importance of soil ecoenzyme stoichiometry for efficient polycyclic aromatic hydrocarbon biodegradation, *Chemosphere* 359, 142348.

12 Remediation Approaches and Cases for EOC-Contaminated Soils

Yuchi Zhong, Dong Zhang, and Ming Zhang

12.1 INTRODUCTION: POSSIBLE TECHNIQUES FOR REMEDIATION OF EOC-CONTAMINATED SOILS

Emerging organic contaminants (EOCs), including pharmaceutical and personal care products (PPCPs), per- and polyfluoroalkyl substances (PFASs), antibiotics, and endocrine disruptive compounds (EDCs), have been of great concern, owing to their flourishing consumption, poor management, contamination for water and soils, and great ecological risks (Mahinroosta and Senevirathna 2020; Gaballah et al. 2021; Ismanto et al. 2022; Fang et al. 2023; Alidoosti et al. 2024; Guedes et al. 2024). As the main reservoir, soils accumulate these emerging contaminants through manure application, wastewater irrigation/discharge, and direct application, posing a great threat to human health and our ecosystems (Horta et al. 2015; Kuppusamy et al. 2018; Shi et al. 2019; Toms et al. 2019). Several remediation approaches have been developed to eliminate EOC-contaminated soil, such as bioremediation, phytoremediation, immobilization, soil washing/flushing, electrochemical remediation, thermal approaches, earthworm-based technology, chemical oxidation etc. (Zhou et al. 2019; Mahinroosta and Senevirathna 2020; Guedes et al. 2021; Kumar et al. 2021a; Gu et al. 2022; Kewalramani et al. 2022; Bolan et al. 2023; Dong et al. 2023; Wu et al. 2023; Sun et al. 2024). The feasibility, efficiency, and scope of application of these technologies are highly dependent on the conditions of the contaminated soil, especially the characteristics of the pollutants. There is a wide variety of EOCs, and they differ greatly in their structural properties. To streamline our discussion, we focus specifically on PFAS – recently prioritized contaminants that rank among the most recalcitratnt to remediate – while reviewing soil remediation technologies for contaminated sites.

Generally, PFAS are a specific sub-group of fluorinated substances that include a perfluoroalkyl moiety. They typically contain a hydrophobic tail with fluorinated aliphatic chain (representing as C_nF_{2n+1}) and a polar hydrophilic head, which makes them behave as surfactants and environmental persistence. They have an incredible range of applications in many industrial and daily life scenarios, such as firefighting foams, nonstick pans, clothing, carpets, makeup, and packaging, making these compounds ubiquitous in the environment. Their specific molecular structures allow them to strongly sorb/partition on soil organic matter, to be easily mobile in groundwater,

 DOI: 10.1201/9781003509523-12

and difficult to be degraded biologically and chemically, known as "forever chemicals." Furthermore, variable partitioning coefficients for octanol and carbon across the PFAS spectrum complicate their behaviors in soil matrix. Therefore, the remediation of PFAS-contaminated soil is extremely challenging.

Eliminating these forever chemicals from the soil and/or even destroying them in a molecular level are difficult to achieve. The update of health advisory levels for PFAS (specifically perfluorooctanoic acid (PFOA) and perfluorooctane sulfonate (PFOS)) has been released by the U.S. Environmental Protection Agency (EPA) around 2016 (USEPA 2016a–e, 2017). On the other hand, a vast number of sites around the world are contaminated and affected by PFAS. In the United States alone, over 57,000 sites have been long-term impacted by PFAS emissions, potentially jeopardizing the drinking water safety of 45%–60% of the U.S. population (Andrews and Naidenko 2020; Salvatore et al. 2022; Smalling et al. 2023). In Europe, more than 17,000 locations have already detected PFAS contamination, with an additional 21,500 sites likely to be affected (Dagorn et al. 2023). To mitigate the hazards and risks associated with PFAS pollution, the United States has planned to invest over 220 billion dollars in remediation efforts (Scaggs 2024). The environmental, ecological, economic and healthy concerns triggered the exploration of related remediation technologies.

The EPA Victoria, Australia, released a technical guide for validation of PFAS soil treatment technologies on April 2024 (EPA Victoria 2024). The guide provides a series of available techniques/technologies for PFAS-contaminated soils, dividing destructive and nondestructive technologies. The former includes thermal desorption, incineration, smouldering/combustion, oxidation-reduction, ball milling, and bioremediation; while the latter only reduces the total and/or available concentrations of PFAS in soil without destroying or changing their structure, such as immobilization, soil washing, electrokinetic remediation, phytoremediation (specially phytoextraction), and containment. A brief summary of the treatment mechanisms is also illustrated in Table 12.1. Best available techniques or technologies and optimal operation conditions should be carefully evaluated and selected, which is truly hard, largely depending on the soil physical-chemical properties as well as types of PFAS.

Soil Properties Influencing Feasibility for Technique Selection: Soil particle size, moisture content, pH, organic matter content, soil heterogeneity, and permeability (soil texture) may be important considerations affecting the selection of suitable techniques to treat EOCs contaminated soil. For examples, big lumps, high heterogeneity, and high moisture content may affect the heat transfer and thus require more energy consumption (higher cost) in a destructive thermal desorption and degradation process. High organic carbon content (f_{oc}) can impede the effectiveness of soil washing due to the strong sorption of hydrophobic EOCs by organic matter. Low permeability of soils is considered to be one of the most difficult situations to overcome in soil remediation, almost for all conventional techniques. Pretreatment, therefore, is required before the selection and application of conventional techniques.

TABLE 12.1
Summary of Treatment Mechanisms for EOC-Contaminated Soil

Category	Technologies	Mechanisms
Destructive	Chemical oxidation-reduction	Transformation of EOCs by hydroxyl radicals (oxidation) or solvated electrons (reduction), such as cleaving C–F bonds
	Thermal desorption-degradation/ incineration	Thermal desorption of EOCs from soil matrix and destruction of EOCs via thermal-assisted oxidation
	Mechano-chemical destruction (e.g., ball milling)	Chemical destroy the EOCs molecules based on mechanical energy
	Biological degradation	Microbial enzyme related degradation of EOCs
	Electrochemical oxidation	Cleaving bonds (e.g., C–C, C–F, C–Cl) in EOCs based on electrochemical reactions
Non-destructive	Immobilization	Sorption of EOCs and reducing its mobilization by sorbents such as biochar, activated carbon, minerals, organoclays, and resins
	Stabilization/solidification	Encapsulation of EOCs to form a solid
	Phytoremediation (phytoextraction)	Use of plant to uptake and/or extract EOCs from soil to plant aboveground tissues
	Containment	Contain EOCs in soil by using of physically engineered barriers
	Separation (e.g., washing)	Use of washing agents (such as surfactants, solvents) to separate EOCs from soil particles into effluents; usually includes further treatment of washing effluents
	Electrokinetic treatment	Basically use of electronic fields for mobilizing EOC molecules in the soil to receive EOC concentration/ separation

Operation Parameter Optimization Based on Chemical Properties: After a specific remediation technique has been determined, it is necessary to select some key process parameters, such as desorption temperature, types and dosages of amendments and soil washing agents, microbial agents, and plant species. For example, carbon-based amendments (e.g., biochar, activated carbon) are considered to be more effective for immobilizing hydrophobic EOCs (e.g., long-chain PFAS, tetrabromobisphenol A), whose hydrophobicity (log K_{ow}) tends to be high; whereas for those chemicals with moderate hydrophobicity (e.g., PFAS with short chains), clay-based materials (e.g., organoclay, mineral composites) show higher immobilization effectiveness (Kabiri and McLaughlin 2021; Sleep and Juhasz 2021; Yu et al. 2023). The halogen amount/ratio (e.g., fluorocarbon, chlorine) and total content of EOCs in soil should be considered to determine the optimal temperature for a thermal treatment system, because the halogen (fluorine) content may significantly affect the volatilization/desorption and thermal degradation efficacy. For example, it is believed that the PFAS desorb/volatilize from the soil matrix at temperatures in a range of 350°C–650°C; thermal destruction occurs at the temperatures greater than 850°C and generally 1100°C is required (Ross et al. 2018; Barranco et al. 2020; Sörengård et al. 2020; Bolan et al. 2021; ITRC 2022; Weber et al. 2023).

Due to the persistent nature of the forever chemicals, various and complicated properties of EOCs, and the complexity of EOCs in the soil matrix, most remediation techniques and/or technologies are still in laboratory/batch scales and limited in small-scale testing. In this chapter, we will discuss some promising technologies with the potential to remediate EOC-contaminated soil and groundwater at a pilot scale or full scale, and some recent practical cases.

12.2 IMMOBILIZATION TREATMENT FOR CONTAMINATED SOILS

In most cases, the toxicity and ecological risks of soil contaminants largely depend on the mobility of these pollutants. Based on this principle, "immobilization" technology has been developed to alleviate the contamination of various toxic compounds in soil. In soil remediation, "immobilization" refers to a process where contaminants are trapped or bound within the soil matrix, preventing their movement and potential leaching into groundwater, essentially "fixing" them in place using various physical and chemical methods, often involving the addition of binding agents to the soil. The "fixing" efficiency is mainly determined by pollutant types/property and their interaction/binding ability with soil matrix and/or immobilizing agents, as well as the soil property (Palansooriya et al. 2020). A wide range of novel stabilization materials, such as biochar, resins, activated carbons, ferrous sulfate, layered double hydroxides, clay minerals, and their modified products, are currently being developed and applied for this purpose (Shen et al. 2019; Huang et al. 2023). Immobilization technologies can be applied either *ex situ* or *in situ*. The *in situ* applications are much preferred (Wuana and Okieimen 2011), due to many advantages, including (i) relatively low costs, depending on choice of immobilizing agents. For example, the price of granular activated carbon (GAC) with 0.07 mmol PFOS/g sorption capacity is $14.60/kg (Das et al. 2015). (ii) Low site disturbance. (iii) High public acceptability. (iv) Easy operation. (v) Wide ranges of targeted contaminants, such as PPCPs and PFAS (Chen et al. 2022; Keerthanan et al. 2023; Wen et al. 2023). (vi) Combination with other techniques (e.g., bioremediation, phytoremediation) owing to release of nutrients and/or concentrating effects for pollutants. For example, agricultural waste-derived biochar enhanced the growth of bacteria (e.g., *Burkholderia* sp., *Bacillus* sp.) and promoted the yield of tomato (Tripti et al. 2017). A commercial immobilizing agent (BAM) developed by ORIN Technology can be used solely as an immobilizer to trap organic pollutants, and it can also allow microbes to access and degrade contaminants sorbed on BAM. The details of BAM-assisted bioremediation and phytoremediation will be discussed in the Sections 12.4.1 and 12.5, respectively.

A case study. A commercial immobilizing agent RemBind 100 was developed by an Australia-based company RemBind®. The agent was used to stabilize PFAS in the PFAS-chromium co-contaminated soil in a foundry and electroplating facility located in New Plymouth, New Zealand (RemBind n.d.). The vulcanic ash soil was contaminated with PFAS to a depth of approximately 3 m, originating from the demisting additive of heat-resistant surfactant of PFOS, with a contamination level of 58 mg/kg. A bench experiment was carried out to test the effectiveness of the immobilizing agent RemBind, with 1%, 2%, 4%, and 8% addition rates. Results showed the concentration of PFAS in the leachate was under detection (<0.01 μg/L). Based on the bench results, a full-scale immobilization remediation with 300 tons of PFOS-contaminated soil was employed by using a conventional rotary hoe with the addition of RemBind at 1% dosage and a further 0.7%. The total PFAS in soil samples and leachates were analyzed by a third-party testing company (Eurofins Environmental Testing Australia Pty Ltd in Brisbane) followed the Australian Standard Leaching Procedure tests (Australian Standard AS4439). The leachates were conducted under the "worst case" of pH 5 leaching fluid. Results indicated that on-site immobilization of PFAS in soil by RemBind was successfully achieved. The concentrations of PFAS in leachates at all test positions with 1% RemBind treatment were <0.01 μg/L, compared with concentrations of 0.61–1.35 μg/L of PFAS in the leachates of untreated soil.

Finally, there is one point that needs our attention. Although laboratory experiments and some case studies have demonstrated that commonly used immobilizing agents (e.g., biochar, GAC, organoclays) or commercial agents (e.g., RemBind, BAM) show good immobilization effects on various pollutants, so far, researchers and engineers cannot guarantee that this solution will remain effective over the long term. In other words, it is uncertain whether these methods irreversibly bind EOCs in the soil and stop future transfer to groundwater; therefore, the potential solubilization and slow release of the immobilized contaminants necessitates long-term monitoring, pontetially spanning decades (Palansooriya et al. 2020).

12.3 THERMAL TREATMENT

Thermal treatment is considered to be the possible method to totally remove PFAS from soils using a high temperature, breaking down PFAS compounds. This technique includes processes such as pyrolysis, thermal air oxidation, thermal desorption, and smoldering (Crownover et al. 2019; Sörengård et al. 2019; Duchesne et al. 2020; Alinezhad et al. 2022; Yao et al. 2022; Sun et al. 2024). Figure 12.1 provides a typical process and possible changes in PFAS structure in a thermal treatment of PFAS-contaminated soil. Thermal-based techniques have exhibited promise potential against recalcitrant PFAS, and the effectiveness depends on the temperatures, processing time, and the turbulence in the combustion chamber (Crownover et al. 2019; Duchesne et al. 2020; Alinezhad et al. 2022; EU 2019). Thermal treatment is widely recognized as a high-cost technology, with the primary expenses attributed to the high temperatures required to break the bonds of PFAS (which have a high melting point), as well as the excavation and transport needed for the ex situ (off site)

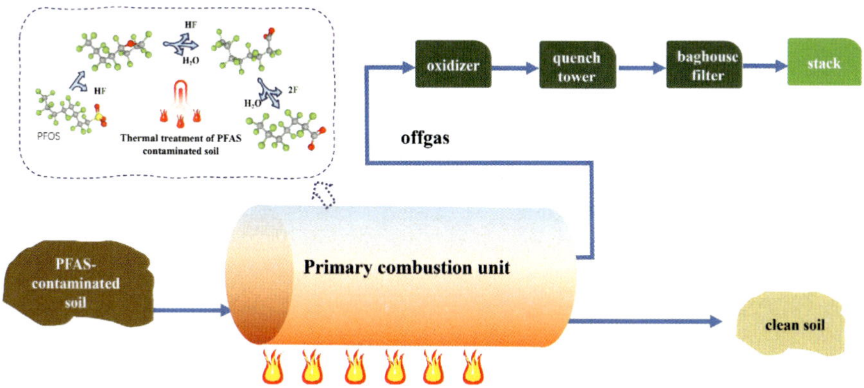

FIGURE 12.1 Diagrammatic sketch of thermal treatment of PFAS contaminated soil. (The decomposition pathway was modified from Longendyke et al. 2022.)

treatment. In this section, we will talk about a field-scale remediation using thermal treatment for PFAS-contaminated soils.

In 2019, Organic Incineration Technology Inc. (OIT) started and completed a commercial operation of the thermal treatment of PFAS-contaminated soils (OIT 2020), mentioned as the Moose Creek Facility case study, after the completion of a preliminary test trial (2017) and a second test trial (2018) with positive results. The Moose Creek Facility, where contaminated soils were treated, was established by OIT in 1990. The thermal treatment system comprises a refractory-lined rotary kiln designed for primary combustion. This unit subjects PFAS-contaminated soils to elevated temperatures, which serves to break down certain organic compounds and desorb residual contaminants from the inorganic matrix being processed and sanitized for recycling and subsequent utilization. The contaminated soil is introduced into the kiln via a hopper. Post-treatment, the products are discharged from the kiln and undergo rehydration in an open water bath. They are then transferred to a designated storage area for clean waste, where they undergo final testing prior to being cleared from the treatment zone for reuse. Another crucial consideration is the monitoring and treatment of gases generated during the thermal treatment process. Gases emanating from the kiln are channeled to a refractory-lined secondary combustion unit, known as the oxidizer. Here, they are subjected to temperatures high enough to destroy and demineralize the contaminants present in the exhaust stream. The exhaust gases from the oxidizer are then directed through a quench tower, where both the gases and the particulate matter carried within the exhaust stream are cooled. Subsequently, the treated gases are passed through a baghouse filter, where any remaining particulate matter is removed before being discharged through a stack that is approximately 60 feet tall.

In the 2018 test trial, approximately 89.7 tons of PFAS-contaminated soil derived from the Eielson Air Force Base (EAFB) were used. The initial concentration of total PFOA and PFOS (28 samples, highest detected) were 0.0765 and 7.24 mg/kg, respectively. These soil samples definitely exceeded the target cleanup levels (CULs) of the Alaska Department of Environmental Conservation (ADEC), which are 0.0017 mg/

kg for PFOA and 0.0030 mg/kg for PFOS. After thermal treatment, all PFOA in all the soil and baghouses were not detected. PFOS was detected at concentrations of less than 0.00220 mg/kg (ranging from 0.00030 to 0.00220 mg/kg), which complies with the CULs of ADEC. The results indicated that the thermal remediation, as mentioned in this trial, is effective for remediating PFAS-contaminated soils.

The 2019 commercial operation soil remediation. Around 200 tons of soil contaminated with PFAS were thermal-treated in this particular case. The PFAS-contaminated soil originating from construction activities was used as "grave fill" cover material for a utilidor system completed at EAFB in fall 2018. The contamination was confirmed with PFOS concentrations ranging from 0.000640 to 0.010 mg/kg, in which five of the seven samples were above the ADEC CULs (0.0030 mg/kg). After treatment, the 200 tons of treated soil was subsequently divided into four separate stockpiles, each containing 50 tons of post-treatment soil. A single grab sample was extracted from each of these stockpiles, amounting to a total of four samples collected. These samples were subjected to analysis for the presence of six PFAS compounds that are proposed for regulation by ADEC. PFOS was detected in two of the samples, albeit at levels below the proposed regulatory thresholds (0.00023 and 0.00028 mg/kg). No other PFAS compounds were detected in any of the samples. ADEC has confirmed that the treated soil satisfies the applicable cleanup standards and is no longer deemed contaminated. After the successful treatment process, the party responsible for the soil was issued a certificate of destruction, serving as evidence that the soil had been effectively treated. Consequently, the soil was deemed appropriate for use as clean construction fill.

Although specific operating parameters cannot be obtained from publicly released reports and databases, it is clear that factors such as soil moisture, contaminant levels, rotary kiln temperature, and feed rate are important and have been optimized and selected from two pretest trials. Additionally, we can see that this particular case illustrates that remediation projects are highly complex and involve multiple layers. In this case, only PFOS contamination was detected in the soil. However, to ensure that the Moose Creek Facility can handle more complex contamination scenarios in the future and to assist ADEC in better optimizing contaminant standards, the project tested for six types of PFAS for which ADEC currently has standards (namely PFBS, PFHpA, PFHxS, PFNA, PFOS, and PFOA). Only after all these indicators meet the standards can subsequent appropriate reuse be carried out. On the other hand, in addition to the detection of PFAS in the soil, during the remediation process, the concentration of PFAS in the exhaust gas was continuously monitored, and the PFAS content in the filters used to capture the gas and the water used for cooling was also tested. Furthermore, related pollutants such as HF, total fluoride, and particulate matter in flue gas were also detected. All these steps are performed to meet the standards of soil remediation, air emission, and soil recycling and re-use, as well as to establish or optimize these standards.

Sun et al. (2024) recently demonstrated that high-frequency heating (HFH) enabled ultra-rapid and highly efficient degradation of diverse PFAS (e.g., long-chain, short chain, emerging PFAS, and precursors) in contaminated soil. Remarkably, their system achieved near-complete degradation (>99.6%) within 60–120 seconds of treatment. They tested the effects of soil contamination levels (two orders of magnitude) and soil types (natural soil, clay, and peat) on the thermal degradation efficiency and claimed negligible effects. The phenomenon occurred probably because of the ultrafast degradation rate of PFAS under high temperature (845°C) with a high heating rate (up to 14.5°C/s). The HFH technology emerges as a promising thermal treatment for PFAS-contaminated soils, while simultaneously addressing key limitations of conventional thermal methods including reactor corrosion risks and insufficient mechanistic understanding.

12.4 BIOREMEDIATION

12.4.1 Bacteria-Based Bioremediation

As mentioned in the Chapter 11, bioremediation has been recognized as a promising technology for remediating emerging organic pollutant-contaminated soils with lots of benefits such as low cost, ease of operation, adaptability, and low production of hazardous by-products (Guo et al. 2016; Kucharzyk et al. 2017; Arora 2018; Aguilar-Romero et al. 2022; Bala et al. 2022; Ismanto et al. 2022; Leung et al. 2022; Li et al. 2022; Roy et al. 2024). The roles of microorganisms and biointerfacial processes (cell surface sorption, transmembrane process, and intracellular enzymatic degradation) in the bioremediation of emerging organic compounds have been discussed in Chapter 11. However, there are still a number of drawbacks, such as time consuming, instability of the culture state, ongoing requirement for oxygen and nutrients, and incomplete degradation for some specific compounds (Chaudhari et al. 2017). For example, Bolong et al. (2009) believed that biological remediation could be effective for those polar EDCs, whereas the biodegradation of EDCs such as steroids and estrogens was usually incomplete (Ismanto et al. 2022). Therefore, strategies should be developed to overcome these constrains. One of the key issues is the presence of appropriate degradative enzyme systems and the maintenance of suitable environmental conditions. It is also possible for the bioremediation of PFAS-contaminated soil and groundwater, if there are suitable microbes along with right nutrition and oxygen conditions.

Aerobic and anaerobic bacteria isolated from diverse contaminated environments – including wastewater, contaminated soil, and groundwater – have demonstrated significant PFAS degradation capabilities. Notable examples include *Pseudomonas aeruginosa*, *Acidimicrobium* sp. *Desulfococcus* genus, which have been successfully characterized for their PFAS transformation potential. Their targeted compounds include PFOS, 6:2 fluorotelomer alcohol (FTOH), 8:2 FTOH, 6:2 fluorotelomer sulfonate, *N*-ethyl perfluorooctane sulfonamidoethanol, and perfluorohexanoate (Key et al. 1998; Rhoads et al. 2008; Wang et al. 2009; Liu et al. 2010; Kwon et al. 2014; Huang and Jaffé 2019; O'Carroll et al. 2020; Wang et al. 2011). Depending on the structure of targeted chemicals and enzymes involved (e.g., cytochrome P450 monooxygenase, transaminase system), the breakdown mechanisms vary, such as partial cleavage of C–C bonds, C–F bond cleavage, and removal of multiple -CF_2- groups (Key et al. 1998; Wang et al. 2009; Liu et al. 2010; Kwon et al. 2014; Huang and Jaffé 2019; Wackett 2022). It is hypothesized that bioremediation of PFAS-contaminated soil remains a largely unexplored solution, although microbes with the ability of breaking down PFSA bonds exist. However, a pilot study conducted in the field of Wisconsin airport derived from fire truck rinse water containing fire-fighting foam may offer surprising results. The pilot study was jointly employed by US-based Orin Technologies and Canada-based Fixed Earth Innovations. Two treatment strategies were carried out: (i) injecting PFAS-degrading bacteria only (bioaugmentation), and (ii) PFAS-degrading bacteria addition, along with BAM amendment and low-voltage electricity-induced aerobic conditions (bioaugmentation + biostimulation). After 26 weeks, the two strategies showed similar results in which both short- and long-chain PFAS reached nondetectable concentrations. However, there are differences in the process. For the BAM with microbes strategy, the total concentrations of PFAS sharply decreased to nondetectable levels; while for the only PFAS-degrading bacteria strategy, total PFAS concentrations rebounded during the weeks 2–9 (due to the generation of short-chain PFAS from long-chain PFAS breakdown) and then decreased to nondetectable levels.

The pilot offers us some insights. (i) First and foremost, bioremediation technologies can effectively degrade PFAS when used in conjunction with appropriate complementary techniques; this is possible in soil and potentially in groundwater as well. (ii) Bioaugmentation only can often achieve relatively good results in a relatively simple environment (where there is no significant competitive pressure, as in this case). However, there may be some risks. For example, metabolic products may introduce new risks (in this case, the increase in total PFAS during the 2–9 weeks was due to the generation of short-chain PFAS). This poses higher demands on the degrading bacterial community (e.g., capable of efficiently degrading both parent compounds and metabolic products). (iii) The combined use of bioaugmentation and biostimulation can avoid these issues. The BAM adsorbent can adsorb PFAS (both parent and metabolic products) onto its surface and also promote the growth and enrichment of degrading bacteria. While promoting bacterial growth, it also avoids the risk of potential metabolic product overflow (see Figure 12.2). While this technology holds great promise, it remains in the nascent phase of field trials and necessitates further testing before it can be implemented on a large scale.

12.4.2 Fungal Bioremediation and Fungi-Assisted ECOHR

Fungi are capable of degrading complex organic compounds, such as PFAS, EDCs, and PPCPs, and have considerable adaptability to changing environmental conditions, making them promising and effective agents in bioremediation (Rodarte-Morales et al. 2011; Křesinová et al. 2017; Merino et al. 2018). For example, *Trametes hirsute* MTCC-1711 could effectively degrade chlorpyrifos with more than 95% removal efficiency (Patil et al. 2021). In another study, acetaminophen and bisphenol A were effectively removed by laccase derived from a fungal strain *Bjerkandera* spp. TBB-03, with remarkable stability under a wide range of pH and temperature conditions (Kang et al. 2021). Tseng et al. (2014) investigated and found that the 6:2 fluorotelomer alcohol could be biotransformed by a wood-rotting fungal strain, *Phanerochaete chrysosporium*. In addition to various enzymes for EOC degradation, fungal hyphae tend to be useful "friends" and "collaborators" for the co-degradation of organic pollutants in soil. As mentioned in Chapter 11, fungi's hyphae could act as

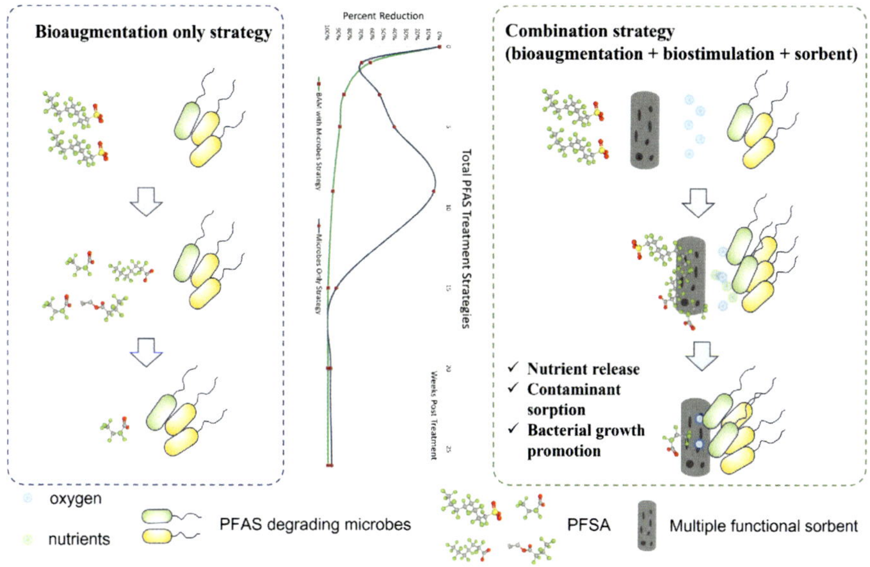

FIGURE 12.2 Bioremediation of PFAS contaminated environment. (Data were collected from the reports provided by Orin Technologies 2022.)

a "highway" for bacteria capable of degrading EOCs to reach the mission site more easily to eliminate target compounds (Wick et al. 2007; Furuno et al. 2012; Banitz et al. 2013; Otto et al. 2016). Worrich et al. (2016) reported that the soil-borne bacterial strain *Pseudomonas putida* KT2440 could efficiently degrade benzoate at consistently high percentages (average of 70.7% ± 7.8%) along "fungal highways" under different strength conditions. However, in some cases, fungal degradation ability for EOCs can be influenced by soil conditions, such as pH, carbon and nitrogen sources, and competition with fast-grown bacteria (Barrios-Estrada et al. 2017; Jureczko et al. 2020; Patil et al. 2021; Ismanto et al. 2022).

Additionally, fungi have special pathways to achieve the degradation/decomposition of complex inert compounds (e.g., PFOA) through enzyme-catalysed oxidative humification reactions (ECOHR). The ECOHR can be carried out by the extracellular enzymes, including lignin peroxidases (LiP), manganese peroxidases (MnP), laccases (LaC), xylanases, cellulases, and hemicellulases, secreted by mainly two fungi: white-rot fungi *Phanerochaete chrysosporium* and brown-rot fungi *Aspergillus niger* (Colosi et al. 2009; Tseng 2012; Weber and Huang 2003). In the ECOHR, the enzymes mentioned above can first convert compounds in the soil into reactive radicals and then react with complex target pollutants. The ECOHR enzymes are ubiquitously present in soil systems and may facilitate the oxidative decomposition of diverse hydrophobic organic compounds (e.g., PFOA, PAHs) during the humification processes. A concise mechanism of PFOA degradation via ECOHRs can be proposed as follows (Luo 2015). The reaction commenced with the transfer of an electron from the carboxylate head group of POFA to a 1-hydroxybenzotrizole (HBT) radical, which then underwent Kolbe decarboxylation to produce a perfluoroheptyl (C_7F_{15}) radical.

This C_7F_{15} radical can further undergo hydrolysis, simultaneously releasing an HF molecule and a fluoride ion (F^-), resulting in the formation of $C_6F_{13}COOH$. The Kolbe decarboxylation process is widely acknowledged as the mechanism for the removal of a CF_2unit from perfluorocarboxylic acids (PFCAs), ultimately yielding a perfluoroalkyl acid with one fewer CF_2unit (Vecitis et al. 2009; Lin et al. 2012). Continuous Kolbe decarboxylation process coupled with corresponding perfluoroalkyl radical generation progressively reduced the carbon chain length as well as the number of fluorine substitutions in PFOA, thereby yielding shorter-chain PFOA. Additionally, shorter-chain perfluoroalkyl radicals can be directly formed from longer-chain radicals through free radical rearrangement. HBT radicals can also transform other non-fluorinated organic compounds in the solution into free radicals. These free radicals cross-couple with perfluoroalkyl radicals to form products that are either partially fluorinated or contain a perfluoroalkyl group, which is a common feature in reactions mediated by laccase (Lu et al. 2009; Sangha et al. 2012). It's important to recognize that laccase substrates containing phenolic functional groups are prevalent in natural organic matter and could act as mediators for ECOHRs (Cañas et al. 2007; Luo et al. 2018). While these natural processes might only lead to very slow PFOA degradation due to the low PFOA concentrations and the fluctuating humification enzyme activities in the natural environment, it is feasible to amplify the humification reactions via an engineered approach for remediation goals.

12.5 PHYTOREMEDIATION

Phytoremediation, actually a bioremediation technology, basically refers to the use of plants and associated soil microbes to eliminate the concentrations and/or adverse impacts (e.g., soil health, ecological risk, plant growth, soil microbial functions) of contaminants in soil and groundwater. It has a relatively long history to in situ soil and groundwater contamination (typically <5 m depth), focusing on various contaminants such as polychlorinated biphenyls, polycyclic aromatic hydrocarbons, chlorinated solvents, explosive compounds, and heavy metals (Schnoor et al. 1995; Campbell et al. 2002; McCutcheon and Schnoor 2003; Panz and Miksch 2012; Passatore et al. 2014; Khandare and Govindwar 2015; Wang et al. 2017a). These contaminants are treated by plants followed by different mechanisms (see Figure 12.3), including uptake and phytoaccumulation, phytoextraction, phytodegradation, phytovolatilization, phytostabilization, and associated microbial degradation (rhizodegradation) (van Aken 2008; Wang et al. 2017b; Wu et al. 2021). For example, root exudates secreted by the plant can facilitate the growth of rhizospheric microorganisms, induce microbial enzymes responsible for pollutant degradation, release nutrients in turn that help plant grow (Didier et al. 2012; Upadhyay et al. 2022), and also can shape the rhizosphere microbiota (Hu et al. 2018). Ding et al. (2024) reported *Arabidopsis* with co-expressed two genes (*tfd*B and *bph*C) could efficiently remove PCB28 from the soil through the uptake and degrade PCB28 within plants, therefore alleviating PCB toxicity. Nevertheless, the phytoremediation of emerging organic compounds such as PPCP, PFAS, and EDCs in soil and groundwater is a relatively novel approach, and the pollutant treatment efficiency and specific mechanisms are still not fully understood (Kurade et al. 2021).

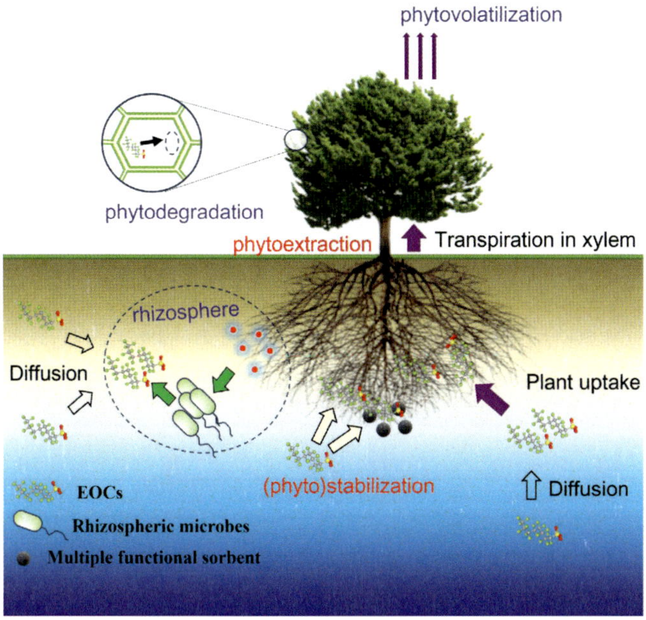

FIGURE 12.3 Proposed phytoremediation processes and mechanisms. (Modified from Kurade et al. 2021; Wu et al. 2021; Mayakaduwage et al. 2022; Orin Technologies 2024.)

Furthermore, the long remediation period is another concern for phytoremediation of contaminated soil and groundwater. For example, an integrated phytoremediation/greenbelt project was carried out since 1986 for remediating a retired field of Union Pacific Railroad Laramie Tie Plant (Wyoming, USA), which was heavily contaminated with zinc, DNAPL (e.g., creosote), LNAPL (e.g., pentachlorophenol), and mobile oil. Specific trees, legumes, and grasses were used to reduce contaminant levels in soil and groundwater. Several mechanisms were involved. Plants enhanced the degradation of organic pollutants by increasing microbial activity and aeration in the rhizosphere. Plants also may uptake and metabolize or fix contaminants. In addition, evapotranspiration drives plume interception and control. After the remedy action started for near three decades, the site has been delisted from the National Priorities List; Plants are growing well in Laramie; the Greenbelt is beneficial to local communities; and the phytoremediation project is progressing. The good removal efficacy for pollutants, ecological and community benefits, and economic friendliness make phytoremediation a competitive candidate for EOC-contaminated soil and groundwater remediation. However, the long-term operational requirements significantly limit its potential for large-scale application.

In recent years, laboratory research and pilot-scale studies on using phytoremediation to treat soils contaminated with emerging pollutants have been increasing. There have also been some field-scale trials in engineering projects. Here is a brief introduction. The uptake and transport of EOCs by plants have been systematically and mechanistically studied. Plants can uptake pollutants from the air through their leaves and from the soil or groundwater through their roots. However, for EOCs (e.g., PPCPs, EDCs), it is generally believed that the uptake of pollutants through the plant roots is the primary pathway, as many EOCs have relatively low volatility (Zhang et al. 2017). Similar to other conventional pollutants, the uptake of EOCs by plant roots occurs through two modes: active transport and passive uptake. Similarly, for the vast majority of EOCs, passive uptake remains the predominant pathway; whereas some hormone-like chemicals (e.g., phenoxy acid herbicides) can only be taken by active transport (Collins et al. 2006; Madikizela et al. 2018). Pollutants diffuse from the soil solution or groundwater to the root surface, enter the xylem and phloem through the apoplastic pathway (with a portion also passing through the symplastic pathway), and then are transported to the above-ground parts, ultimately reaching the stems, leaves, and fruits driven by phytotranspiration (Miller et al. 2016; Kurade et al. 2021). The uptake potential and upward transport ability, also known as BCF (bioaccumulation factor), depends on EOCs properties such as molecular weight, water solubility, pK_a, hydrophobicity (octanol-water partitioning coefficient K_{ow} or D_{ow}, octanol-air partitioning coefficient K_{oa}), soil organic carbon contents, soil water chemistry, and plant types and physiology (Lin et al. 2007; Boonsaner and Hawker 2010, 2012; Goldstein et al. 2014; Cui et al. 2017; Madikizela et al. 2018; Mordechay et al. 2018; Grifoni et al. 2020; Wu et al. 2021). The phytodegradation of ECOs and related enzymes and micro-processes have been also well studied. This topic has already been well summarized and published (Kurade et al. 2021), so we will not go into specific details here.

A field pilot on PFAS phytoremediation has been done in a former tannery site in Northeast Michigan, USA, by ORIN Technologies. The soil texture in the field is mainly silty sands. The PFAS concentrations in the groundwater ranged from 2200 to 6500 ng/L, and the contamination plum distributed 4–12 feet below ground. The concentrations of PFAS in background leaf tissues were detected and were more than 80 ng/g. The integrated phytoremediation strategy was carried out to remediate PFAS-affected groundwater, employing a blend of proven remediation techniques (see Figure 12.3). Specific tree species were planted. BAM (multiple functional biochar sorbent developed by ORIN Technology) was incorporated at a volume ratio of 8% with soil into 8-foot-wide trenches that reached the saturated zone. A specialized sleeve was used to isolate the rhizosphere from the surrounding vadose zone soils and rainwater infiltration, effectively creating a controlled bioreactor for groundwater uptake. PFAS-degrading microbes were also introduced into this bioreactor setup. A control tree (BKG-1) was similarly isolated but did not receive BAM and microbial treatments. Groundwater and tree leaf tissues were analyzed periodically. The control tree BKG-1 (without additional PFAS-degrading microbes) demonstrated that plants can bioaccumulate PFAS but not enough to satisfy regulatory levels. After 1 month's treatment by the integrated strategy (phytoremediation plus BAM and PFAS-degrading microbes), groundwater samples achieved regulatory concentration reductions within 1 month in areas of the site with a pH of 6.5–7. The plants and BAM provide a complementary to facilitate PFAS bioattenuation within the rhizosphere. BAM acts as an immobilizer to sequestrate PFAS and reduce the flux to the rhizosphere, while the trees capture the more mobile short-chain compounds ($C<4$). Both long-chain and short-chain compounds become bioavailable for PFAS microbial degradation. The success of this pilot study offers validated green treatment methodology and framework for integrated treatment systems (plant, microbes, sorbents) under multiple mechanisms.

12.6 SOIL WASHING

Soil flushing and washing are also promising technologies for soils contaminated with EOCs. Contrary to the immobilization approach (see Section 12.2), the primary purpose of these mobilizing techniques (flushing and washing) is to mobilize pollutants from the soil matrix to the aqueous phase. In the cases of mobilization strategies, soil-trapped contaminants can be desorbed by the application of mobilizing amendments, such as surfactants, organic acids, organic solvents, cyclodextrin, and chelating agents, and the mobilized pollutants are susceptible to leaching (Barbafieri et al. 2017; Fatin-Rouge 2020; Rinklebe et al. 2020; Bolan et al. 2021; Kumar et al. 2021b, 2022; Bika et al. 2024). As a consequence, active plant uptake or continuous extraction by recovery well are necessary; otherwise, it would lead to groundwater contamination and toxicity to soil plants and biota (Bolan et al. 2021). Therefore, *ex situ* soil washing is widely accepted and applied, owing to flexible application, simple operation, short period, low cost and high removal efficiency, without subsequent regular monitoring (Chen et al. 2021; Lee and Son 2021; Liu et al. 2021; Fazle Bari et al. 2022; Gu et al. 2022; Tran et al. 2022). Nevertheless, the washing method requires large quantities of solvents to wash EOCs from soil. It should also be noted that soil washing does not degrade EOCs but instead transfers them into the washing water, which then needs to be treated. The large quantities of washing effluents, which usually contain high concentrations of pollutants and washing agents, are of potential high risk and are needed to be properly treat. In addition, the recovery of washing agents can decrease the cost and be environmentally friendly. Therefore, a typical soil washing plant includes at least four sections, soil sorting and attrition system, washing system (core system), washing effluents recovery system, and wastewater treatment system, as illustrated in Figure 12.4.

Surfactant-aided soil washing has been widely recognized to be a very efficient approach in the treatment of organic polluted soils. For example, completed removal of oxyfluorfen (a diphenyl-ether herbicide) from contaminated soil was achieved by using anionic surfactant sodium dodecyl sulfate (SDS) at a dosage below 5 g/kg (dos Santos et al. 2016). Surfactants have been successfully applied in the removal of PFAS from contaminated soils in the process of soil washing, capable of increasing water solubility, reducing sorption to soil, and mobilizing the transport of PFAS (Pan et al. 2009; Guelfo and Higgins 2013). The soil washing efficiency is governed by several factors, including surfactant concentration and property (e.g., hydrophilic-lipophilic balance value), soil properties (e.g., pH, dissolved organic matter, and salinity), and pollutants' property (e.g., hydrophobicity) (Lamichhane et al. 2017). For example, the mobilization effect of PFAS from soil by surfactant depends on surfactant concentration. Basically, high concentrations of surfactant (e.g., multiple critical micelle concentrations, CMC) leads to mobility increase; whereas low concentrations of surfactant (e.g., less than CMC, or 1~2 CMC) can immobilize PFAS due to the extra sorption site of sorbed surfactants (Pan et al. 2009; Guelfo and Higgins 2013). Another anionic surfactant, sodium dodecyl-benzene sulfonate (SDBS), decreased

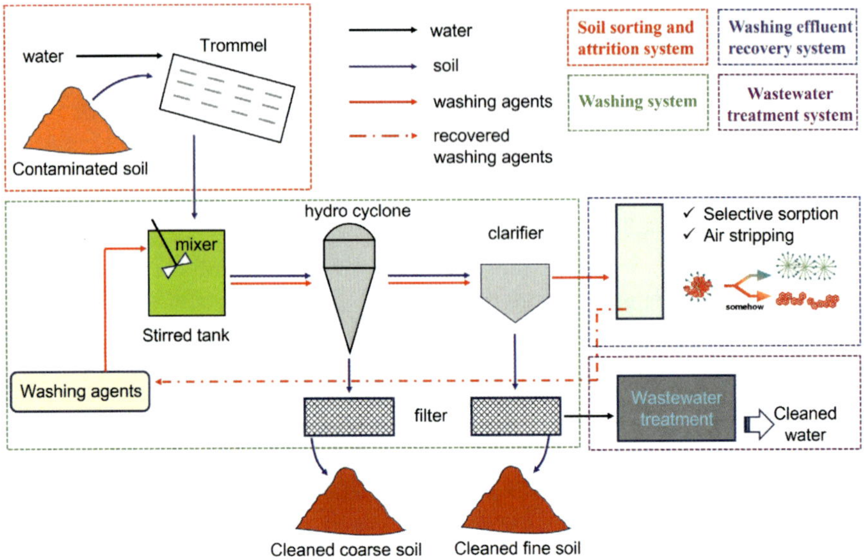

FIGURE 12.4 Soil washing procedure. (Modified from FRTR 2020.)

the sorption of PFOS on soil when the addition dosage was greater than 21.7 mg/L, and enhanced the sorption of PFOS with SDBS concentrations lower than 4.34 mg/L (Pan et al. 2009). Organic solvents and organic acids can also be effective pollutant extracting agents. Munoz et al. (2018) reported that methanol with ammonium acetate effectively improved the desorption of most cationic and zwitterionic PFASs from soil. Oxalate and root exudates could increase the desorption of PFOS from soil, with an enhancement of nearly 20 times (ranging from 1.43 to 17.14 folds) (Tang et al. 2017).

The soil washing strategy has been also applied in the real contaminated soil remediation. Grimison et al. (2023) reported a full-scale soil washing remediation for high clay content (61%), PFAS-contaminated soil. The soil washing plant, known as SourceZone®, was established by Sydney-based Venita Services Pty Ltd and Nova Scotia-based CleanEarth Technologies Inc. The remediation site was a fire training area in South Australia with a long history of aqueous film-forming foams usage. The total soil amount for treatment was approximately 2200 tons and was treated in 25 batches with 90 tons per batch. The total concentration of PFAS in contaminated soil was 261,000 μg/kg. The concentration profile of PFAS in the soil before treatment was analyzed as following: 0.2–15,000 μg/kg for eight perfluorosulfonic acids (PFSAs, C_3-C_{10}), 0.2–420 μg/kg for six PFCAs (C_3-C_8), and 0.6–12 μg/kg for perfluorooctane sulfonamide (POSA) and *N*-methyl perfluorooctane sulfonamide (*N*-MeFOSA). PFAS removal efficiencies after soil washing (average) were listed as follows: 94.4%–97.1% (±<8% standard deviation) for PFCAs (C_4-C_7), 85.7%–94.9% (±<8% standard deviation) for PFSAs (C_4-C_8), 51.4% ± 44% for C_9 PFSAs, and 13.6% ± 74% for C_{10} PFSAs, respectively. Results indicated that the soil washing removal efficiency depended on the chemistry properties of PFAS (e.g., carbon chain length, hydrophilic head group, hydrophobicity), rather than total PFAS concentrations. Leachable PFAS from washing treated soil was substantially decreased compared with untreated soil. For PFOS, contributing for more than 79% of measured PFASs in soil, a ~90% reduction was reached from 116 μg/L of untreated soil to 11 μg/L of washing soil. For all other PFASs, the leachable concentration after soil washing were below 1 μg/L. The combination use of GAC and ion-exchange resin (IXR) as washing effluent treatment also exhibited excellent performance. After treatment, the PFAS concentrations in the wash solution were 0.16 ± 0.13 μg/L. GAC (estimated sorption capacity of 1.1–1.3 g PFAS/kg GAC) played the major role in PFAS sorption without breakthrough, whereas IXR acted as backup sorbents. Overall, the soil washing plant for PFAS-contaminated soil treatment is effective, cost affordable, and climate friendly. The throughput of clay soil reached 11.1 tons soil/h, indicative treatment cost reached AUD $400–500/ton soil treated, and the carbon dioxide generation was approximately 18.8 kg CO_2 equivalent/ton soil.

Both lab scale studies and field applications demonstrate soil washing technology remains a viable and promising remediation strategy for soils contaminated with hydrophobic organic compounds.

12.7 SUMMARY

Soil contaminated with various EOCs, including PPCPs, EDCs, PFAS, and antibiotics, poses huge challenges to our health and ecological security. To this end, many remediation technologies, such as immobilization, bioremediation, phytoremediation, thermal treatment, soil washing, and oxidation approaches, have been introduced to deal with soil pollution. However, due to the complex and heterogeneous nature of soil, as well as the diversity of pollutants, there are not many field remediation cases for these technologies to remediate EOC-contaminated soils, and they are currently still mainly at the laboratory research stage. Fortunately, many researchers and companies have already started pilot experiments, bringing us hope. For example, "Sorbent+" technologies (e.g., sorbent + phytoremediation, sorbent + bioremediation), thermal treatment technologies, and others have all had successful cases. There are still some challenges at present, such as the recent discovery of large-area contamination of farmland soil by PFAS due to biosolids and PFAS-bearing wastewater irrigation, which also brings great challenges to the safe use and efficient remediation of farmland soil.

REFERENCES

Aguilar-Romero I., van Dillewijn P., Nesme J., Sørensen S.J., Nogales R., Delgado-Moreno L., Romero E., 2022, A novel and affordable bioaugmentation strategy with microbial extracts to accelerate the biodegradation of emerging contaminants in different media, *Science of the Total Environment* 834, 155234.

Alidoosti F., Giyahchi M., Moien S., Moghimi H., 2024, Unlocking the potential of soil microbial communities for bioremediation of emerging organic contaminants: omics-based approaches, *Microbial Cell Factories* 23, 210.

Alinezhad A., Sasi P.C., Zhang P., Yao B., Kubátová A., Golovko S.A., Golovko M.Y., Xiao F., 2022, An investigation of thermal air degradation and pyrolysis of per- and polyfluoroalkyl substances and aqueous film-forming foams in soil, *ACS Environmental Science & Technology Engineering* 2, 198–209.

Andrews D.Q., Naidenko O.V., 2020, Population-wide exposure to per- and polyfluoroalkyl substances from drinking water in the United States, *Environmental Science & Technology Letter* 7, 931–936.

Arora N.K., 2018, Bioremediation: a green approach for restoration of polluted ecosystems, *Environmental Sustainability* 1, 305–307.

Bala S., Garg D., Thirumalesh B.V., Sharma M., Sridhar K., Inbaraj B.S., Tripathi M., 2022, Recent strategies for bioremediation of emerging pollutants: a review for a green and sustainable environment, *Toxics* 10, 484.

Banitz T., Johst K., Wick L.Y., Schamfuß S., Harms H., Frank K., 2013, Highways versus pipelines: contributions of two fungal transport mechanisms to efficient bioremediation, *Environmental Microbiology Reports* 5, 211–218.

Barbafieri M., Pedron F., Petruzzelli G., Rosellini I., Franchi E., Bagatin R., Vocciante M., 2017, Assisted phytoremediation of a multi-contaminated soil: investigation on arsenic and lead combined mobilization and removal, *Journal of Environmental Management* 203, 316–329.

Barranco F., Caprio P., Hay G., 2020, *Evaluation of Indirect Thermal Desorption Coupled with Thermal Oxidation Technology to Treat Solid PFAS-Impacted Investigation-Derived Waste*. SERDP Project ER18-1572. Department of Defense Strategic Environmental Research and Development Program (SERDP), Environmental Security Technology Certification Program (ESTCP), Alexandria, VA.

Barrios-Estrada C., Rostro-Alanis M., Parra A., Belleville M.P., Sanchez Marcano J., Iqbal H., Parra R., 2017, Potentialities of active membranes with immobilized laccase for bisphenol A degradation, *International Journal of Biological Macromolecules* 108, 837–844.

Bika S.H., Adeniji A.O., Ojemaye M.O., Elsheikh E.A.E., Anyanwu B.O., Okoh A.I., Okoh O.O., 2024, Synthesis of EDTA-cysteine-*β*-cyclodextrin for the removal of organophosphate flame retardants (OPFR) from sediments and soil samples from the Buffalo River Estuary, Eastern Cap of South Africa, *Heliyon* 10, e24951.

Bolan N., Sarkar B., Yan Y., Li Q., Wijesekara H., Kannan K., Tsang D.C.W., Schauerte M., Bosch J., Noll H., Ok Y.S., Scheckel K., Kumpiene J., Gobindal K., Kah M., Sperry J., Kirhham M.B., Wang H., Tsang Y.F., Hou D., Rinklebe J., 2021, Remediation of poly- and perfluoroalkyl substances (PFAS) contaminated soils: to mobilise or to immobilise or to degrade? *Journal of Hazardous Materials* 401, 123892.

Bolan S., Hou D.Y., Wang L.W., Hale L., Egamberdieva D., Tammeorg P., Li R., Wang B., Xu J.P., Wang T., Sun H.W., Padhye L.P., Wang H.L., Siddique K.H.M., Rinklebe J., Kirkham M.B., Bolan N., 2023, The potential of biochar as a microbial carrier for agricultural and environmental applications, *Science of the Total Environment* 886, 163968.

Bolong N., Ismail A.F., Salim M.R., Matsuura T., 2009, A review of the effects of emerging contaminants in wastewater and options for their removal, *Desalination* 239, 229–246.

Boonsaner M., Hawker D.W., 2010, Accumulation of oxytetracycline and norfloxacin from saline soil by soybeans, *Science of the Total Environment* 408, 1731–1737.

Boonsaner M., Hawker D.W., 2012, Investigation of the mechanism of uptake and accumulation of zwitterionic tetracyclines by rice (*Oryza sativa* L.), *Ecotoxicology and Environmental Safety* 78, 142–147.

Campbell S., Paquin D., Awaya J.D., Li Q.X., 2002, Remediation of benzo[a]pyrene and chrysene-contaminated soil with industrial hemp (*Cannabis sativa*), *International Journal of Phytoremediation* 4, 157–168.

Cañas A.I., Alcalde M., Plou F., Martínez M.J., Camarero S., 2007, Transformation of polycyclic aromatic hydrocarbons by laccase is strongly enhanced by phenolic compounds present in soil, *Environmental Science & Technology* 41, 2964–2971.

Chaudhari A.U., Paul D., Dhotre D., Kodam K.M., 2017, Effective biotransformation and detoxification of anthraquinone dye reactive blue 4 by using aerobic bacterial granules, *Water Research* 122, 604–613.

Chen W.X., Zhang H.Y., Zhang M., Shen X.F., Zhang X.Y., Wu F., Hu J., Wang B., Wang X.L., 2021, Removal of PAHs at high concentrations in a soil washing solution containing TX-100 via simultaneous sorption and biodegradation processes by immobilized degrading bacteria in PVA-SA hydrogel beads, *Journal of Hazardous Materials* 410, 124533.

Chen X., Lin H., Dong Y., Li B., Liu C., Yin T., 2022, Mechanisms underlying enhanced bioremediation of sulfamethoxazole and zinc(II) by *Bacillus* sp. SDB4 immobilized on biochar, *Journal of Cleaner Production* 370, 133483.

Crownover E., Oberle D., Kluger M., Heron G., 2019, Perfluoroalkyl and polyfluoroalkyl substances thermal desorption evaluation, *Remediation Journal* 29, 77–81.

Collins C., Fryer M., Grosso A., 2006, Plant uptake of non ionic organic chemicals, *Environmental Science & Technology* 40, 45–52.

Colosi L.M., Pinto R.A., Huang Q., Weber W.J., 2009, Peroxidase-mediated degradation of perfluorooctanoic acid, *Environmental Toxicology and Chemistry* 28, 264–271.

Cui H., de Angelis M.H., Schröder P., 2017, Iopromide exposure in *Typha latifolia* L.: evaluation of uptake, translocation and different transformation mechanisms in planta, *Water Research* 122, 290–298.

Dagorn G., Aubert R., Horel S., Martnon L., Steffe T., 2023, "Forever pollution": explore the map of Europe's PFAS contamination, Le Monde. https://www.northeastern.edu/environmentalhealth/pfas-project-labs-methodology-used-by-european-investigative-journalists/

Das P., Megharaj M., Naidu R., 2015, Perfluorooctane sulfonate release pattern from soils of fire training areas in Australia and its bioaccumulation potential in the earthworm *Eisenia fetida*, *Environmental Science and Pollution Research* 22, 8902–8910.

Didier T., Philippe L.G., Sonia H., Amar B., Claudia M.C., Marielle D., Jairo F., 2012, Prospects of *Miscanthus x giganteus* for PAH phytoremediation: a microcosm study, *Industrial Crops and Productions* 36, 276–281.

Ding Y.Z., Ren H.J., Hao X.Y., Zhang R.N., Hao J.J., Liu J.L., Pan H.Y., Wang Y., 2024, Enhanced phytoremediation of PCBs-contaminated soil by co-expressing *tfd*B and *bph*C in Arabidopsis aiding in metabolism and improving toxicity tolerance, *Environmental and Experimental Botany* 217, 105548.

Dong M., Sun N., Liu C.G., 2023, Bromide ion enhancing the phytodegradation of emerging phenolic pollutants and its mechanisms mediating wheat resistance to phenolic pollutants stress, *Journal of Cleaner Production* 411, 137295.

dos Santos E.V., Sáez C., Martínez-Huitle C.A., Cañizares P., Rodrigo M.A., 2016, Removal of oxyfluorfen from ex-situ soil washing fluids using electrolysis with diamond anodes, *Journal of Environmental Management* 171, 260–266.

Duchesne A.L., Brown J.K., Patch D.J., Major D., Weber K.P., Gerhard J.I., 2020, Remediation of PFAS-contaminated soil and granular activated carbon by smoldering combustion, *Environmental Science & Technology* 54, 12631–12640.

EPA Victoria, 2024, *Technical Guide: Validation of PFAS Soil Treatment Technologies*, Victorian Government, Melbourne, Australia.

EU, 2019, Regulation (EU) 2019/1021 of the European Parliament and of the Council of 20 June 2019 on persistent organic pollutants (recast) (Text with EEA relevance) (OJ L 169, 25.6.2019, pp. 45–77).

Fang L.F., Chen C.Y., Li S.Y., Ye P.P., Shi Y.J., Sharma G., Sarkar B., Shaheen S.M., Lee S.S., Xiao R., Chen X.P., 2023, A comprehensive and global evaluation of residual antibiotics in agricultural soils: accumulation, potential ecological risks, and attenuation strategies, *Ecotoxicology and Environmental Safety* 262, 115175.

Fatin-Rouge N., 2020, Contaminant mobilization from polluted soils: behavior and reuse of leaching solutions. In E. van Hullebusch, D. Huguenot, Y. Pechaud, M.-O. Simonnot, S. Colombano (eds.), *Environmental Soil Remediation and Rehabilitation*, Springer, pp. 1–59.

Fazle Bari A.S.M., Lamb D., MacFarlane G.R., Rahman M.M., 2022, Soil washing of arsenic from mixed contaminated abandoned mine soils and fate of arsenic after washing, *Chemosphere* 296, 134053.

FRTR, 2020, Soil washing, Federal Remediation Technology Roundtable. https://www.frtr.gov/matrix/Soil-Washing/#

Furuno S., Foss S., Wild E., Jones K.C., Semple K.T., Harms H., Wick L.Y., 2012, Mycelia promote active transport and spatial dispersion of polycyclic aromatic hydrocarbons, *Environmental Science & Technology* 46, 5463–5470.

Gaballah M.S., Guo J., Sun H., Aboagye D., Sobhi M., Muhmood A., Dong R., 2021, A review targeting veterinary antibiotics removal from livestock manure management systems and future outlook, *Bioresource Technology* 333, 125069.

Goldstein M., Shenker M., Chefetz B., 2014, Insights into the uptake processes of wastewater-borne pharmaceuticals by vegetables, *Environmental Science & Technology* 48, 5593–5600.

Grifoni M., Rosellini I., Angelini P., Petruzzelli G., Pezzarossa B., 2020, The effect of residual hydrocarbons in soil following oil spillages on the growth of *Zea mays* plants, *Environmental Pollution* 265, 114950.

Grimison C., Knight E.R., Nguyen T.M.H., Nagle N., Kabiri S., Bräunig J., Navarro D.A., Kookana R.S., Higgins C.P., McLaughlin M.J., Mueller J.F., 2023, The efficacy of soil washing for the remediation of per- and poly-fluoroalkyl substances (PFASs) in the field, *Journal of Hazardous Materials* 445, 130441.

Gu F., Zhang J.P., Shen Z.Q., Li Y., Ji R.T., Li W., Zhang L.J., Han J., Xue J.M., Cheng H., 2022, A review for recent advances on soil washing remediation technologies, *Bulletin of Environmental Contamination and Toxicology* 109, 651–658.

Guedes P., Dionísio J., Couto N., Mateus E.P., Pereira C.S., Ribeiro A.B., 2021, Electro-bioremediation of a mixture of structurally different contaminants of emerging concern: uncovering electrokinetic contribution, *Journal of Hazardous Materials* 406, 124304.

Guedes P., Mateus E.P., Couto N., Rodríguez Y., Ribeiro A.B., 2014, Electrokinetic remediation of six emerging organic contaminants from soil, *Chemosphere* 117, 124–131.

Guelfo J.L., Higgins C.P., 2013, Subsurface transport potential of perfluoroalkyl acids at aqueous film-forming foam (AFFF)-impacted sites, *Environmental Science & Technology* 47, 4164–4171.

Guo Q., Yan J., Wen J.J., Hu Y.Y., Chen Y.B., Wu W.J., 2016, Rhamnolipid-enhanced aerobic biodegradation of triclosan (TCS) by indigenous microorganisms in water-sediment systems, *Science of the Total Environment* 571, 1304–1311.

Horta A., Malone B., Stockmann U., Minasny B., Bishop T.F.A., McBratney A.B., Pallasser R., Pozza L., 2015, Potential of integrated field spectroscopy and spatial analysis for enhanced assessment of soil contamination: a prospective review, *Geoderma* 241–242, 180–209.

Hu L.F., Robert C.A.M., Cadot S., Zhang X., Ye M., LI B.B., Manzo D., Chervet N., Steinger T., van der Heijden M.G.A., Schlaeppi K., Erb M., 2018, Root exudate metabolites drive plant-soil feedbacks on growth and defense by shaping the rhizosphere microbiota, *Nature Communications* 9, 2738.

Huang J., Tan X., Ali I., Duan Z.P., Naz I., Cao J., Ruan Y.L., Wang Y.M., 2023, More effective application of biochar-based immobilization technology in the environment: understanding the role of biochar, *Science of the Total Environment* 872, 162021.

Huang S., Jaffé P.R., 2019, Defluorination of perfluorooctanoic acid (PFOA) and perfluorooctane sulfonate (PFOS) by *Acidimicrobium* sp. strain A6, *Environmental Science & Technology* 53, 11410–11419.

Ismanto A., Hadibarata T., Kristanti R.A., Maslukah L., Safinatunnajah N., Kusumastuti W., 2022, Endocrine disrupting chemicals (EDCs) in environmental matrices: occurrence, fate, health impact, physico-chemical and bioremediation technology, *Environmental Pollution* 302, 119061.

ITRC, 2022, *PFAS Technical and Regulatory Guidance Document*, Interstate Technology Regulatory Council (ITRC), Washington, DC.

Jureczko M., Przystaś W., Krawczyk T., Gonciarz W., Rudnicka K., 2020, White-rot fungi-mediated biodegradation of cytostatic drugs-bleomycin and vincristine, *Journal of Hazardous Materials* 407, 124632.

Kabiri S., McLaughlin M.J., 2021, Durability of sorption of per- and polyfluorinated alkyl substances in soils immobilized using common adsorbents: 2. Effects of repeated leaching, temperature extremes, ionic strength and competing ions, *Science of the Total Environment* 766, 144718.

Kang B.R., Kim S.Y., Kang M., Lee T.K., 2021, Removal of pharmaceuticals and personal care products using native fungal enzymes extracted during the ligninolytic process, *Environmental Research* 195, 110878.

Keerthanan S., Rajapaksha A.U., Mašek O., Vithanage M., 2023, Plant uptake of personal care products and biochar-assisted immobilization in soil: an appraisal, *Journal of Soils and Sediments* 23, 2669–2684.

Kewalramani J.A., Wang B.R., Marsh R.W., Meegoda J.N., Freire L.R., 2022, Coupled high and low-frequency ultrasound remediation of PFAS-contaminated soils, *Ultrasonics Sonochemistry* 88, 106063.

Key B.D., Howell R.D., Criddle C.S., 1998, Defluorination of organofluorine sulfur compounds by *Pseudomonas* sp. strain D2, *Environmental Science & Technology* 32, 2288–2287.

Khandare R.V., Govindwar S.P., 2015, Phytoremediation of textile dyes and effluents: current scenario and future prospects, *Biotechnology Advances* 33, 1697–1714.

Křesinová Z., Linhartová L., Filipová A., Ezechiáš M., Mašín P., Cajthaml T., 2017, Biodegradation of endocrine disruptors in urban wastewater using *Pleurotus ostreatus* bioreactor, *New Biotechnology* 43, 53–61.

Kucharzyk K.H., Darlington R., Benotti M., Deeb R., Hawley E., 2017, Novel treatment technologies for PFAS compounds: a critical review, *Journal of Environmental Management* 204, 757–764.

Kumar A., Yadv A.N., Mondal R., Kour D., Subrahmanyam G., Shabnam A.A., Khan S.A., Yadav K.K., Sharma G.K., Cabral-Pinto M., Fagodiya R.K., Gupta D.K., Hota S., Malyan S.K., 2021a, Myco-remediation: a mechanistic understanding of contaminants alleviation from natural environment and future prospect, *Chemosphere* 284, 131325.

Kumar M., Bolan N.S., Hoang S.A., Sawarkar A.D., Jasemizad T., Gao B., Keerthanan S., Padhye L.P., Singh L., Kumar S., Vithanage M., Li Y., Zhang M., Kirkham M.B., Vinu A., Rinklebe J., 2021b, Remediation of soils and sediments polluted with polycyclic aromatic hydrocarbons: to immobilize, mobilize, or degrade? *Journal of Hazardous Materials* 420, 126534.

Kumar M., Bolan N., Jasemizad T., Padhye L.P., Sridharan S., Singh L., Bolan S., O'Connor J., Zhao H.C., Shaheen S.M., Song H., Siddique K.H.M., Wang H.L., Kirkham M.B., Rinklebe J., 2022, Mobilization of contaminants: potential for soil remediation and unintended consequences, *Science of the Total Environment* 839, 156373.

Kuppusamy S., Kakarla D., Kadiyala V., Mallavarapu M., Young-Eun Y., Yong B.L., 2018, Veterinary antibiotics (VAs) contamination as a global agro-ecological issue: a critical review, *Agriculture, Ecosystems & Environment* 257, 47–59.

Kurade M.B., Ha Y.H., Xiong J.Q., Govindwar S.P., Jang M., Jeon B.H., 2021, Phytoremediation as a green biotechnology tool for emerging environmental pollution: a step forward towards sustainable rehabilitation of the environment, *Chemical Engineering Journal* 415, 129040.

Kwon B.G., Lim H.J., Na S.H., Choi B.I., Shin D.S., Chung S.Y., 2014, Biodegradation of perfluorooctanesulfonate (PFOS) as an emerging contaminant, *Chemosphere* 109, 221–225.

Lamichhane S., Bal Krishna K.C., Sarukkalige R., 2017, Surfactant-enhanced remediation of polycyclic aromatic hydrocarbons: a review, *Journal of Environmental Management* 199, 46–61.

Lee D., Son Y., 2021, Ultrasound-assisted soil washing processes using organic solvents for the remediation of PCBs-contaminated soils, *Ultrasonics Sonochemistry* 80, 105825.

Leung S.C.E., Shukla P., Chen D.C., Eftekhari E., An H.J., Zare F., Ghasemi N., Zhang D.K., Nguyen N.T., Li Q., 2022, Emerging technologies for PFOS/PFOA degradation and removal: a review, *Science of the Total Environment* 827, 153669.

Li J.H., Li X.H., Da Y.B., Yu J.L., Long B., Zhang P., Bakker C., McCarl B.A., Yuan J.S., Dai S.Y., 2022, Sustainable environmental remediation via biomimetic multifunctional lignocellulosic nano-framework, *Nature Communication* 13, 4368.

Lin H., Niu J., Ding S., Zhang L., 2012, Electrochemical degradation of perfluorooctanoic acid (PFOA) by Ti/SnO_2-Sb, Ti/SnO_2-Sb/PbO_2 and Ti/SnO_2-Sb/MnO_2 anodes, *Water Research* 46, 2281–2289.

Lin H., Tao S., Zuo Q., Coveney R., 2007, Uptake of polycyclic aromatic hydrocarbons by maize plants, *Environmental Pollution* 148, 614–619.

Liu J., Wang N., Szostek B., Buck R.C., Panciroli P.K., Folsom P.W., Sulecki L.M., Bellin C.A., 2010, 6–2 fluorotelomer alcohol aerobic biodegradation in soil and mixed bacterial culture, *Chemosphere* 78, 437–444.

Liu J., Zhao L., Liu Q., Li J., Qiao Z., Sun P., Yang Y., 2021, A critical review on soil washing during soil remediation for heavy metals and organic pollutants, *International Journal of Environmental Science and Technology* 19, 601–624.

Longendyke G.K., Katel S., Wang Y.X., 2022, PFAS fate and destruction mechanisms during thermal treatment: a comprehensive review, *Environmental Science: Processes & Impacts* 24, 196–208.

Lu J., Huang Q., Mao L., 2009, Removal of acetaminophen using enzyme-mediated oxidative coupling processes: I. Reaction rates and pathways, *Environmental Science & Technology* 43, 7062–7067.

Luo Q., 2015, Degradation of perfluoroalkyl acids by enzyme catalyzed oxidative humification reactions, Docter's Thesis, University of Georgia, Athens, GA.

Luo Q., Liang S.T., Huang Q.G., 2018, Laccase induced degradation of perfluorooctanoic acid in a soil slurry, *Journal of Hazardous Materials* 359, 241–247.

Madikizela L.M., Ncube S., Chimuka L., 2018, Uptake of pharmaceuticals by plants grown under hydroponic conditions and natural occurring plant species: a review, *Science of the Total Environment* 636, 477–486.

Mahinroosta R., Senevirathna L., 2020, A review of the emerging treatment technologies for PFAS contaminated soils, *Journal of Environmental Management* 255, 109896.

Mayakaduwage S., Ekanayake A., Kurwadkar S., Rajapaksha A.U., Vithanage M., 2022, Phytoremediation prospects of per- and polyfluoroalkyl substances: a review, *Environmental Research* 212, 113311.

McCutcheon S.C., Schnoor J.L., 2003, *Phytoremediation: Transformation and Control of Contaminants*, John Wiley & Sons, Inc., Hoboken, NJ.

Merino N., Wang M., Ambrocio R., Mak K., O'Connor E., Gao A., Hawley E.L., Deeb R.A., Tseng L.Y., Mahendra S., 2018, Fungal biotransformation of 6:2 fluorotelomer alcohol, *Remediation* 28, 59–70.

Miller E.L., Nason S.L., Karthikeyan K.G., Pedersen J.A., 2016, Root uptake of pharmaceuticals and personal care product ingredients, *Environmental Science & Technology* 50, 525–541.

Mordechay B.E., Tarchitzky J., Chen Y., Shenker M., Chefetz B., 2018, Composted biosolids and treated wastewater as source of pharmaceuticals and personal care products for plant uptake: a case study with carbamazepine, *Environmental Pollution* 232, 164–172.

Munoz G., Ray P., Mejia-Avendaño S., Vo Duy S., Tien Do D., Liu J., Sauvé S., 2018, Optimization of extraction methods for comprehensive profiling of perfluoroalkyl and polyfluoroalkyl substances in firefighting foam impacted soils, *Analytica Chimca Acta* 1034, 74–78.

O'Carroll D.M., Jeffries T.C., Lee M.J., Le S.T., Yeung A., Wallace S., Battye N., Patch D.J., Manefield M.J., Weber K.P., 2020, Developing a roadmap to determine per- and polyfluoroalkyl substances-microbial population interactions, *Science of the Total Environment* 712, 135994.

Organic Incineration Technology Inc. (OIT), 2020, *Moose Creek Facility Thermal Remediation of PFAS-Contaminated soil 2019 Case Study*, LLC/US Ecology, Inc., NRC Alaska.

Orin Technologies, 2022, Fire-fighting tank rinse PFAS totes at Airport in MidWest. https://orinrt.com/2024/10/groundbreaking-pfas-treatment-study-combined-technologies-show-superior-results-at-midwest-airport/

Orin Technologies, 2024, Phytoremediation of former Tannery site in Northeast Michigan. https://orinrt.com/2024/10/former-tannery-northeast-michigan-phytoremediation/

Otto S., Banitz T., Thullner M., Harms H., Wick L.Y., 2016, Effects of facilitated bacterial dispersal on the degradation and emission of a desorbing contaminant, *Environmental Science & Technology* 50, 6320–6326.

Palansooriya K.N., Shaheen S.M., Chen S.S., Tsang D.C.W., Hashimoto Y., Hou D.Y., Bolan N.S., Rinklebe J., Ok Y.S., 2020, Soil amendments for immobilization of potentially toxic elements in contaminated soils: a critical review, *Environment International* 134, 105046.

Pan G., Jia C., Zhao D., You C., Chen H., Jiang G., 2009, Effect of cationic and anionic surfactants on the sorption and desorption of perfluorooctane sulfonate (PFOS) on natural sediments, *Environmental Pollution* 157, 325–330.

Panz K., Miksch K., 2012, Phytoremediation of explosives (TNT, RDX, HMX) by wild-type and transgenic plants, *Journal of Environmental Management* 113, 85–92.

Passatore L., Rossetti S., Juwarkar A.A., Massacci A., 2014, Phytoremediation and bioremediation of polychlorinated biphenyls (PCBs): state of knowledge and research perspectives, *Journal of Hazardous Materials*, 278, 189–202.

Patil P.D., Singh A.A., Yadav G.Y., 2021, Biodegradation of organophosphorus insecticide chlorpyrifos into a major fuel additive 2,4-bis(1,1 dimethylethyl) phenol using white-rot fungal strain *Trametes hirsute* MTCC-1711, *Journal of Indian Chemical Society* 98, 100120.

RemBind, n.d., PFAS immobilization at an electroplating site in New Zealand. https://rembind.com/projects/onsite-immobilization-of-pfas-using-rembind-at-an-electroplating-site-in-new-zealand/

Rhoads K.P., Janssen E.M.L., Luthy R.G., Criddle C.S., 2008, Aerobic biotransformation and fate of N-ethyl perfluorooctane sulfaon-amidoethanol (N-EtFOSE) in activated sludge, *Environmental Science & Technology* 42, 2873–2878.

Rinklebe J., Shaheen S.M., El Naggar A., Wang H., Du LLaing G., Alessi D.S., Ok Y.S., 2020, Redox-induced mobilization of Ag, Sb, Sn, and Tl in the dissolved, colloidal and solid phase of a biochar-treated and un-treated mining soil, *Environment International* 140, 105754.

Rodarte-Morales A.I., Feijoo G., Moreira M.T., Lema J.M., 2011, Degradation of selected pharmaceutical and personal care products (PPCPs) by white-rot fungi, *World Journal of Microbiology and Biotechnology* 27, 1839–1846.

Ross I., McDonough J., Miles J., Storch P., Kochunarayanan P.T., Kalve E., Hurst J., Dasgupta S.S., Burdick J., 2018, A review of emerging technologies for remediation of PFASs, *Remediation* 28, 101–126.

Roy A., Dubey P., Srivastava A., Kaur I., Shrivastava A., Vajpayee P., Srivastava S., Srivastava P.K., 2024, Exploring the potential of *Meyerozyma caribbica* and its combined application with bacteria for lindane bioremediation, *Chemosphere* 361, 142413.

Salvatore D., Mok K., Garrett K.K., Poudrier G., Brown P., Birnbaum L.S., Goldenman G., Miller M.F., Patton S., Poehlein M., Varshavsky J., 2022, Presumptive contamination: a new approach to PFAS contamination based on likely sources, *Environmental Science & Technology Letters* 9, 983–990.

Sangha A.K., Parks J.M., Standaert R.F., Ziebell A., Davis M., Smith J.C., 2012, Radical coupling reactions in lignin synthesis: a density functional theory study, *The Journal of Physical Chemistry B* 116, 4760–4768.

Scaggs A., 2024, Anything can be a PFAS-growing industry, *Finacial Times*. https://www.ft.com/content/cf9a2df4-49bd-4c00-8e71-cf192563df0f

Schnoor J.L., Licht L.A., McCutcheon S.C., Wolfe N.L., Carreira L.H., 1995, Phytoremediation of organic and nutrient contaminants, *Environmental Science & Technology* 29, 318A–323A.

Shen Z.T., Jin F., O'Connor D., Hou D.Y., 2019, Solidificatino/stabilization for soil remediation: an old technology with new vitality, *Environmental Science & Technology* 53, 11615–11617.

Shi H., Bai C.H., Luo D.L., Wang W., Zhou C.M., Meharg A.A., Yao L.X., 2019, Degradation of tetracyclines in manure-amended soil and their uptake by litchi (*Litchi chinensis* Sonn.), *Environmental Science and Pollution Research* 26, 6209–6215.

Sleep J.A., Juhasz, A.L., 2021, A review of immobilisation-based remediation of per- and poly-fluoroalkyl substances (PFAS) in soils, *Current Pollution Reports* 7, 524–539.

Smalling K.L., Romanok K.M., Bradley P.M., Morriss M.C., Gray J.L., Kanagy L.K., Gordon S.E., Williams B.M., Breitmeyer S.E., Jones D.K., AdCicco L.A., Eagles-Smith C.A., Wagner T., 2023, Per- and polyfluoroalkyl substances (PFAS) in United States tapwater: comparison of underserved private-well and public-supply exposures and associated health implications, *Environment International* 178, 108033.

Sörengård M., Kleja D.B., Ahrens L., 2019, Stabilization and solidification remediation of soil contaminated with poly- and perfluoroalkyl substances (PFASs), *Journal of Hazardous Materials* 367, 639–646.

Sörengård M., Lindh A.S., Ahrens L., 2020, Thermal desorption as a high removal remediation technique for soils contaminated with per- and polyfluoroalkyl substances (PFASs), *PLoS ONE* 15(6), e0234476.

Sun R.Z., Babalol S., Ni R.C., Dolatabad A.A., Cao J.F., Xiao F., 2024, Efficient and fast remediation of soil contaminated by per- and polyfluoroalkyl substances (PFAS) by high-frequency heating, *Journal of Hazardous Materials* 463, 132660.

Tang J., Zhang Y., Zha Y., Li X., Fan S., 2017, Oxalate enhances desorption of perfluorooctane sulfonate from soils and sediments, *Water, Air, & Soil Pollution* 228, 462.

Toms L.M.L., Bräunig J., Vijayasarathy S., Phillips S., Hobson P., Aylward L.L., Kirk M.D., Mueller J.F., 2019, Per- and polyfluoroalkyl substance (PFAS) in Australia: current levels and estimated population reference values for selected compounds, *International Journal of Hygiene and Environmental Health* 222, 387–394.

Tran H.T., Lin C., Hoang H.G., Bui X.T., Le V.G., Vu C.T., 2022, Soil washing for the remediation of dioxin-contaminated soil: a review, *Journal of Hazardous Materials* 421, 126767.

Tripti, Kumar A., Usmani Z., Kumar V., Anshumali, 2017, Biochar and flyash inoculated with plant growth promoting rhizobacteria act as potential biofertilizer for luxuriant growth and yield of tomato plant, *Journal of Environmental Management* 190, 20–27.

Tseng N., 2012, Feasibility of biodegradation of polyfluoroalkyl and perfluoroalkyl substances, Master's Thesis, UCLA, Los Angeles, CA.

Tseng N., Wang N., Szostek B., Mahendra S., 2014, Biotransformation of 6:2 fluorotelomer alcohol (6:2 FTOH) by a wood-rotting fungus, *Environmental Science & Technology* 48, 4012–4020.

Upadhyay S.K., Srivastava A.K., Rajput V.D., Chauhan P.K., Bhojiya A.A., Jain D., Chaubey G., Dwivedi P., Sharma B., Minkina T., 2022, Root exudates: mechanistic insight of plant growth promoting rhizobacteria for sustainable crop production, *Frontiers in Microbiology* 13, 916488.

USEPA, 2016a, Health effects support document for perfluorooctane sulfonate (PFOS), EPA 822-R-16-002.

USEPA, 2016b, Health effects support document for perfluorooctanoic acid (PFOA), EPA 822-R-16-003.

USEPA, 2016c, Drinking water health advisory for perfluorooctane sulfonate (PFOS), EPA 822-R-16-004.

USEPA, 2016d, Drinking water health advisory for perfluorooctanoic acid (PFOA), EPA 822-R-16-005.

USEPA, 2016e, Risk management for per- and polyfluoroalkyl substances (PFASs) under TSCA, https://www.epa.gov/assessing-and-managing-chemicals-under-tsca/risk-management-and-polyfluoroalkyl-substances-pfas

USEPA, 2017, Technical fact sheet – perfluorooctane sulfonate (PFOS) and perfluorooctanoic acid (PFOA), EPA 505-F-17-001.

van Aken B., 2008, Transgenic plants for phytoremediation: helping nature to clean up environmental pollution, *Trends in Biotechnology* 26, 225–227.

Vecitis C., Park H., Cheng J., Mader B., Hoffmann M., 2009, Treatment technologies for aqueous perfluorooctanesulfonate (PFOS) and perfluorooctanoate (PFOA), *Frontiers of Environmental Science & Engineering in China* 3, 129–151.

Wackett L.P., 2022, Strategies for the biodegradation of polyfluorinated compounds, *Microorganisms* 10, 1664.

Wang L., Ji B., Hu Y.H., Liu R.Q., Sun W., 2017a, A review on in situ phytoremediation of mine tailings, *Chemosphere* 184, 594–600.

Wang M., Zhang D.Q., Dong J.W., Tan S.K., 2017b, Constructed wetlands for wastewater treatment in cold climate: a review, *Journal of Environmental Sciences-China* 57, 294–311.

Wang N., Liu J., Buck R.C., Korzeniowski S.H., Wolstenholme B.W., Folsom P.W., Sulecki L.M., 2011, 6:2 Fluorotelomer sulfonate aerobic biotransformation in activated sludge of waste water treatment plants, *Chemosphere* 82, 853–858.

Wang N., Szostek B., Buck R.C., Folsom P.W., Sulecki L.M., Gannon J.T., 2009, 8-2 Fluorotelomer alcohol aerobic soil biodegradation: pathways, metabolites and metabolite yields, *Chemosphere* 75, 1089–1096.

Weber N.H., Delva C.S., Stockenhuber S.P., Grimison C.C., Lucas J.A., Mackie J.C., Stockenhuber M., Kennedy E.M., 2023, Thermal mineralization of perfluorooctanesulfonic acid (PFOS) to HF, CO_2, and SO_2, *Industrial & Engineering Chemistry Research* 62, 881–892.

Weber W.J., Huang Q., 2003, Inclusion of persistent organic pollutants in humification processes: direct chemical incorporation of phenanthrene via oxidative coupling, *Environmental Science & Technology* 37, 4221–4227.

Wen C., Liu T., Wang D., Wang Y., Chen H., Luo G., Zhou Z., Li C., Xu M., 2023, Biochar as the effective adsorbent to combustion gaseous pollutants: preparation, activation, functionalization and the adsorption mechanisms, *Progress in Energy and Combustion Science* 99, 101098.

Wick L.Y., Remer R., Würz B., Reichenbach J., Braun S., Schäfer F., Harms H., 2007, Effect of fungal hyphae on the access of bacteria to phenanthrene in soil, *Environmental Science & Technology* 41, 500–505.

Worrich A., Köbug S., Banitz T., Centler F., Frank K., Thullner M., Harms H., Miltner A., Wick L.Y., Kästner M., 2016, Bacterial dispersal promotes biodegradation in heterogeneous systems exposed to osmotic stress, *Frontiers in Microbiology* 7, 1214.

Wu X.Y., Zhu Y., Yang M.R., Zhang J.Y., Lin D.H., 2023, Earthworms enhance the bioremediation of tris(2-butoxyetheyl) phosphate-contaminated soil by releasing degrading microbes, *Journal of Hazardous Materials* 452, 131303.

Wu Y.D., Trejo H.X., Chen G., Li S.M., 2021, Phytoremediation of contaminants of emerging concern from soil with industrial hemp (*Cannabis sativa* L.): a review, *Environment, Development and Sustainability* 23, 14405–14435.

Wuana R.A., Okieimen F.E., 2011, Heavy metals in contaminated soils: a review of sources, chemistry, risks and best available strategies for remediation, *Communications in Soil Science and Plant Analysis* 42, 111–122.

Yao B., Sun R.Z., Alinezhad A., Kubátová A., Simcik M.F., Guan X.H., Xiao F., 2022, The first quantitative investigation of compounds generated from PFAS, PFAS-containing aqueous film-forming foams and commercial fluorosurfactants in pyrolytic processes, *Journal of Hazardous Materials* 436, 129313.

Yu Y., Guo H., Zhong Z., Lu Z., Zhu X., Li Z., Chang Z., 2023, Enhanced removal of tetrabromobisphenol A by *Burkholderia cepacian* Y17 immobilized on biochar, *Ecotoxicology and Environmental Safety* 249, 114450.

Zhang C., Feng Y., Liu Y.W., Chang H.Q., Li Z.J., Xue J.M., 2017, Uptake and translocation of organic pollutants in plants: a review, *Journal of Integrative Agriculture* 16, 1659–1668.

Zhou Z., Liu X.T., Sun K., Lin C.Y., Ma J., He M.C., Ouyang W., 2019, Persulfate-based advanced oxidation processes (AOPs) for organic-contaminated soil remediation: a review, *Chemical Engineering Journal* 372, 836–851.

Index

Note: **Bold** page numbers refer to tables and *italic* page numbers refer to figures.

A

Abiotic degradation, 112, 113, 115
Acetaminophen, **61**, 257–258, 276
Activated carbon, 21, 121, 202, 209, 220, 229, **270**–271
Additives, 3, 7–8, 10, 32, 35–36, 39, 57, 59, 60, 63, 69, 92–93, 107, 109, **118**, **183–184**, 233
Adsorption rates, 58
Advanced Fenton oxidation, 222
Advanced nanomaterials, 21
Advanced oxidation processes, 21–23, 119, 222–223, 226
Advection, 91, 110, 113
Agricultural runoff, 5, 49, 121
Agroindustry, 6
Analytical models, 114
Anthropogenic activities, 1, 52, 58, 69, 71, 74, 110, 123
Antibiotic resistance genes, 160, 162–163
Antibiotic-resistant bacteria, 160, 162–163
Antibiotics and hormones, 5
Apoplastic, 133, 280
Aquaculture, 5, **2**, 8
Aquatic ecosystems, 6, 53
Artificial intelligence, 24
Atmospheric deposition, 33, 40, 42–43, 45, 59, 67, 74, 110–111

B

Bacterial transport, 243–245, 247–249, *250*, 259
Basel Convention, 77, 177–178
Bioaccessibility, 133, 204
Bioaccumulation, 3, 7, 9, 16, 25, 58, 67, 77, 94, 132, 140, 149, 152–154, 156, 200, 280
Bioaccumulation factor, 154, 280
Bioaccumulative potential, 154
Bioadsorption, 132
Bioaugmentation, 21, 122, **202**, 240, 242–243, 259, 276
Bioavailability, 91, 93–94, 97–99, 102, 133, 141, 155, 226–228, 231, 235, 242–*243*, 249, 253
Biochar, 21, 115, 119, 121, 199, 207–**208**, 210–212, 220, 228, 234, **270**–272, 280
Biochar technology, 199, 207, 210
Biodegradability, 8, 111, **117**
Biodegradable, 22, 226, 229, 230
Biodegradation, 94, 97, 110, 112, 113, **117**, 119, **120**, 122, 132–133, 135, 206, 241–242, 249–252, 254–255, 257–259, 275
Biogeochemical interactions, 112
Biological interfacial processes, 240
Biological treatments, 22–23, 58, 122
Biomagnification, 7, 13, 149, 153–154, 156
Biosensors, 19
Biosolids, 5, 7–10, 34–35, 60, 67, 75, 110, 151, 159, 199, 283
Biosorption, 101–102, 140, 250–251
Biostimulation, 21, 122, 240, 259, 276
Biosurfactants, 139, 140, 229, 240, 245, 254
Biota sediment accumulation factor, 151
Biotransformation, 24, 110, 112–113, 250, 255, 257
Biphenyls, **2**, 7–8, 17, 57, 63, 96, 101, **117**, 139, **152**, **161**, 170, **174**, **190**, 204, 220, 278
Bisphenol A, **12**, 17, 40, 48, 50, **62**, 75–**76**, 98, 109, 148, **161**, 228, 251, **257**–258, 276
Brominated retardants, 15
Buffering capacity, 98–99

C

Cancer risk, 14–15
Carbamazepine, **12**, 50, **62**, 95, 98, 101, **117**, 122, 135–137, 151, 154, 251, 255–257
Carcinogenicity, 15–17, 138
Carcinogens, 15, 171
Cation exchange capacity, 98, 100
Cell tumors, 15
Cell surface hydrophobicity, 244
Chelating agents, 205, 228, 281
Chelation, 228
Chemical degradation, 94, 150
Chemical oxidation, 220, 226, 268
Chemical reduction, 224
Chlorinated paraffins, 57, 69, **174**, **184**, **191**–192, 194
Chlorination, 14, 222
Chronic toxicity, 157
Coagulation-flocculation, 121
Combined remediation, 212
Combined treatment methods, 22
Complexation, 98, 100

Consumer products, 14
Containment, 269, **270**
Contaminant transport models, 113
Controlled emerging contaminants, 180, 192
Conventional persistent pollutants, 24
Compost, 34, 36, 163
Cosmetic products, 60
Congeners, 10, 15–17, 63–65, **70–73**, **191**

D

Debye length, 248
Defluorination, 258
Degradation pathways, 1, 4
Deposition, 7–8, 33, 40–49, 59, 67, 74, 110–111, 132, 198, 243–244, 247–248
Desorption, 58, 110–111, 135, 199–**201**, 203–206, 212, 230, 241, 243, 246–247, 269, **270**–272, 282
Detection methods, 19, 20
Development disorders, 16
Diclofenac, 49, **61**, 92, 95, 101, **117**, 122, 151, 154, 228, 255, 257
Diethyl phthalate, **12**
Discharge plasma, 204, 232
Dissolution, 95, 101, 133, 227, 231
Dissolved organic carbon, 98, 136
Doxycycline, 63, 92

E

Eco-friendly, 21, 24, 34, 52, 124, 170, **173**, 205–206, 230
Ecological risks, 103, 121, 159, 160, 268, 271
Ecological systems, 1, 19, 32, 52, 102, 160, 171, 179
Ecosystem resources, 15
Ecosystem safety, 3
Efficient treatment systems, 12
Electrochemical degradation, 226–227, 235
Electrokinetic, 199, 269, **270**
Electrostatic attraction, 50, 98, 100, 210, 243, 251, 253
Electrostatic interaction, 94, 100–101, 113, 247, 251
Endochrine disrupting chemicals, 95
Endocrine disruptors, 5, 17, 96, 98, 130, 138, 148, 157, 159, 169, 171
Endocrine receptors, 16
Environmental degradation pathways, 4
Environmental hazards, 9
Environmental toxicity, 12
EOC migration, 44, 51, 110–111
Evapotranspiration, 136, 279

F

Fenton's reagent, 222
Fisheries, 8
Flame retardants, **2**, 10, 32, 39, 42, **63**, 69, 71, 93, 109, 110, 130, 137–138, 148, **152**, 157, **161**, 192–194, 198, 200
Fluoxetine, 60, **62**, 137, 151, 154, 255

G

Genetic engineering, 8, 21
Genotoxic effects, 15, 159
Genotoxicity, 17–18
Green livers, 131
Green oxidant, 222
Green solvents, 207
Green technology, 142
Groundwater contamination, 122, 278, 281

H

Health risks, 13–14, 18, 32, 40, 74, 93, 115, 123, 142, 177, 180, 195
Herbicides, 5, 90, 96, 101, 109, 132, 140, 157, 226, 233, 280
Hexachlorobutadiene, **174**, **185**, 192–194
Hormonal interference, 15
Hormones, 1, **2**, 5, 8–9, 17, 32, 60, **62**, 90, 96, 109, **117**, 140
Hybrid models, 114
Hydraulic conductivity, 114–115, 150
Hydrogen bonding, 94, 101, 135, 210, 251
Hydrolysis, 91, 113, 131, 140, 241, 278
Hydrophobic interactions, 50, 210, 244–246, 251, 253
Hydrophobicity, 6, 97, 100, 102, 133–134, 142, 152, 244–246, 249, 251, 259, 270, 280–281, 283
Hydroxyl radical, 122, 220–222, 224, 228–229

I

Ibuprofen, 49, 50, **61**, 92, 98, 123, 136, 138, 251, **256**–258
Immobilization, 211, 227–228, 268–272, 281, 283
Industrial effluents, 4, 207
Industrial growth, 8–9
Ingestion, 9–10, 58, 60, 138, 153
Innovative technologies, 24
Integrated approach, 123, 234–235
Ionic strength, 50, 94, 96, 98, 133, 243

L

Landfills, 5, 10, 40–41, 110, **117**, 148
Leaching, 7, 8, 41, 44, 48, 50, 77, 97–99, 111, 121, 135, 142, 150, 271–272, 281

M

Machine learning, 115
Membrane filtration, 22
Membrane technologies, 122
Micelle, 230–231, 245–246, 281
Microbial activity, 50–51, 103, 115, 122, 141, 150, 155, 206, 279
Microbial degradation, 67, 103, 112–113, **118**, 121, 123, 130, 278, 280
Microbiological diversity, 149
Microplastics, 11, 35–43, 49–51, 90, 94, 96, 98, 159, 169, 194–195, 223, 231
Microwave-assisted remediation, 199
Mineralization, 95, 150, 155, 158, 207
Molecular techniques, 19, 119
Monitoring, 14, 37, 39, 57, 65, 77–78, 107, 115–**117**, 121, 124, 163, 169, 170, 177, 180, 202, 212, 272–273, 281
Mortality, 14–15, 154, 159–**161**
Multidisciplinary approach, 103, 116
Mutagenic effects, 18
Mutations, 18

N

Nano-biochar, 231
Nano-catalysts, 232
Nano-enhanced remediation, 231
Nanomaterials, 3, 5, 6, 21, 57, 107, 109, 121, 153, 228, 231, 232, 235
Nano-phytoremediation, 231
Nanoplastics, 38, 94, 162
Naproxen, 58, **61**, 257
Natural attenuation, 115, 119, 121, 123
Nonionic surfactants, 230, 246, 248
Nonpersistent EOCs, 109
Numerical models, 114

O

Organochlorine pesticides, 7, 57, **117**, 192–194
Organophosphate esters, 35, 57
Ozonation, 23, 122, 222

P

Partition coefficient, 11–**12**, 49, 95, 101, 134, 154
Partitioning, 12, 63, 69, 94–96, 101, 114, 135, 203, 269, 280
Passive transport, 132, 142
Perfluoroalkyl sulfonate, 148
Permeability, **117**, 132, 153, 225, 248, 254, 269
Persistent EOCs, 108–109
Persistent organic pollutants, **2**, 21, 42, 92, 109, 169, **191**–192, 228
Personal care products, 1, **2**, 3, 5, 8, 13, 32, 37, 39, 57, 59, 60, 90, 92, 107, 109–111, 114, 116, **117–118**, 121, 130, 148, 157–158, 162, 198, 203, 225, 240, 268
Photocatalysis, 22, 122, 205
Photolysis, 113
Phthalate esters, 7, 59
Phthalates, **2**, 7–8, 17, 35, 37, 39–41, 57–60, 75, **76**–77, 96, 98, 138, 152, 200, 229, 231
Phthalic acid esters, 36
Phytodegradation, 139, 140, 278, 280
Phytoextraction, 21, 139–141, 269, **270**, 278
Phytoremediation, 21, 122, 130, 138–142, 199, **201**, 211, 230, 234, 268–269, **270**–271, 278–280, 283
Phytostimulation, 141
Phytovolatilization, 21, 139, 141, 278
Plant growth-promoting bacteria, 140, 211, 234
Plant physiology, 136, 142
Plant uptake, 67, 132, 139, 198, 200, 281
Polarity, 11, 13, 45–46, 95, 97, 101, 135, 142, 206, 230, 253
Policymakers, 12, 24, 53, 123, 169
Polyaromatic hydrocarbons, 152
Polybrominated diphenyl ethers, 10, 40, 93, **152**, **174**, 229
Polychlorinated biphenyls, **2**, 7, 57, 63, 96, 101, **117**, 139, **152**, **161**, 170, **174**, **190**, 204, 220, 278
Pore filling, 94, 101, 210
Predictive models, 115
Public awareness, 25, 77, 121, 123, 180

R

Reclaimed water, 13, 98
Rhizodegradation, 139, 141, 278
Risk factor, 14
Rotterdam Convention, 177–178
Removal efficiency, 21–22, 122, **201**, 205, **208**, 229, 276, 281–282
Redox potential, 98–99, 113

S

Salinity, 96, 98, 232, 281
Semi volatile organic compounds, 92
Separation technologies, 24
Septic systems, 5, 35
Sewage sludge, 3, 5–8, 10, 34–36, 77, 92–93, 148
Soil ecosystem, 3, 6, 9, 12, 18, 24–25, 41, 44, 93, 149, 150, 153–154, 163, 200

Soil health, 13, 15, 35, 38, 58, 157, 159, 278
Soil horizons, 7–8
Soil invertebrate, 156–158
Soil microbiome, 155, 162–163
Soil microorganisms, 13, 51, 93, 137, 150, 154–155, 234
Soil organic matter, 50, 91, 97, 111, 114, **117**, 135, 268
Soil porosity, 114–115
Soil vapor extraction, 199, 200, **201**
Soil washing, 199–**201**, 205–206, 210, 212, 228–229, 235, 268–270, 281–283
Soil-groundwater systems, 113, 115–116, 119, 122
Sorption capacity, 44, 46, 101–102, 110, 113, 133, 229, 271, 283
Sorption coefficient, 97–98, 114, 151, 255
Sorption of EOCs, 101–102, 241, 251, **270**
Stabilization, 35, 139, 210, 227, **270**–271
Stockholm Convention, 75, 169, 170–171, **172–174**, 177, **191**
Strategies, 13, 21, 32–34, 40–41, 102, 113, 116, 119, 123, 158, **187–188**, 195, 211–212, 243, 246, 250, 257, 259, 275–276, 281
Sulfamethoxazole, **61**, 63, 96, 122, 135–137, 155, 229, 251, **256**, 258
Supercritical fluid extraction, 199
Supercritical fluids, 207, 209
Surface sorption, 94, 101, 250, 255, 259, 275
Surfactant-enhanced remediation, 230
Surfactants, 32, 90, 92–94, 96, 110, 130, 205, 206, 222, 225, 228, 230–231, 242, 244–246, 248, 268, **270**, 281
Sustainability, 37, 53, 102, 122–123, **172**, 211–212, 229
Symplastic, 133, 280
Synergistic effects, 198, 211

T

Thermal desorption, 199, 200, 203–205, 212, 269, 272
Thermal treatment, 204, 270, 272–275, 283
Toxins, 5–6, 21, 99, 138, 231
Translocation, 131–136, 249
Transmembrane transport, *252*, 252–255, 259
Transport pathways, 108, 113, 116
Transportation mechanism, 110, 132
Treatment processes, 5, 20, 22–23, 35, 37–38
Triclosan, **2**, 58, **61**, 63, 95, 101, 136, 148, 151, 240, **256**
TSCF, 131

U

Urban soils, 34, 39, 40, 60, **64**–65

V

Volatilization, *33*, 40, 47, 51, 67, 95, 97, 101, 130, 202, 234, 271

W

Washing agents, 205, 229, 270, 281
Wastewater irrigation, 38, 48, 67, 74, 90, 110, 268, 283
Wet deposition, 7–8, 42–43, 46

X

Xenobiotics, 159

Z

Zero-valent iron, 220

Made in the USA
Las Vegas, NV
19 September 2026

35357f1e-94b2-4c2b-8cee-907047029276R01